LISTENING TO A CONTINENT SING

DRAWINGS BY NANCY HAVER

LISTENING TO A CONTINENT SING

Birdsong by Bicycle from the Atlantic to the Pacific

DONALD KROODSMA

PRINCETON UNIVERSITY PRESS
PRINCETON & OXFORD

Published by Princeton University Press,
41 William Street, Princeton, New Jersey 08540

In the United Kingdom: Princeton University Press,
6 Oxford Street, Woodstock, Oxfordshire OX20 1TW

press.princeton.edu

Jacket images courtesy of Shutterstock. Warbler courtesy of
Nancy Haver

ISBN 978–0–691–16681–0
Library of Congress Control Number: 2015950553
British Library Cataloging-in-Publication Data is available
This book has been composed in Arno Pro and Lydian BT
Printed on acid-free paper. ∞
Printed in the United States of America
1 3 5 7 9 10 8 6 4 2

For the Birds, and David

HOW TO LISTEN USING THE QR CODES

Here's how to listen to the 381 recordings referenced by number in the text. On page 2 of the text, for example, on the left side of the page I refer to a recording of an American robin and provide a brief description of what can be heard in that recording. Above the description is a QR code (essentially a two-dimensional barcode), and in parentheses is the number and duration of the recording (1, 2:00), indicating that this is recording #1 of 381 and it is exactly two minutes in duration.

Using a QR code reader (such an app on a smart phone), you can play this robin recording by accessing the correct page on ListeningToAContinentSing.com. Scan the QR code for the wood thrush on page 2 and you can listen to this Virginia maestro. You can use your QR code reader throughout the book for all 381 sounds, or alternatively, once on the internet, you can easily navigate around the website ListeningToAContinentSing.com to find whatever you wish (see below).

LEARN SO MUCH MORE AT LISTENINGTOACONTINENTSING.COM

At this website, detailed notes supplement the book's text so that you learn how to listen to more than 200 different species, with multiple recordings from many of them. Also, you can download each of the sounds so that you can view and study them in a program such as free Raven-lite (http://www.birds.cornell.edu/brp/raven/RavenOverview.html). Seeing the sounds dance across a computer screen as you listen is an excellent way to learn how to listen to birds in general. As I have learned over the last 40 years, learning to hear with your eyes will give your ears a big boost.

Also, at this website, besides choosing to listen by number from 1 to 381 as referenced in the book's text, you can choose where you want to listen, anywhere from Virginia to Oregon. Or you can choose to listen from a list of over 200 species placed in the usual field guide order (just look at all those thrushes, or warblers, or sparrows!). Or surf the dawn wave of birdsong and first light from coast to coast, listening to the special recordings that show how birds greet the day during the dawn chorus.

CONTENTS

How to Listen Using the QR Codes vii
Acknowledgments xi
Prologue xiii

1. Beginnings *1*
2. Peace, and War *12*
3. Lemonade *20*
4. Blue Ridge Dawn *30*
5. A Virginia High *44*
6. Appalachia *56*
7. Boone Country *68*
8. A Ride in Heaven *74*
9. Laid Up *82*
10. On the Road Again *86*
11. Dawn Sweeps the Shawnee *96*
12. The Ozarks *106*
13. A Prairie Gem *115*
14. Kansas Oceans *125*
15. Shortgrass Prairie *136*
16. Western Birds *146*
17. Riding the Rockies *153*
18. Sage and Song *163*

19. Hello, Wyoming *173*

20. The Oregon Trail *179*

21. Grand Tetons *190*

22. Into the Fire *199*

23. Caterpillars Marching *207*

24. Chief Joseph Pass *217*

25. Lewis and Clark *225*

26. Pacific Islands Incoming *233*

27. Ascending into Oregon *242*

28. Geological Chaos *251*

29. Over the Cascades *261*

30. A Homecoming *270*

31. Land's End *280*

Epilogue—Where Are They Now? *287*

Notes 289
References 295
Index 299

ACKNOWLEDGMENTS

Life would be infinitely poorer without my fine riding companion and son David. Thank you, David, for being you and helping make this trip happen, for your flair in the kitchen, for riding early in the morning with me, and for teaching me more than a few things about life.

A bicycle trip across the country will renew your faith in the human spirit, I was told early on, and so it was. Here's a hearty thank you to the many individuals and groups and towns who welcomed us to stay overnight at "their place" along the way: in Virginia, Jim Kirchner and family in Mechanicsville, the Volunteer Fire Department in Mineral, (the late) June Curry in Afton, Elizabeth Brown Memorial Park in Wytheville, and United Methodist Church in Rosedale; in Kentucky, Jeff Kline in Virgie, Booneville Presbyterian Church in Booneville, Lincoln Homestead State Park in Springfield, LaRue County Park in Hodgenville, and Elementary School Park in Utica; in Illinois, Cole Memorial Park in Chester; in Missouri, the courthouse lawn in Hartville, and City Park in Ash Grove; in Kansas, the Environmental Education Center at Quivira National Wildlife Refuge, and the Biking-across-Kansas crowd in Tribune; in Wyoming, Jim and Carol States in Saratoga, City Park in Lander, and Dave and Jo-An Martin in Dubois; and in Oregon, Presbyterian Church in Dayville, Geoff Keller in North Bend, and Sue and Dick Ulbricht in Beaverton.

Thank you to the friendly folk who took time to talk with me in their native tongues through Virginia and Kentucky, as mentioned in the book; and to the late Carolyn Jensen for taking some of these interviews, as well as other TransAm sounds, into a "Radio Expeditions" program (*http://www.npr.org/templates/story/story.php?storyId=4699207*).

Riding companions enrich a journey in many ways. Thanks to Kris and Kes, John and Mark, Al, Stefan, Keith, and especially to Tom Hunt and to Ed and Ev Wengrofsky.

Thanks to those along the way who made a big difference: Heidi and Cycle Ed repairing a bike in Ashland, VA, on our third day out; Jeff Kline for his hospitality in Virgie, KY; the family in Harvey County East Park who provided a fine dinner; Lee for reeling us in out of the rain in Ness City, KS; medical professionals in Lexington, VA, Bardstown, KY, and Eads, CO, who cared for my ailing leg; the townspeople of Kremmling, CO; and countless others along the way who gave a friendly honk or wave.

Our bicycle journey occurred during 2003, but during 2008–2010, I crossed the country again in three segments as I sought out recordings to accompany the text of this book. Many individuals told me of favorite places where it just might be quiet enough to record without undue man-made noises: in Virginia, the late Curt Adkisson, Roger and Lynda Mayhorn, Ryan Mays, and Aubrey Neas; in Kentucky, John Omer; in Illinois, Rod McClanahan; in Missouri, John Faaborg, Brad Jacobs, Mark Robbins; in Kansas, Mark Robbins; in Colorado, Ken Behrens, Pam Bilbeisi, Nathan Pieplow, Jim Sedgwick; in Wyoming, Terry McEneaney, Jim and Carol States, Jack States; in Idaho, Fred and Melly Zeilemaker, who kindly hosted me at their Xylem Acres; in Oregon, Sallie Gentry, Dave Herr, Geoff Keller, Kim Nelson. I know there are others, but my memory and records fail me. Thanks to all!

Others helped move this project along in significant ways. Wil Hershberger and especially Lang Elliott offered generous help with sound-editing software. Greg Budney and Bill McQuay at Cornell's Macaulay Library (of Natural Sounds) were indispensable in preparing the sounds. And you're the best, Nancy Haver, for producing such fine drawings, and in so timely a fashion.

I owe a special thanks to Janet Grenzke for recording many of the sounds that accompany the text and for studying the prose word by word, offering her (always strong) opinions about what does and doesn't work. Other readers were Zane Kotker, Ed Wengrofsky, Kenda Kroodsma, and Greg Oates.

A huge thank you to the consummate professional, literary agent Regina Ryan, who believes in birdsong (and me), and to Robert Kirk at Princeton University Press, for seeing promise in this project. Thanks also to Jodi Beder for her superb copyediting, and to all those at Princeton University Press (especially Ellen Foos and Amanda Weiss) who saw this book through to publication.

PROLOGUE

Birdsong coast to coast, from seaside sparrows and laughing gulls on the Atlantic to wrentits and western gulls on the Pacific! I'm surfing the wave of first light and dawn song across the continent, sweeping the Appalachians, hurdling the mighty Mississippi, roller-coastering the Ozarks, bursting out of the eastern forests and into the Great Plains; halfway now, I hit the Rocky Mountain wall and charge north up its very backbone, into the Tetons and through Yellowstone, joining Lewis and Clark, dipping into Hells Canyon and climbing out into the geological wonderland that is now Oregon, listening to birds and all else that "sings" along the way, all from the best seat imaginable, all from the seat of a bicycle.

We are pioneers, embarking on a transcontinental journey unlike any before us. Others have traversed the continent, to be sure, including birders who have listed all of the species they've identified. Not us. We will reach more deeply, more intimately, far beyond such lists as we listen in on the personal lives of individual birds, using their songs and calls as a window into their minds. For 35 years as a scientist I have relentlessly studied birdsong, and here is the grand payoff, listening my way across the continent. No one before has experienced the continent as we will, because no one before has listened as I know how.

MAY 3: AMHERST, MASSACHUSETTS, TO YORKTOWN, VIRGINIA, BY CAR

Daydreaming, I am already on my bicycle, heading west, when I realize that he's laughing. In his sleep! Mid-afternoon on this most glorious third of May, deep into Virginia on Interstate 95, my son sleeps as I drive and daydream from Amherst, Massachusetts, to Yorktown, at the southeastern tip of Virginia. There we'll jettison the rental car, mount our bikes, and pedal west. I'm eager to listen to this entire continent sing; David, well, he'll be doing whatever he pleases. Maybe we'll both figure out what to do with the rest of our lives.

Though we have similar builds, as our nearly identical bikes hanging from the back of the car attest, our worldviews differ. I'm fifty-six years old and have worked hard, intensely even, all my life. As a kid, I toiled in the hot summer sun on my hands and knees, bunching radishes, topping onions, picking or weeding beans or cucumbers and the like, earning a few pennies to eke out some spending money. I studied relentlessly, racing directly from high school to college to graduate school to a postdoctoral position and then a university professorship, supporting a growing family, with never any significant time off. I need a break.

He's twenty-four. I'd like to think that my hard work has given him the opportunity to flourish. He's an athlete, a musician, a scholar, with a wonderful combination of talents, full of life and a joy to be around. Best of all, he's my son, and he's journeying with me on the ultimate bicycle tour. He finished his master's degree at Stanford one term early so that we could start in early spring rather than wait until midsummer after the peak of the birdsong season. He also knows how to cook.

≈

Reflecting on life as we drive to Virginia, I realize that for me this bicycle journey began five years ago in an Amherst bicycle shop. Making what I had intended to be light conversation with John, the owner, I mentioned offhandedly that someday I'd like to ride beyond the confines of western Massachusetts. With a bit of a twinkle in his eye, he looked at me and said, *"Every year you don't do it, it's less likely you ever will."* I chuckled politely, then realized the enormity of his comment, not just about bicycling but about life in general.

The next summer I was in Colorado, "Riding the Rockies" during a weeklong event with two thousand other cyclists. The following year it was "Cycle

Utah," this time in Zion and Bryce Canyon National Parks. I then began two years of preparation for the Big One, clearing my slate so that I could spend an entire spring and summer on the road, away from work and home.

I planned furiously (David's plan was "just go do it"). The route we eventually settled on was the popular TransAmerica route, established in 1976 for the "Bikecentennial" of our country, when over 4000 cyclists celebrated our nation's two-hundredth birthday by biking coast to coast. Using detailed maps now available from the Adventure Cycling Association, we would begin where the British surrendered in Yorktown, Virginia, and over the next two to three months revel in the history and natural wonders of this country as only a cyclist can. I reviewed the birds we could expect to encounter and searched out the best campgrounds, all to guarantee the best birdsong experiences. We bought our bikes, solid steel-frame, go-anywhere Bruce Gordon touring bikes with 26-inch wheels; I chose a red frame, David black. Both were outfitted with front and rear racks and four red panniers in which we would carry all our gear. My 16-digit credit card number was burned into memory as I chased a host of supplies.

One decision remained to be made: Do we bike from east to west or west to east? From Virginia to Oregon or Oregon to Virginia? Most cyclists start at the Pacific to take advantage of the prevailing westerly winds, but that most popular direction increasingly made little sense to me. I listed the reasons:

1. North America was settled (at least in recent times) from east to west, and it's the way we easterners naturally think of experiencing the continent. The sun also sweeps from east to west, as does the dawn chorus of birdsong. How could we buck the settling pioneers, the sun, *and* the birds?

2. When biking to the west, the morning sun will be at our backs, not in our eyes. What a difference that makes in how one perceives the landscape, and what a difference in safety, too, as motorists approaching us from behind will not be blinded by the sun.

3. Because most cyclists are eager to take advantage of the prevailing winds, they ride from west to east, so we'll meet far more bicyclists than if we were all heading in the same direction, and meeting other cyclists on the road is one of the joys of bicycle touring.

4. And the trump card, of course, was *birdsong*! To navigate the western mountain passes, one must wait until June to start biking from the

> Pacific. That would put us in hot and humid Kentucky and Virginia during July or August, maybe even into September, well past the season of birdsong. How much better to begin on the Atlantic in May and experience Virginia and Kentucky in springtime, one of the greatest shows on earth. When we hit the cool, dry Rockies, we'll head up and north, thereby prolonging spring, the birds singing with us all across the continent.

It's a no-brainer: East to west is the only way to go. Thoreau concurred: "Eastward I go only by force; but westward I go free." My sentiments exactly.

The planning now completed, here we are in Virginia driving to our campground at the Newport News City Park, where we'll unload our gear. I'll return the rental car and mail the car's bike rack back home. We'll eat dinner, fuss over last-minute details, and get a good night's sleep. Well before sunrise we'll be up, eat breakfast, break camp, and bike the few miles through the woods to see and hear the sun rise at the Yorktown Victory Monument, the starting point of the 4200-mile TransAm Bicycle Trail.

1.
BEGINNINGS

DAY 1, MAY 4: YORKTOWN TO JAMESTOWN, VIRGINIA

Rain. Why, when we are about to embark on the journey of a lifetime, why must it rain? Restless, anxious, hoping for a break, I've done my best to enjoy the soothing patter on the tent's fly over the last few hours, but now it's nearly 5 a.m. and time to get moving. The muffled words I catch from deep in David's sleeping bag are ". . . rain . . . sleep more . . . won't miss anything. . . ."

Reluctantly, I accept. So much for nearly two years of my planning and imagining a Grand Beginning to this Journey, at the towering Monument with birds singing through sunrise.

From the comfort of our sleeping bags, Plan B comes all too easily, but for me sleep does not. I lie alert, thinking through the preparations for the trip

and wondering what I am doing here. Over four thousand miles of biking lie ahead, from the Atlantic to the Pacific, with all kinds of challenging terrain and weather. I had better get used to Plan B, I lecture myself, for who knows what we'll encounter in the coming weeks and months.

American robin. Low carols followed by a high *hisselly*. (1, 2:00) See p. vii for how to listen.

A robin begins to sing, 5:34 a.m. according to my watch, about half an hour before sunrise. His low, sweet carols drop from above one by one, *cheerily, cheer-up, cheerio, cheerily,* and I am soon silently singing with him, three to five carols over a few seconds, then a brief pause, then a few more carols, and another pause. I feel his tempo, counting the number of carols in the next package and pausing, counting and pausing, his initial measured pace calming. I try to stretch each quarter-second carol into a second or more, slowing his performance, relishing the varying patterns in pitch and rhythm, listening and watching as miniature musical scores float through my mind. He accelerates now, adding a single high screechy note, a *hisselly,* after each caroled series, but soon there will be two or more such high, exclamatory notes. I know how to listen to the patterns in his singing, how he combines sequences of different caroled and *hisselly* notes to express all that is on his mind, sometimes even singing the two contrasting notes simultaneously with a low carol from his left voice box and a high *hisselly* from his right, but for now the effort of deep listening is too much like work. Instead, I curl up in the sleeping bag, drifting along on a robin's song, floating, a broad smile creeping over my face, this robin having reminded me why I'm here.

Wood thrush. Five different *ee-oh-lay* phrases in the rain-songs of this Virginia maestro. (2, 2:02)

A wood thrush joins in. He awakes with sharp *whit whit* calls, as if a bit peeved, then gradually calms to softer *bup bup* notes, and soon he's in full song, so rich and melodious, the stuff of boundless superlatives, one of the wonders of eastern woodlands. I sing with him, too, acknowledging the soft *bup bup bup* notes at the beginning of each song, then gather in the low, rich, flute-like, *ee-oh-lay* prelude, then smile at what sounds to my human ears like a harsh and percussive terminal flourish. At first I simply dissect each song into its prelude and flourish, marking the contrast between the two, but I'm soon sketching each prelude in my mind as he sings. Emerging are five different half-second masterpieces of rising and falling, rich, pure notes, delivered just slowly enough that I can detect the overall patterns. And the flourishes—what a pity that I cannot slow them down now and hear the pure magic in the way the thrush must hear it, with his precision breathing through his two voice boxes producing the most extraordinary harmonies imaginable.

Wood thrush

In my mind's eye, I see a grand evolutionary tree, the massive trunk emerging from the primordial soup, the branches and twigs sufficient to accommodate every lineage and every living creature that ever was and is. At the very tips of three twigs in this grand array are the robin, the thrush, and me. Trace each of our lineages back in recent time and we each find two parents, four grandparents, then eight great-grandparents, and so on. Climbing down this tree, some tens of millions of years back in time the robin and thrush meet at a branching point where they have the same ancestor, where they are one. The robin and thrush now travel back in time together in search of their roots, meeting up with me some hundreds of millions of years ago, when we all had the same ancestor, when we were one. We belong to an extended family, each of us an extraordinary success story, each of us with an unbroken string of successful ancestors dating back to the beginning of time. The robin, the thrush, and I are equals: "*Mitakuye oyasin,*" the Sioux would say as they end a prayer, "*all my relations.*"

A chickadee sings now, too, a Carolina chickadee. Song after whistled song pierces the air, each sharp and sure. He sings the common high-low-high-low pattern, *fee-bee-fee-bay,* four whistles alternating from high to low frequency. But then he's on to another pattern, this one high-low-low-high, as in *fee-bee-bee-fee;* and soon he sings *fee-bee-bay-fee-bee,* three different song patterns, now leaping excitedly among them, successive songs always different. What frenzied singing, as if he's eager to show off all that he knows, eager in this pre-sunrise chorus to challenge other males and to impress listening females that he's *the one.* How different from what I can expect in an hour or so, when he'll repeat one of his songs many times before switching to another, perhaps well after attentive females have made their mating decisions. I listen for a neighboring male chickadee, hoping to hear a dialogue between them, but hear none.

Carolina chickadee. Excitedly singing a variety of songs, interspersed with calls. (3, 2:02)

But I do hear a conversation among the tufted titmice. The nearby male sings *peter peter peter,* and two other males in the distance, each on his own territory, echo with the identical song, songs that they've learned from each other. Back and forth and around they go as they answer each other; I note the time, 5:51 a.m., knowing that this *peter peter* discussion could go on for a while, as each male can sing 500 or more

Tufted titmouse

Tufted titmice. Neighboring males dueling with identical songs. (4, 1:21)

renditions of a particular song before they all switch, almost in unison, to a different song in their repertoires.

The robin, the wood thrush, the chickadee, the titmice . . . *Yes, I know why I'm here,* and I'm not even out of the sleeping bag on the first day. Disjointed thoughts surface with jumbled words that do no justice to the certainty of purpose . . . to celebrate life, David's and mine, and the lives of other creatures along the way . . . to hear this continent sing, not only the birds but also the people, flowers and trees, rocks and rivers, mountains and prairies, clouds and sky, all that is . . . to discover America all over again, from the seat of a bicycle . . . to embrace reality, leaving behind the insanity of a workplace gone amuck . . . to simply *be,* to strip life to its bare essentials and discover what emerges . . . and in the process, perhaps find my future . . . by listening to birds!

The rain having abated, David stirs, and we agree it's time to get going. "Best not to get up before the sun," David offers as he squeezes out of the tent. I sense something rather profound in that statement, something about awaking *after* the birds' finest hour. What opposites we are, as I cherish dawn and he dusk, our preferred waking hours a half day out of sync. We should address this issue head-on, but I choose to just smile and wonder quietly how this will play out over the coming weeks.

We dress for the cool, wet weather, staying dry and warm in our bright yellow rain gear. Sleeping bags are soon stuffed into their sacks, sleeping pads rolled and tied, the wet tent collapsed and stuffed into its sack. David warms water over his homemade soda-can stove, and we refuel on a breakfast of oatmeal loaded with brown sugar and raisins, mixed with powdered milk for extra protein. Continuing what will be the routine for the next two to three months, we rinse dishes and utensils, return miscellaneous items to their places in the panniers, fill the water bottles, and load the panniers onto the bikes, strapping sleeping bags and pads and tent to the bikes' racks. Somewhat mystified at this point, I stare at the loaded bikes, such a pretty sight, astonished that all of the gear we had spread out last night is now neatly tucked away, ready to ride.

"Ready?" "Yep!" After zeroing our odometers for the day, we set out through the campground and then onto an unmarked trail into the nearby woods. We do our best to follow the directions of a park attendant, but are rather uncertain in these first minutes of our journey how this path will lead us to Yorktown's Victory Monument, from where our maps will guide us to the Pacific Ocean.

Though we may be uncertain of the trails, I know what I'm hearing: It is springtime in Virginia and migrants abound, many of them probably having arrived overnight. The treetops are alive with every male bird trying his best to impress. "David, listen to all this! There are robins and Baltimore orioles and scarlet tanagers and wood thrushes and great crested flycatchers and Carolina chickadees and tufted titmice and brown-headed cowbirds and song sparrows and red-eyed vireos . . . and warblers . . . blue-winged and black-throated green and prairie and yellow and chestnut-sided warblers, northern parulas, just for starters, and many of them have plastic, wavering songs, showing that they're still learning them."

David smiles, nods. *Too much,* he seems to be saying, but maybe in the coming weeks he'll also come to love birds for all they have to say. Or perhaps he's wondering how he can turn his interest in carbon cycles and climate science into a lifetime of exploration, much as I've done with birdsong. I've heard him say "It's a chance to get to know my father better . . . but we could have chosen a more adventurous trip than one that's all mapped out from coast to coast." I relish time with him, too, but I look at the road ahead as uncharted, as fresh and unexplored, for we will be listening to the world pass by in ways that no one else ever has.

Heading generally northeast along woodland trails and secondary roads, we enter a clearing where, a sign informs us, George Washington had his headquarters during late 1781. The American army was camped just to the east, our allies the French to the north, and about three miles to the northeast were the besieged British along the York River. Two miles to the east we ride into Surrender Field, where British General Cornwallis and his thousands of troops gave up their arms on October 19, 1781, effectively ending the American Revolution.

We bike on, through the battlefields, past the earthen redoubts where the Americans and French stormed the British positions in a surprise night attack. Cannon are strategically placed throughout the landscape, and in mock battle, David dismounts his iron steed and mans one of the cannon, then scrambles up the earthen redoubt. With the visitor center still closed, we continue on through Yorktown to the Victory Monument itself, a column of Maine granite almost 100 feet high with Lady Victory herself standing tall at the top, proclaiming proudly that this nation "of the people, by the people, and for the people" stands united, strong, and *independent.* Here, amid all the symbolism commemorating the defeat of the British and the birth of a nation, here is the official beginning of the 1976 Bikecentennial route that we'll follow across the country.

Carolina wren

Carolina wrens. An escalated singing interaction, two males countering each other with matching songs. (5, 4:06)

VIC-to-ry! VIC-to-ry! VIC-to-ry! Or perhaps it is heard as *LIB-er-ty! LIB-er-ty! LIB-er-ty!* How appropriate these mnemonics for the Carolina wren's song that explodes from the bushes at the edge of the clearing just beyond the monument. His challenges are answered almost immediately by three other males whose songs ripple into the distance. I listen intently to the pitch and rhythm of the responses; they're all different, each of the males, at least for now, choosing to sing what the others are not. But every five seconds each male chooses among several singing options, all made possible because each male has about 30 different renditions of this *VIC-to-ry!* song, with most of them learned from and therefore identical to those of his neighbors. The default choice for each male is to continue with the current version of his *VIC-to-ry!* song, and if they continue singing and relations remain peaceful, eventually each will switch to another song that none of his neighbors is singing at the moment. I listen for tensions to escalate, when neighbors are more likely to address each other with identical songs, but for now calm prevails.

I'm jarred from my listening by the crinkling of food wrappers beside me. What does it take to stoke a twenty-four-year-old across the country? I wonder, but I'm hungry, too. Over fig bars, a bagel, and some cheese, I explain to David how to listen to the wrens. "That's great, Pops." He seemed to be listening attentively, though judging from his tone of insincerity, he must be wondering how this will all play out over the coming weeks.

Far more unceremoniously than I had imagined, we take a last look around, check the map one more time, and mount our bikes, heading west down Yorktown's Main Street. Within 50 yards we turn right on Compte de Grasse Street, coast down about 200 yards to the York River itself at Cornwallis Cove, then take a left onto Water Street. In a flash, the entire trip unfolds before me, one road after another leading us beyond Yorktown all the way to the Pacific as we follow our Adventure Cycling maps through ten states, over four thousand miles in two to three months. For now, though, we've chosen an easy first day, a shake-down ride to test our bodies and bikes on the essentially flat coastal plain of Virginia. Within a mile we're on the Colonial National Historical Parkway, a 23-mile stretch of road that connects the battlefields of Yorktown with colonial Williamsburg and historic Jamestown.

"We cheated," announces David, more than half serious I sense, as he stops at a pull-off beside the York River. "Should have started on the Atlantic, not a river that dumps into the ocean." Given his rude assessment, with the word "tragedy" slipped in there somewhere, and faced with the stone wall we'd have to traverse with loaded bikes to reach the river, we forgo the usual cross-country ritual of dipping the rear tire into the water. I chuckle at the symbolism, the dipping of tires, how meaningless, but deep down a little voice nags at me that we didn't do this quite right. The little voice swells as I imagine standing on the Atlantic shore, listening to all the birds in the saltwater marshes there. Get over it, I advise myself.

Seaside sparrow. Beside the Atlantic, wheezing two subtly different songs. (6, 2:06)

Fish crows and laughing gulls are suddenly everywhere. From the crows flying all about it's an outright laugh, a nasal *caa-ha, caa-ha,* or often a simple *caa*. The gulls laugh from fields beside the road and from high in the trees above the river—*HA-a HA-a HA-a,* every once in a while letting rip a wild, prolonged *ha-ha-ha-ha-ha-hah-haah-haaah*. It's an omen, I decide, the birds smiling on us and providing a hearty send-off for the journey of a lifetime. Yes, that's what I hear, and *Thank you,* I find myself saying, *and I wish you well, too*.

A CELEBRATORY SEND-OFF

Fish crow. Nasal, "laughing" *caa-ha, caa-ha*. (7, 1:15)

Laughing gull. *U-ah,* laughing halfheartedly. (8, 0:39)

The road leaves the river and we enter a forest of oaks and tulip trees and sweet gums and loblolly pines, the trees arching over the road, a tunnel inviting us deeper into a magical world at a pace befitting a cross-country bicycle trip, now escorted by songbirds all around. So many sounds, so many stories, all from family. I try to acknowledge each voice, each brown thrasher and indigo bunting and eastern towhee and prairie warbler and tufted titmouse and Baltimore oriole and scarlet tanager and great crested flycatcher and common grackle and . . . wow, listen to the blue jays here—how different their calls from the birds back home. Dialects, the oral traditions, yes, the culture of the jays here is different from the culture elsewhere. All these voices and more—it's dizzying, my mind racing among them as I float westward, the tires barely touching the pavement.

SO MANY VOICES!

Eastern towhee. Two different *drink-your-tea* songs. (9, 1:00)

Blue jays. A variety of calls from the local dialect in eastern Virginia. (10, 1:40)

David rides on ahead, giving me space, and I soon find myself focusing on the soothing songs of red-eyed vireos. They're packed in here, a male singing over the road every hundred yards or so, one or two singers always within earshot. I pick out the next bird, perhaps 50 yards ahead—he sings,

Blue jay

just a quarter second burst of energy, pauses a second, then sings again. The nearer I approach the sharper his songs, each clearly different from the one before. Beneath him I ride and then beyond, while he sings what seems a never-ending series of different songs: *Here I am . . . over here . . . vireo . . . listen now . . . believe me . . . that's right. . . .*

I check my speedometer: 12 miles per hour. Too fast. At that speed, I have only 15 to 20 seconds with each vireo, but I need more time. A minute would be good, so I slow to four miles per hour, then three for good measure, challenging my balance as I now take a full minute to pass each singing vireo. I listen intently, trying to pick out from the next male an especially distinctive song, the handle by which I can get to know him better. The third bird obliges, singing as I approach what sounds like an imitation of a goldfinch's call, a thin, rising *twweeeee.* With his next song, I begin counting: 1, 2, 3, 4, 5 songs, but none of them the goldfinch imitation, 6-7-8-9-10, 11-12-13-14-15, 16-17, and just after passing underneath him, there is the *twweeeee* again, at 18. I count again, 5 . . . 10 . . . 15, and I circle back to keep him in earshot. . . . There it is again, at 20. Nice. Given that a red-eyed vireo tends to sing most of his songs before repeating himself, I can hear that he has about 19 different songs at his command.

Red-eyed vireo. Distinctive songs occur periodically throughout his performance. (11, 5:38)

Turning west again toward Williamsburg, I speed up to catch David. Continuing with the vireos, I listen to each in turn, but now I'm content to just identify a distinctive song from each without lingering long enough to hear him repeat it. None of these birds has the distinctive *twweeeee* of that first bird, telling me another part of the red-eye's story, that each singer is unique, the songs in his repertoire an identifiable voiceprint. How different these vireos are from the tufted titmice, the Carolina wrens, and the Carolina chickadees, neighboring males of which learn each other's songs and share nearly identical repertoires.

Interstate 64. How jarring. Trucks and buses and cars thunder by just overhead. Where is everyone going so fast? What's the hurry? With windows closed, the air being conditioned, the radios no doubt blaring, the real world is shut out. But just 24 hours ago we were there, too, I remind myself. I take a few deep breaths, happy with the choice that we have made for these next few months. The trees again envelop me, the roar of traffic soon replaced by the songs of birds, peace restored.

At 13 miles, Williamsburg! And there's David, stretched out on the grass beside the road, waiting for me. "Sorry—got tangled up in some birds," I confess. He looks happy, relaxed, and, more importantly, tolerant of my delay.

"Stay away from that huge, sprawling visitors' center," he advises as he points across the packed parking lot. "People everywhere. There's extra parking for $37, according to the big flat-screen TVs hanging from the ceiling. Feels more like Disney World than a historic site."

It's Sunday, and we walk our bikes among the throngs of people in the streets of the historic area. It's big history, as here, during the 1700s, was the thriving capital of Virginia, the most influential American colony. We sit on the grass beneath a small maple tree and break out the bagels and cheese, the bananas, and the fig bars, adding a Snickers bar for dessert. Just across the pathway from us, tourists pair off and pose to have their pictures taken with heads and arms secured in the stocks. Soon we're there, too, held at the neck and wrists, heads and hands protruding, smiling, giddy even, our picture taken by a willing passerby.

All about us is authentic Williamsburg . . . *chirrup chirrup . . . chirrup chirrup* . . . except for those house sparrows—they weren't here in the 1700s. Ironically, they seem to flourish best within the restored area. They appear to thrive on the forage in the horse dung, and the abundant nesting opportunities in the nooks and crannies of the buildings suit them well. But these birds don't belong in a restored Williamsburg, because it was much later, in 1851, when the first North American *chirrup* was heard in Brooklyn, New York, where the birds were introduced from England, and our singing continent was forever changed. In 150 years, these sparrows have taken the continent by storm, their *chirrups* and *cheeps* to accompany us from coast to coast. These much-maligned birds have grown on me, I confess, as I have come to appreciate the subtle richness in their incessant calling.

House sparrows. The ruckus from a Virginia flock. (12, 2:01)

It's just another 11 miles on the Colonial Parkway to Jamestown. I take the red-eyed vireos at full speed. *Drink-your-teeee,* eastern towhees encourage from roadside bushes; in the canopy are scarlet tanagers and Baltimore orioles, a delight to eye and ear, but now only heard, none seen. *BOB WHITE!* I melt at the sound, just off the road to the left, for what could be more Dixie than the call of the bobwhite, whose Latin name *Colinus virginianus* announces it as the "Quail of Virginia." Closer now, I hear the full call, a striking *oh BOB WHITE!*; he begins soft and low with the *oh*, pauses a quarter second, rises to the louder *BOB*, pauses half a second, then slides up the scale with the ringing *WHITE*. Yes, this is Virginia in May! I'm smiling, beyond happy. How far my head has come in the last 24 hours.

We soon exit the forest and cycle along the James River, another spreading estuary of brackish water just like the York River. It wasn't always so, I remind myself. I imagine the scene here during the Pleistocene ice ages, over the last two and a half million years, when the oceans were lower because so much of the earth's water was locked up in ice. During that time powerful rivers raged here, draining the mountains to the west and gouging out the broad valleys that are now drowned by the higher oceans.

With about thirty easy miles of biking for the day, we roll into Jamestown Beach Campground. It's empty on this Sunday night, and we choose the best site, perched on the bank above the river. Our gourmet dinner is from cans purchased at the camp store and cooked over our camp stove. Though the sunrise this morning escaped us, the sunset does not. The thin, orange fabric of the tent is soon aglow, matching the glow of the sun setting between two cypress trees out in the river. A pileated woodpecker works the cypress for a few last morsels before going to roost, and above it an osprey settles onto its nest for the night.

"I don't get it," says David, looking up from his journal. "I just feel lost when you start telling me about birdsong. What's the big deal?"

"Oh, where do I begin? . . . Thirty-five years ago, I suppose . . . in graduate school, 11 years before you were born. . . . I've come to know these birds, perhaps better than I know most of my human friends. Most important, I think, is that I hear each bird not as a species to be identified and listed, which is a rather limited endgame, but as an individual with something to say, much as I listen to any human individual with something to say, not just someone to be identified. And when a bird sings or calls, it tells what is on its mind, which varies from moment to moment, so that every listen is new and different and interesting. And with each bird heard, a lifetime of wonderful experiences and connections cascades through my mind, each new listen building on others, the entire soundscape richer with each passing day."

"Yeah," responds David, maybe understanding some. "You've got a few years' head start on me. I'll give it another try in the morning."

Knowing that morning will come soon and that we hope to be up by 5 a.m., a good hour before sunrise, we turn in, crawling into our lightweight two-man tent, carefully chosen so that without the rain fly we will have views through the mesh in all directions. Settling in, we confirm the views, the western horizon aglow, Jupiter emerging a little west of overhead, silhouettes of cypress against the horizon and the oaks overhead. Yes, it's all good, in a setting so tranquil now that it is difficult to imagine the massive conflicts with Native Americans here, difficult to imagine the first African slaves introduced here into the future United States.

How long I sleep I am not sure, but sometime in the hour before midnight, with a quarter moon hanging above the river to the west, I hear her call: *who-cooks-for-you, who-cooks-for-you-alllllll.* A barred owl! It's the vibrato in the "*alllllll*" that gives her away; it's an extended waver, typical of the female's call, as if she can't easily let go. From farther up river comes her mate's reply, *who-cooks-for-you, who-cooks-for-you-all?,* the simpler "*all*" the clue that it's a male.

I doze, but down by the cypress trees before the moon has fully set, they have at it. It all begins simply enough with one of his calls, *who-who-who-who-who-all,* she immediately responding in kind with her *who who-who-who-who-alllllll.* With the two calls back-to-back, I can now hear that his call is richer and more mellow, her call higher pitched, and the vibrato in her *alllllll* contrasts sharply with his simple *all.* They exchange these calls one more time before his wailing begins, the *who* now more of a *wha,* as if his mouth were now wide open and he were baring his teeth, though he has none, of course. He continues his uncouth wailing for ten to 15 seconds, though it seems forever, she now accompanying him with a simple *who-allllllllll,* the vibrato extended and exaggerated. What an extraordinary sound, like goblins in the night, or caterwauling cougars, or dueling demons, in reality two barred owls simply declaring "I'm yours." And just as suddenly as it began, it's over. Silence.

Barred owls

Barred owls. Mates caterwauling, the extra vibrato hers. (13, 1:12)

I smile, but out of respect for the hour and the slumbering son who slept through it all, I manage to contain myself. *Thank you, thank you,* I find myself whispering, *and a very good night.*

2.
PEACE, AND WAR

DAY 2, MAY 5: JAMESTOWN TO MECHANICSVILLE, VIA MALVERN HILL, VIRGINIA

"The time is May 5th, at 5:41 p. . . . a.m. We're at Jamestown campground," David announces haltingly into the microphone. I'm impressed. Sunrise is still half an hour away, yet here he is, Mr. Night-Life, up so early and making a valiant effort to appreciate my enthusiasm for the dawn and all who sing now. "We have what appears to be a lot of robins. . . ." With the little mini-disc recorder in his left hand, he points the shotgun microphone in his right at one of the robins who sings from a nearby fence post. And sing he does, a nonstop performance, but only briefly. "OK, that robin has flown away, so we will walk elsewhere. . . ." Good, I say to myself—he's voicing useful notes into the recorder.

A Carolina wren begins singing above us, an unmistakable *LIB-er-ty! LIB-er-ty! LIB-er-ty!*, and David points the microphone at him. This kid has promise as a field recordist, but I wonder if he knows what he's recording. We stand side by side, motionless, with David recording, and I silently try to follow along with the wren. He alternates two different songs, then introduces a third, drops back to the second, on and on, a song every four seconds as he explores five different songs to express whatever is on his mind during the next four minutes. I've never heard such a wonderful jumble of songs from a Carolina wren, as usually he'd repeat one song many times, up to 500 times even, before switching. He's clearly excited about something.

"What an incredible variety of Carolina wren songs," I announce into David's microphone after the bird has flown off. "That was a Carolina wren? I thought it was a cardinal," replies David. I try to explain the difference between the two, how the cardinal sings with slurred whistles, the songs often two-parted, but the wren's song is explosive, a vigorous chant of three-parted phrases. Hoping our field guide will help, I open it and read aloud:

Cardinal, "bright clear whistles like *what-cheer, what-cheer,* or *teew teew teew*; many variations" and then for the wren, "rollicking, full-toned chant, *liberty-liberty-liberty-whew*. Many variations."

"It's the 'many variations' that are the problem. Each cardinal has a dozen songs and each wren 30 or more, and both have local dialects as well, with songs changing from place to place, yielding a seemingly infinite variety of songs for each species. I think if you just listen long enough, you'll get a feel for the difference."

"Maybe. It's like learning a new language, with so many words to learn. Remembering just the birds' names is hard enough, but then keeping all their songs separate is really hard. If I could make the sounds myself, like I can with a new language, I think it'd be easier. 'Total immersion' is supposed to help with a language, just living the language in a foreign country for a few months—maybe that's what I'm into here for the coming weeks."

Chipping sparrows. Nearly identical dawn songs from our two campground birds. Bird 1 (14, 0:59) Bird 2 (15, 1:01)

Out toward the campground entrance, two birds duel on the gravel driveway, just a few hops apart. "Chipping sparrows," I whisper. "They sing like this before sunrise, spitting short songs at each other from the ground. Very few birders, even the experts, ever hear this. Later they'll sing much longer songs up in the trees, well separated, each on his own territory—the birds, that is, not the birders." We edge closer, with David recording first one chipping sparrow, then the other.

"Ah, listen carefully—this is special. Hear how one bird sings and the other immediately echoes an almost identical song? . . . Each bird has only one song, but there's so much variety among males that rarely do neighbors have similar songs like this. That happens when a young male learns his song from an adult and then stays beside that adult on an adjacent territory." David nods, quietly acknowledging that he hears the unique relationship between these two singers before us.

"Quick, get that chickadee." David swings the microphone, pointing it up into the small maple nearby. Song after song the chickadee whistles, just as sharp and piercing as the bird I heard from the tent yesterday morning, and this male also cycles rapidly among three different songs. The chickadee flies off, leaving David to make his announcement on the recorder: "Carolina wren . . . no, what is this? . . . oh yeah, Carolina chickadee, I knew that."

What a bonanza, this first dawn on the road! Comparing the pine warbler and chipping sparrow overhead now is irresistible, their songs nearly indistinguishable to even the most experienced human listener, but after a brief session with them, I realize that I've been pressing David pretty hard this morning; in his 24 years, he has never listened to birds with me like this, and he is giving it his best shot at learning what it's all about. Given his preferred waking time, though, I can't imagine this kind of predawn effort happening very often.

Pileated woodpecker. Such power in these seven drumrolls. (16, 2:34)

Off to the east, the thin layer of clouds on the horizon glows a flaming orange, giving way to a cooler orange above, then to a light blue that shades to a deep blue overhead. Down by the river a pileated woodpecker drums and then calls, a wild *kuk-kuk-kukkukkukkukkuk-kuk-kuk*, both the drumroll and the call so distinctively accelerating and then trailing off. Why these woodpeckers arise so late I don't know, but there's no better natural alarm for sunrise than the sounds of the pileated woodpecker.

White-throated sparrow. Nearby, a young migrant practices his song. (17, 1:09)

We retreat to the campsite and eat a late breakfast as we pack up. In the nearby bushes, a young, migrant white-throated sparrow warms us with his bungled attempts at song. He falters and stutters, his voice breaking and crackling over whistled notes that will eventually be pure and steady. As if undecided about the form his single song will eventually take, he tries starting on a high frequency and ending low, but in the next song he begins low and ends high; each whistle wavers and the entire effort lacks the fine rhythm of the adult's song, a masterful *Ohhhh Sweeeet Canada Canada Canada,* two or three long whistles followed by several whistled doublets or triplets. He's practicing as he readies his journey, and within a few weeks he'll be the virtuoso whistling his perfected song over some pristine landscape far to the north.

After an hour of packing and eating, we're on the road, heading west. David guides us with the map, which tells us that Route 5, otherwise known as the John Tyler Memorial Highway, oozes with history. There's Sherwood Forest Plantation, former home of two presidents, William Henry Harrison and John Tyler, the house itself dating back to about 1730. In the next few miles the map prepares us for more of Old Virginia, with Evelynton and Berkeley and Shirley Plantations. The owner of one plantation fired the first shot at Fort Sumter in the Civil War, and another claims to be the site of the first official Thanksgiving in America, in 1619. Another is the oldest family-owned business (a farm) in North America and the site of a Civil War battle, and an owner signed the Declaration of Independence. Today, though, with birds singing and flowers lining the way, there's little hint of all that has transpired here on this peaceful road in the last four hundred years.

We enter Charles City, county seat of Charles City County, except we find no city, only a Citgo gas station and country store, with *two* purple martin houses on 15-foot poles, at opposite ends of the parking lot. I wheel up to one, the unfazed martins clearly accustomed to plenty of activity. Tallying the nesting holes, I count three rows of six facing me, 18 more on the other side, another 36 across the parking lot, 72 homes altogether for these spectacular swallows.

This close, I can see the martin hierarchy. Perched on their condo are three adult males, at least two years old; they're handsome, a uniform bluish black that iridesces in the sunlight. Older females accompany them—each has small patches of iridescent blue on the crown and back, but they're a smudgy gray below, with a gray collar and forehead. A few yearlings fly about, drabber versions of the adults. Given our somewhat bumbling use of maps and compass and timepieces and other aids, it's humbling to realize that these birds have been to southern Brazil and back since last summer, the older birds having made at least one previous round trip, all flying on their own power with navigational skills honed over eons of time.

Purple martin

How they *cher* and *chortle* and *croak* as they swarm about the box. They seem to greet each other with a simple *cher cher cher*. A little more exuberance leads to more of a *chortle*, and the males sing a rapid jumble of gurgled notes, ending in a harsh, grating *croak*.

Charles Haupt. A love affair with purple martins, as told in the local dialect of this Virginia gentleman. (18, 5:34)

A gentleman in clerk's overalls emerges from the store and joins me beneath the martins. "Charles Haupt. H-a-u-p-t," he spells his name for me. Of "Haupt's Country Store" I quickly realize, glancing back at the large sign over the store's entrance. He continues: "We've had this store here in our family continuously operating since 1893. . . . We got a pretty good amount of birds this year, and they're happy. . . . My grandfather's family was born and raised at the courthouse itself, which is a mile above us. They had purple martins there the turn of the century [1900], and when they built this complex in 1938 here, he brought a martin house here and we've had purple martins ever since. . . . They are very, very pleasurable birds to be around. They'll make you happy."

I could listen all day to Charles, his easy manner spilling out his vast knowledge of his martins, how he worked to bring his colony back after losing them in a storm one year, how he protects them ("Well, to be honest with you, I'm constantly shooting starlings"). In Charles's voice I hear not only his love for purple martins but also the voice of a local man with roots here, cultured with the local dialect. As we head west, up and over the Appalachians and into Kentucky, I'll seek out more locals like Charles. I want to hear their voices, to hear how those voices and the voices of birds change as we sweep across the continent.

David having emerged from this sell-everything country store, we continue west, buoyed by the sights and sounds of the Virginia countryside in May. In another 15 miles, we turn right onto Willis Church Road, toward the Civil War battlegrounds outside Richmond where some of the most intense fighting occurred. It was on this very road where troops and artillery maneuvered during the infamous Seven Days Battles, and it was at Willis Church itself, just a few miles down the road, where Robert E. Lee had his headquarters. In another mile we crest the steepest hill we have yet climbed, Malvern Hill, now preserved as part of the Richmond National Battlefield. Here the last of six intense battles was fought over that seven-day period in 1862, the clash between 80,000 Union and 80,000 Confederate troops.

Pausing on the hilltop, we straddle our bicycles, gazing down upon the open field to the north. On July 1, 141 years ago, we would have seen shocks of wheat dotting the slope, awaiting harvest, with the Confederate army amassing just a quarter mile away in the trees. To our right is Western Run, to our left Crew's Run, roughly a mile and a half apart, the steep terrain of these two ravines providing natural boundaries that funnel the Confederate army up this hill into the Union infantry. Lining this ridge are 29 Union cannon; another 70 rim the hill, and still another 150 are held in reserve nearby. The cannon from this hilltop lob solid shot into the distant enemy, but as gray-clad troops waving their battle flags emerge from the trees, the Union artillery switches to canister, now each blast like a giant shotgun blowing gaping holes in the Confederate lines. By nightfall, at the battle's end, no Southerner had reached the hilltop, and thousands lay on the gentle slope below. The following daybreak, as reported by a Union officer, "Over five thousand dead and wounded men were on the ground, but enough were alive and moving to give the field a singular crawling effect."

Coasting down to the Confederate position at the base of the hill, we learn that Stonewall Jackson commanded here, and Magruder's forces were just to the west; Longstreet and A. P. Hill were nearby—all familiar names from Civil War history. We look up the hill into the Union artillery and marvel at the madness of such a frontal attack. I close my eyes, trying to imagine the horrific scene: explosions of Union cannon, suicide marches up the slope into such fire power, the slaughter of more than 8000 men on that last day, bringing the casualties to some 35,000 Union and Confederate soldiers over the entire seven days.

"I could have been here," reflects David soberly, "charging up this very hill. Now it all seems so, well, ridiculous that they killed each other like this, but according to that book *Cold Mountain*, had I been born in the South, I'd have had no choice. Be shot for desertion or avoiding the army, or charge up a hill and be killed like this."

Field sparrow

Emerging from the insanity of this scene is the requiem now offered by a field sparrow nearby. His gentle whistles accelerate, sliding down the scale, each whistle a little shorter and lower than the one before, a two-second lament for all who suffered here. Every few seconds he repeats his mournful song, over and over, never-ending. In the distance I hear two others, each with a unique cadence, each offering his own comment on the scene that his ancestors some hundred generations ago would have witnessed here.

VOICES FROM THE CIVIL WAR BATTLEFIELD AT MALVERN HILL

Field sparrows. Each (lamenting) with his own unique voice.
Bird 1 (19, 1:03)
Bird 2 (20, 1:02)
Bird 3 (21, 1:09)

Upslope two yellowthroats argue, each in turn offering his own two-second rendition of the song we know as *wich-i-ty wich-i-ty wich-i-ty*, a three-syllable phrase repeated three or four times. Though they're a good 50 yards from us, they glow a brilliant yellow in the sunlight, and I see them as clearly as the battle scene before me. Back and forth, song and countersong, each declares his territory, their vocal diplomacy seeming so much more reasonable than the mortal combat between human forces on that horrific day nearly a century and a half ago.

Other voices add to the now tranquil scene. From high up on the hillside, not far from the Union line, the plaintive whistles of an eastern meadowlark drift down to us, the five

Common yellowthroats. Two males counter each other with song. (22, 1:20)

Summer tanager. The rolling, musical carols of a summer tanager. (23, 2:22)

Blue-gray gnatcatchers. The warning wheezes of two birds overlooking the battlefield. Bird 1 (24, 1:47) Bird 2 (25, 4:03)

or so fluty whistles slurred together, gently descending. From the trees lining Carters Mill Road, which the Confederates crossed on their charge up the hill, sing an indigo bunting . . . a red-eyed vireo . . . a cardinal . . . a summer tanager now just above us, his rich, robin-like song such a contrast to the wheezes of the gnatcatchers nearby. In neutral blue-gray dress and as if hoarse from their heroic effort, the blue-gray gnatcatchers wheeze incessantly, *no no no no don't go, no no no don't go,* the same thought for minutes on end, no doubt the words of many a Confederate soldier's last thoughts here.

In the few minutes it's taken me to gather in this soundscape, David has taken some pictures, of the battleground itself and of the explanatory signs. We listen some together, reflecting on past and present, numb from the immensity of the conflict that was once here, such a contrast to the peaceful setting before us now.

In silence, we bike on . . . past the Willis Church and Lee's old headquarters, through Glendale, and in another 14 miles cross the Chickahominy River. In just two more miles we come to Cold Harbor National Cemetery, burial ground for 2000 Union soldiers—*only* 2000 of the more than 60,000 Union soldiers lost during Grant's drive on Richmond during the spring of 1864. Sixty thousand, and that's only the Union losses, as the Confederate losses are unknown.

Sobered, we bike on to Mechanicsville, just north of Richmond. Our second stop, after braking impulsively at the Waffle House, is the Pedal Power bike shop, to buy simple "platform pedals" for David's bike. His ankle hurts from a severe sprain suffered during an Ultimate Frisbee match, and he needs to adjust his pedaling motion. Instead of clipping his shoe to the pedal at the ball of his foot, he needs to move his foot forward on the pedal so that he pushes more on his heel. He'll lose the leverage and more efficient fluid motion of the foot while pedaling, but he willingly gives that up to bike on.

Talk is easy in a bike shop, and one of the customers, Jim, invites us to camp in his backyard overnight. "That campground where you were headed is at the end of the airport runway. It'll be noisy there," he explains. I had been told that biking cross-country would renew one's faith in humanity, and here is good evidence. From the gentle smile in Jim's face, too, there is

a bit of that faraway look, a clear yearning to join us, to hit the open road, to see and hear America, and it doesn't take long for it to come out: "I'd like to do that someday." Thinking back to advice offered to me in a bike shop a few years ago, I nod and smile: "You know, Jim, every year you don't do it, it's less likely you ever will."

That night, in Jim's backyard, with seventy-some miles behind us for the day, we enthuse about biking the peaceful Virginia countryside, shuddering at the mayhem of that day on Malvern Hill back in 1862. We fall asleep to the sounds of airplanes and trains and, on nearby Interstate 295, cars and trucks and blaring sirens, the sounds of urban America, even the loudest of sounds eventually swallowed by the dead of night and the roar of heavy rain pummeling our tent.

3.
LEMONADE

DAY 3, MAY 6: MECHANICSVILLE TO MINERAL, VIRGINIA

5:15 a.m.—my watch alarm alerts me 55 minutes before sunrise. It's still raining, but given Jim's kind offer of breakfast this morning, there would have been no early departure anyway, so half of us chose to sleep in. In this cozy suburban backyard walled in by dogwoods and sweet gums and tulip trees and holly and oaks and all things southern, I want to lie here in my sleeping bag as the world awakes around me. I need to feel first light sweep by, to hear the dawn wave of light and song pass, sweeping westward to where we'll listen in the days and weeks to come.

There, accompanied by the gentle breathing of David who sleeps beside me . . . a robin, the *American* robin, at 5:29. He awakes softly with a few *tut* calls, then complains sharply with loud *piik* calls, but soon he's fully energized and singing. I follow along, a string of three rich caroled phrases followed by a single, high squeaky *hisselly*, then after a brief pause four carols and another *hisselly*, then three and a *hisselly*. Such a contrast, these two types of song phrases, one rich and low and caroled, the other screeched on so high a note, the reasons for two and the functions of each a mystery. My robin routine is disrupted by a cardinal, then a Carolina wren and a wood thrush, all within a minute of each other, the song of each a magic carpet whisking me into his world.

Four songbirds now sing from the trees nearby, and I listen as each tells his own story and that of his ancestors as well. They are kin, all having inherited their ancestors' special songbird brains and intricate voice boxes that enable them to imitate and sing complex songs, but how differently these four lineages have come to express themselves. A young male wren and a young cardinal copy the songs of adult neighbors, and as a result their songs occur in dialects and males often duel with identical songs; but male robins and wood thrushes are more creative, with each male improvising many unique songs on his own, so no dialects can exist. The wren and cardinal also typically repeat a particular song many times before switching to another, but the robin and thrush always race from one song to the next, with successive songs always different.

Northern cardinal

Northern cardinals. Two males match each other's songs during the dawn chorus. (26, 2:44)

How much they have to say also differs: The cardinal knows perhaps 12 songs, the wren 30 to 40, the robin about 15 to 20 carols and many more *hissellys* (perhaps 100—no one has counted), the thrush half a dozen preludes and a dozen flourishes that he rearranges to produce 30 to 40 different songs. The cardinal uses his two voice boxes in rapid succession to create what sounds like a single whistle spanning a broad frequency range, but the thrush delivers a high- and low-frequency virtuoso duet by simultaneously engaging both voice boxes; the robin uses both techniques, usually singing in succession low carols with his left voice and high *hissellys* with his right, but under some circumstances sings the two very different sounds simultaneously; the wren most likely uses his two voices successively, rarely if ever simultaneously. Female thrushes and robins don't sing, but the female cardinal sings much like her mate, and the female wren has a buzzy chatter that she uses like a song.

And what a treasured variety of sound quality I hear: the rich carols and squeaky *hissellys* of the robin, the contrasting flute-like *ee-oo-lays* and percussive trills of the thrush, the sliding whistles of the cardinal, the ringing chants of the wren.

A fifth singer now joins them, a great crested flycatcher. Over the last two days I've heard the raucous *wheeep* notes of these flycatchers from high in the trees, but how they greet the dawn is special. There's an emphatic *wheeee-up*, followed by another, then a faint low, buzzy note, barely audible, the pattern repeating. I can hear how successive *wheeee-up* songs are different, as he plays with duration, frequency, emphasis, tonal quality, and more—like snowflakes, no two ever quite alike. How different he is from the four songbirds who still sing nearby. This flycatcher is not a "true songbird" but instead belongs to a sister lineage in which the brain is not equipped to imitate songs. The voice box is also simpler, and the songs are somehow encoded directly in the genes and rooted deep in the DNA; as a result, flycatchers everywhere sing the same relatively simple songs, and with no song learning, no local dialects exist. Different from the songbirds, yes, but no less a success, with about 400 flycatcher species throughout the New World.

Great crested flycatcher. Methodical dawn song, such a contrast to raucous daytime outbursts. (27, 2:01)

Five birds sing in the light rain, and I sing along. I mark when the wren and cardinal switch to a series of another song, I race with the thrush and robin as they run through all that they know, and I smile at the deliberate pace and simpler tune of the flycatcher. The singing gradually subsides and I drift off, smiling at the birdy thoughts of Emily Dickinson, such as "I hope you love birds, too. It is economical. It saves going to Heaven." Yes, heaven is now.

JAY JAY JAY JAY! I'm awakened by a gang of blue jays invading the yard, though David sleeps on. There are at least four of them, at 6:10, just after the sun would have peeked over the horizon had it not been for the heavy cloud cover. Each jay is on message, each giving the same version of their harsh, screaming *JAY* call. I listen for one of them to break ranks, changing to a different *JAY* call, or to a more musical *queedle-queedle,* pairs of high and low pure notes. If one switches, will the others follow, and in what pattern? They learn their calls from one another and often match each other's calls, much like the cardinal and Carolina wren match each other's songs. The jays tell yet another success story—they're songbirds without an obvious song, using instead what we think of as "calls." They depart the yard as they arrived, still screaming *JAY JAY JAY JAY!*

"Best not to get up before the sun" were his words, and he's true to them. "I had a wonderful listen over the last hour," I tell my waking son, but I spare him the details. Breakfast would be served at 6:30, Jim had told us, so David awakes just in time for us to dress and work our way out of the tent to Jim's back door.

What Jim has waiting for us is a real treat: a bacon and pancake breakfast, with lots of butter and syrup, prepared as if he knew what he'd like on his cross-country trip. The food is great, the talk genuine, but we feel their rush to meet a daily schedule, as Jim and his wife Cathy both must get to work, and daughters Michelle and Courtney must race off to school. How familiar those tensions, yet how foreign once we abandoned our rental car and entered this new life.

By nine o'clock we're on the road, dressed in our raincoats and rain pants, navigating our way out of suburbia. Our bright yellow rain covers are over the panniers, too, as it continues to rain lightly, though brighter skies give hope for a nice day.

"Titmouse?" David asks of the bird singing up to the right.

"No, pine warbler." I chuckle at what had to be a wild guess on his part, but at least he heard the bird. I like that. I think back to 35 years ago when I first truly listened to a singing bird. Just after graduating from college, I took two field ornithology classes ("Baby Birds" and "Big Birds") simultaneously at the University of Michigan field station; I studied every song clue relentlessly back then, taking endless notes. It was tough at first, I remember, but the more I listened the more I heard, and a third of a century later it's not the identification that I find challenging, but instead trying to understand what the bird is trying to communicate.

In Ashland, just out of town past Randolph Macon College, two gray catbirds sing from lilac bushes in full bloom on opposite sides of the road as we bike the gauntlet between them. *Faster faster faster* they seem to sing with their hurried squeaks and squarks and snarls, nothing repeated, just an endless barrage. Or so it would seem—but I know it's not endless, as I know what's in their heads. Were I to somehow peer into the mind of one, I'd find 200 to 400 different little squeaks, most of which he made up when he was just a few months old and then committed to memory. He's a master improviser, but there is the occasional catbird who is also a master mimic, and I'm listening for him, a Jonathan Livingston Catbird who seemingly sets standards beyond what most catbirds seem to manage.

Beyond the catbirds we bike, past the picture-perfect horse farm with the greenest of pastures and sparkling white rail fences, across the South Anna River and up the steep hill on the other side, where the right pedal falls off David's bicycle. Huh? How could that happen? While we're puzzling over our predicament, the same woman walks by with whom we had exchanged a friendly hello just a few minutes ago. Heidi soon retrieves her pickup and loads David's bike into it, driving him back to Cycle Ed's repair shop in Ashland. I bike back to Ashland on my own steam.

Gray catbird. The master improviser struts his varied songs, and *meows*. (28, 2:05)

Ed's diagnosis is simple: "The pedal has stripped the threads in the crank arm. You'll need a new one." Ed tries to make do with parts he has on hand, but finally concludes it just won't work, so he and David head to Richmond to find a crank arm, leaving me alone . . . thinking, here we are, on our third day out, fixing an ankle problem yesterday, a bicycle today . . . what will it be tomorrow?

As a mockingbird sings nearby, I recall the poster I kept in my university

Northern mockingbird. An extraordinary mimic, telling of flickers, martins, jays, yellowthroats, cardinals, and so many more. (29, 7:25)

office for so many years: With its bright bowl of lemons and frosty pitcher, the poster declared "When life gives you lemons, make lemonade." I soon realize that here is a gift, a forced lingering in Ed's yard with this mockingbird who seems hell-bent on proving something to somebody. "Sings all night, too," Ed had said. A bachelor mockingbird, I concluded, as only they sing in the dead of night, presumably to lure a female from some hapless male within earshot.

Unrolling my camp chair, I insert my air mattress into it and sit comfortably in the yard, notebook in hand, just listening, marveling at what an extraordinary being he is. Five, ten minutes pass, and I begin listing his most obvious mimicry, the sounds I can recognize:

tree swallow calls
northern flicker *klee-yer!* call
tufted titmouse songs
killdeer *killdeer*
purple martin calls
red-bellied woodpecker calls
blue jay calls
northern cardinal song
red-winged blackbird calls
belted kingfisher rattle
wood thrush *whit whit* calls
house sparrow calls
house finch calls
house wren calls
blue-gray gnatcatcher calls
common yellowthroat song
great crested flycatcher calls
eastern bluebird song
American kestrel song
eastern phoebe song
Carolina wren songs

And he sings multiple versions of so many of them, mastering what seems to be the full call repertoire of blue jays and purple martins, and so many different titmouse and cardinal and Carolina wren songs.

I hear a voice—mine, it seems—talking to him: "How many different songs do you know, Mr. Mocker? The record for your kind, you know, is close to two hundred. How do you measure up?" All I need to do is identify when he sings a few of his unique songs, count how many times he switches to another song as he sings (about 15 times per minute, I've already noted), and then do some simple math.

Northern mockingbird

First, I need those few unique sounds that I can easily identify. I look at my list, immediately throwing out the multiple versions of cardinals and wrens and titmice and jays, because

I can't distinguish among those similar versions. I focus instead on the five sounds for which the mocker has only one model. One: The eastern phoebe is a flycatcher, does not learn its songs, and as a result all phoebes sing the same *FEE-bee* song; it's unmistakable, everywhere, always. Two: The wood thrush has a large repertoire of songs, but all wood thrushes use a distinctive *whit whit* call that is not learned, so the mockingbird again has only one model for this imitation. Ditto for the *klee-yer!* call of the northern flicker, the *killdeer* call of the killdeer, and the rattle call of the belted kingfisher. That's enough. I'm set.

I sketch a rough table in my notebook, seven columns wide, start my stopwatch, and begin listening to the first minute. Each time he switches to a new song, I tally one in the rightmost column, and each time he sings one of the five unique songs, I mark the appropriate cell. End of first minute: one flicker *klee-yer!* call, 15 total songs. The second minute: one *killdeer*, another 15 songs. In the third minute he flies to the ground briefly to eat something, and sings none of my five unique songs among a total of eight songs. He continues, on and on, and after 16 minutes flies off to the neighbor's yard.

minute	Phoebe	Wood Thrush	Flicker	Killdeer	King-fisher	TOTAL SONGS
1			1			15
2				1		15
3						8
4		1			1	15
5			1			12
6						15
7			1			13
8		1				15
9						16
10	1		1	1		17
11		1			1	16
12						7
13		1				10
14						14
15						16
16			1			16
TOTALS	1	4	5	2	2	220

For 16 minutes I followed his every tune, and now what do I see? Judging by how many times he sang each sound, he seems to like his flicker *klee-yer!* the most, the phoebe's song the least. The "average" number of times each of these sounds occurred is 1 + 4 + 5 + 2 + 2 = 14, divided by 5, or 2.8. Assuming that these five sounds are representative of the other songs in this mockingbird's repertoire, I can divide 220 by 2.8, thereby estimating that this male sang about 80 different songs during these 16 minutes. Not a record, but still impressive, especially with all the mimicry.

I'm almost sad to see David and Ed return with the parts they're convinced will work. Sure enough, amiable Ed got it right. After a quick tune-up of both our bikes, he sends us on our way, I all the richer for the delay, though a bit poorer in the wallet. "Best thing about the trip is pop's Visa," said David to Ed as they finished.

It's rural Virginia again, past the horse farm and across the South Anna River, all the way up the hill this time and beyond Heidi's house—we're on the way again. David tells me of his adventure with Heidi: "She's an extremely religious woman, or so it appears to my heathen self. She talked about *Focus on the Family*—a radio talk show, and gave me a pamphlet about Jesus. When we arrived at Cycle Ed's she said a prayer with 'Father . . . great how you made the universe . . . Jesus is great . . .' with a few 'let David's Achilles tendon get better' and 'help fix his crank arm' thrown in. She helped us out because it is what Jesus would do."

"Fascinating. This is the South, the Bible Belt," I respond. "Did I ever tell you that I almost spent a summer during college selling Bibles in Kentucky? Some friends of mine were very successful at it, making big bucks. I backed out at the last minute. My prospective boss said I'd be a failure the rest of my life for quitting. 'Watch me' I said."

The miles roll by quickly, past Patrick Henry's home at Scotchtown, then Coatesville, then Bumpass, where David needs a picture of himself flexing his muscles in front of the post office, the large block letters (some needing repair) somehow reflecting the energy he feels:

U ITED STATES POST FFICE
BUMPASS, VA, 23024

In just a few more miles we cross an arm of Lake Anna. We've been pushing hard to reach camp by nightfall, about 45 miles from Ashland to Mineral, but we pause here on the causeway in the early evening, marveling at several princess trees adorned with countless clusters of large purple flowers.

“David, listen . . . on that peninsula across the lake, a brown thrasher.” In the calm of the early evening, the morning rains long gone, the songs of a brown thrasher drift across the lake to our ears. They come mostly in pairs, each phrase sung twice, at a pace inviting words to the tune: *hello hello, I'm here I'm here, listen-up listen-up, twos twos, thrasher thrasher*. We've heard lots of thrashers beside the road, *smack*ing loudly as they scurried into the bushes when we've biked by, but here is the first lusty singer, though at some distance.

Brown thrasher

Brown thrasher. Singing mostly in twos, his mimicry revealing a lifetime of listening. (30, 5:02)

David hears “a brown thrasher” only because I tell him so, but he experiences little in the process; as I hear this bird, cherished memories of thrashers over the past 35 years tumble through my mind. There are memories of a nest I watched from a blind as a budding ornithologist way back in 1968 . . . of baby thrashers we raised, to study how they learned to sing . . . of studying one particular brown thrasher with my mother-in-law and learning that he could sing over 2000 different songs . . . of countless songsters seen hunched in familiar pose, early spring, in the top of a leafless tree . . . of other thrashers out West, especially the sage thrasher. . . . Each encounter stirs old memories that would otherwise have lain dormant and adds to them so that each experience becomes richer than the one before, much like “compound interest,” wrote Louis Halle in *Spring in Washington*. As with all the other birds that so enthrall me on our journey, David and I may hear the same bird but experience it so differently.

What a trio, these birds often called “mimic thrushes”—late morning it was the gray catbirds in Ashland, early afternoon the mockingbird at Cycle Ed's, and now the thrasher. The catbird sings in ones, each squeak uttered only once; the thrasher in twos; the mockingbird in fives or more. In round numbers, a catbird sings about 300 different “songs,” a thrasher 2000, a mockingbird only 100. The mocker is perhaps constrained by his need to mimic, while the catbird and the thrasher improvise freely; the thrasher generates the largest known song repertoire among birds by simply making them up, by improvising on the basics of thrasher themes that are held in their genes.

Leaving Lake Anna and the singing thrasher, we have a short hour of biking to Mineral. The wood thrushes have begun their evening serenade, the bold preludes and ethereal flourishes of these tree-borne spirits saturating the air.

Mineral arrives all too soon, though after a stiff climb, and just after sunset, the lights and cheers from some contest on the athletic field guide us into town. We are to camp behind the volunteer fire station, either in or under the pavilion, the first time we will take advantage of the generous, standing offers from small communities that line our bike route from coast to coast. "Sure, just inside the door to the left, and the kitchen's upstairs—help yourself" is the response to my request about a bathroom.

We work quickly, pitching the tent out in the open, cooking a meal, washing up, and it is ten o'clock by the time we crawl into our sleeping bags.

Mineral's population of 424 is now quiet, too, and I soon realize what kind of night it will be. I should have known, should have anticipated. It had rained all last night, grounding them, delaying them on their relentless flight north, but today was one made in heaven, and the weather continues. Millions upon millions of northbound migrants will be aloft again tonight, and here is our payoff for last night's rains and for that climb up into Mineral just two hours ago.

It began simply, with a split-second *tsck,* dismissed as nothing until it was followed by a *pft,* a *tck,* a *phew,* and so many others, as now overhead flows an enormous river of migrants. "David, up above, flying overhead, listen . . . there. Hear that? . . . Another . . . again . . . some ever so faint, some loud." Yes, he heard, but he is asleep before I have a chance to tell him that I've never heard anything like this. Never. Oh, I've heard the honks of geese flying in the night, and while sleeping at a lake edge in the Adirondacks one September, I heard what must have been a flight of thrushes descending just before dawn, their rich, mellow calls raining down on me. Just last year I studied a CD entitled *Flight Calls of Migratory Birds,* hoping to learn to distinguish some of the more common thrushes from each other.

But nothing could have prepared me for what was now happening overhead. Lying still on my back, facing upward, my feet to the south, head to the north, I listen, and listen, and listen. The variety of *pip*s and *tseep*s and *squeak*s dropping onto our tent is endless. Focus, I remind myself—try to pick one out. I work on the thrushes, knowing some must be up there, as the first week of May is their prime migration time. Is that a veery, or a gray-cheeked? A wood thrush, faint to the south? I think so—he must be approaching, flying overhead by now, *yes,* there he is again, a clear *bzeee* overhead, probably from the same bird. Hurrah!

My veery detector fires, not once, but four times within a second—first overhead, and then within a split second off to the left, to the right, and beyond my feet to the south. Four veeries, four individuals acknowledge each other's presence in the darkness above. I hear four, but what do they hear? There must be others beyond them, perhaps a vast ocean of veeries moving overhead tonight, individuals well spaced but in regular contact with each other, as if assured by their collective navigational skills that they are all on course.

All species combined and weather permitting, the *pips* and *tseeps* and *squeaks* overhead no doubt extend from the Tropics to the northern tip of whatever is inhabitable during early May. And here I lie, quietly, on a hilltop in Virginia that reaches modestly to the heavens, dozing, awaking, drifting off again, the wave of migrants seemingly never ending. Oh to fly with them, beneath the stars with a gentle tailwind, over the whip-poor-will who calls now from the south, over the incessantly singing mockingbird nearby, over the cuckoo calling in the distance to the north, over the sounds of a landscape that sings also by night, drawn to some special place to the north that will be home for the season. Perhaps in my dreams. . . .

4.
BLUE RIDGE DAWN

DAY 5, MAY 8: WHIPPOORWILL LANE TO AFTON, VIRGINIA

The miles add up as we take small daily bites out of the continent. On our first day, Sunday, we biked an easy 33 miles, Monday 73, Tuesday 69. Yesterday it was 70 miles from Mineral to where we pitched out tent at nightfall in the woods beside the Mechums River, along Whippoorwill Lane (though we heard none). It was a tough day, we agreed, partly because of our late start after doing laundry in the morning, which then forced us to bike in the midday heat, and neither of us handles high temperatures well. Dawdling here and there with long breaks along the way didn't help either, nor did the hills.

David's journal summed it up: "Yesterday was a shitty day . . . hot and humid . . . hordes of tourists at Jefferson's Monticello . . . dad and I were both exhausted . . . but there was the University of Virginia in Charlottesville. It had been too long since I'd seen college-age women!"

Today, our fifth day out, we would do things differently, we had decided, by starting earlier and resting more often, but only briefly. Then it rained again overnight and into the morning, giving us yet another late start. So we agree that today will be a "rest day"—just 20 or so miles up to Afton, where we'll spend the night at the Cookie Lady's house. And then, yes, then the next day, I'm eager to bike the Blue Ridge Parkway at dawn. Mid-morning, rather frustrated by our slow progress, we roll out of our campsite.

FEE-bee, FEE-b-bre-be, FEE-bee, FEE-b-bre-be . . . . "David, hear that? Hear the two songs? He alternates his raspy *FEE-bee* with the stuttered *FEE-b-bre-be.* He says his name, phoebe, the eastern phoebe, a flycatcher, and every bird has those two songs. The songs aren't learned so there are no dialects. When he's less excited, he sings a string of raspy *FEE-bee* songs before giving just one stuttered *FEE-b-bre-be.*"

Eastern phoebe

"I knew that," replies David matter-of-factly. Ah, yes, many years ago, when he was about six years old, he had helped raise the baby phoebes back home when I was studying how they acquire their songs. Does he really remember, or think he should, or is he just jerking me around?

Chip-burr. Close by, just above us, the unmistakable call of the scarlet tanager. "David, it's the scarlet tanager! You want to see him?" I did bring a small pair of binoculars, after all, though we have yet to use them. "Who cares?" says David.

Eastern phoebe. With enthusiasm, alternating his two songs, *FEE-bee . . . FEE-b-bre-be.* (31, 1:35)

A squirrel, tail jerking wildly over its body, squeezes between us and an oncoming truck. "I wonder if that was a dare from one of his friends," pipes up David.

Zee-zee-zee-zee-zee-zee-zee-zee-zee, a raspy song climbs the scale, as if he's spelling his name, p-r-a-i-r-i-e, an unmistakable prairie warbler. "David, you gotta see him. He's a brilliant yellow with. . . ."

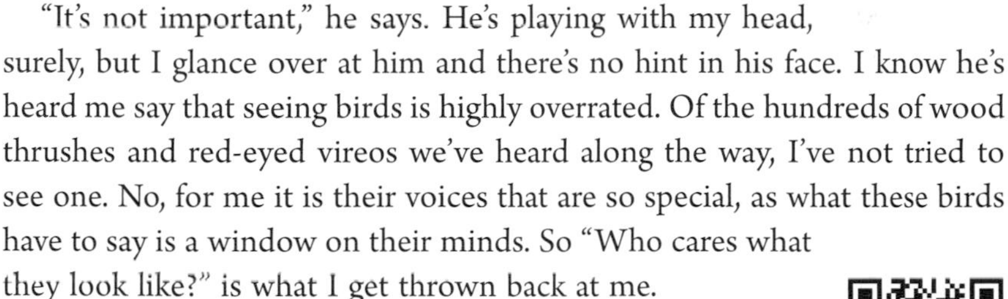

"It's not important," he says. He's playing with my head, surely, but I glance over at him and there's no hint in his face. I know he's heard me say that seeing birds is highly overrated. Of the hundreds of wood thrushes and red-eyed vireos we've heard along the way, I've not tried to see one. No, for me it is their voices that are so special, as what these birds have to say is a window on their minds. So "Who cares what they look like?" is what I get thrown back at me.

Seee-seee-see-see-see-see-se-se. Nice, it's another prairie warbler just ahead, but it's a thin, wiry whistle that now rises the scale. I pull up beside David, explaining: "There's another prairie warbler. They have two different kinds of songs, one used mostly with females—that's the buzzy *zee* song we heard just back there, and this whistled *see* song is the one they use when fighting other males. You can tell the mood of the bird just by listening. We'll hear lots of other warblers doing this kind of thing along the way. Brace yourself!"

Prairie warblers. Raspy, rising mate-attraction song, often sung throughout the day. (32, 0:48) More pure-toned, aggressive song, most predictably heard during dawn chorus. (33, 0:40)

"Mockingbird?" he asks about the bird singing up ahead on the left. "No, brown thrasher. Count them. Remember one for catbird, two for thrasher, five or more repeated phrases for mockingbird." The thrasher continues his couplets—when consciously counted out, so clearly different from the fives and sixes or more of the mockingbird. Perhaps five seconds pass, and I hear a "Mmmmmm," implying that David has grasped the counting game.

"Phoebe! I got it," he proudly proclaims. "Well, close, but that's the mockingbird singing a phoebe song. He's good at it, too. Listen to the tempo—you'll never hear a phoebe singing a string of his *FEE-bee* songs that fast." "You're making this shit up," he decides.

The trees beside us are a rich green and fully leafed out now in late spring, but the Blue Ridge rises in the distance, and up there, perhaps a thousand feet above, the buds are just bursting, the young leaves casting a reddish and golden wash on a mountain in early spring dress. Soon we cross under busy I-64, ride briefly on the Rockfish Gap Turnpike, then turn onto Old Turnpike Road, shifting down, and down, and down, until we are in our lowest gears and gliding smoothly up the mountain. The air feels richer and purer, refined perhaps by the simple thought of leaving the busy valley and rising skyward.

Before we realize it, at 11 a.m. we are in Afton, our destination for the day. But how can we stop so early? We could easily make the 15 miles to Reeds Gap on the parkway by nightfall, camping there at Lake Sherando. It's tempting, but no, we stop. We will rest up and start over tomorrow with a new pattern of biking early in the morning.

Past the post office, on the right side of the road, is the sign: "June Curry. The Cookie Lady." A big red arrow points to the right, and we're soon talking with June herself as if we were longtime friends. She stands next to the garden hose and a sign "Water for Bikers," and wears a friendly smile and her own T-shirt, proclaiming "Afton, Virginia. Home of the Cookie Lady."

"So, why are you called the Cookie Lady?" I ask. "Well, it's because it started when the TransAm bicycle trail started across the country. . . ." We listen to this legend of the TransAm trail tell her own story, how in 1976 she and her father Harold rescued a group of tired Bikecentennial cyclists who thought they could refuel at a small grocery in Afton and continue on the parkway, but the store had closed. June and her father brought out their peanut butter and jelly and whatever else they had, but it was the cookies that were best remembered. Word of their hospitality spread, and since then she's welcomed over 15,000 cyclists to her home from all fifty states and dozens of foreign countries. The "bike house" is

wall-papered with postcards and other memorabilia that cyclists left or sent, and June eagerly provides a tour of each room, telling no doubt the same stories to us that she has to countless others before: "This is the 29th year . . . I'm 83 years old now. . . . I consider the bicyclists my family . . . they remember me. I even have Mother's Day cards, they come back to see me, they write to me." I affirm, "We bicyclists love you, June."

When asked, she tells us about her birds, too. "Because of all the trains and the cars and cats and dogs, you know, so many of the birds we used to have we don't have anymore. . . . Mostly what we get now are blue jays and a woodpecker every once in a while. And these other little birds whatever they are that's whistling up there." One of the featured singers overhead at the moment is a brown thrasher, and I explain how it has thousands of different songs that it can sing. "Oh my goodness!"

(The late) **June Curry, "The Cookie Lady."** Telling all, about her history, the bike house, her love of cyclists, and their love of her. (34, 3:48)

"I read a master's thesis about the TransAmerica Trail last year," David tells me later. "What I remember is that most cyclists are white and rich and male, mostly in their twenties, but all ages do it, and a few hundred complete the trip each year. The Adventure Cycling group in Missoula, Montana, keeps statistics on it."

We eat lunch, do our laundry in the sink, and nap in the fresh air out on the porch. Late afternoon, we're invited into June's house to watch a video that other cyclists have sent and to hear more of her stories. And we eat some more, heading back to our sleeping bags on the porch by eight o'clock. It is our goal to sleep outside every night of this trip, to smell the night air, to hear America by night as well as by day. The alarm is set for 3:55 a.m., about two and a quarter hours before sunrise, giving plenty of time to reach the Blue Ridge Parkway in the dark before the birds begin to sing.

DAY 6, MAY 9: AFTON TO LEXINGTON, VIRGINIA

Throughout the night a mockingbird sings in the distance, a cuckoo now and then. The bright lights of passing cars seem to shine in our faces forever. . . .

Beep beep beep beep beep. My watch alarm sounds off beside my head. Eager for the morning, I look over at a comatose form next to me. "David?" I whisper gently. From somewhere inside the bag comes a muffled "five more minutes." OK. I stuff my sleeping bag into its sack and head to the kitchen to cook the oatmeal.

Minutes later I'm back. "David?" . . . Slowly the bag moves, a head emerges, "OK." Hooray. It's going to happen. This is extreme; just yesterday he said "My favorite time of day to bike is late afternoon, in the early evening light." And I can't get enough of early morning light. Yet he's game, despite his natural body clock, knowing how much biking through the dawn means to me.

By 4:50 we're packed and ready to push off from the Cookie Lady's front door, which is precisely the minute it begins to rain. Rain! *Nooooooo,* I wail silently inside, fearing the morning will be lost. Hoping for a brief passing shower, we dig out our rain gear, covering our bodies and the bike's panniers, and in five minutes we're on the road.

It's a little over a mile to the parkway, but steeply uphill. The narrow beam of my headlight reveals only David up ahead and a short stretch of road between us, the rest of the world having slipped away. Silence, except for the rain pelting my waterproof clothing and the swish of my nylon rain pants as I pedal. How eerie, as if we bike in a small lighted bubble moving through the void. We soon edge onto the normally busy Rockfish Gap Turnpike, quiet now except for a few spooky 18-wheeled behemoths that creep by on the steep upgrade. We take the off-ramp at Rockfish Gap up to the parkway, turn south, and by 5:15 we are all to ourselves. Magically, the rain stops! *Yes! Perfect!*

We cycle south, slipping silently through the dark. How do dogwood flowers manage to glow in the dark? Their brilliant white petals now line the road and show us the way. No, wait, they're not petals or part of the flowers at all, I now recall from one of my favorite college classes—Virginia's showy state flower isn't a flower but instead a formation of four specialized leaves called "bracts," and they're brilliantly white presumably to show pollinating insects the way to the prize, the inconspicuous little flower encircled by the four white bracts. Equal, but different, I find myself saying, as this dogwood tree is a survivor from the beginning of time, just like me and the birds I'm listening for.

The fragrances just after the shower are intense. Is it cherry that predominates? Dogwood has no odor—there's no Yankee Candle fragrance called "Dogwood," but there is "Wild Cherry." So concentrated, so powerful, the air seems at least one part flower fragrance to go with the one part oxygen and four parts nitrogen.

Still, silence . . . until a single, tentative *chewink* of an eastern towhee pierces the quiet. With my right hand, I fumble beneath the sleeves over my left wrist to punch the light button on my watch, leaning over the left handlebar to catch a glimpse: 5:25 a.m., 45 minutes before sunrise. The call is contagious, as *chewinks* now erupt from seemingly every roadside bush. In less than a minute I hear a feeble song, then a louder one, and soon the bushes sing, *drink-your-teeeeeee,* two strongly enunciated introductory notes followed by a rapid series of repeated notes. The birds ease into it, at first repeating one song several times, much as they do later in the morning, but then the warm-up is over and no holds are barred. Up ahead on the left I hear a *drink-your-teeeeeeeee,* a high-low-high sequence; next, from the same bird, is an entirely different pattern, a low-high-low, *drink-your-tea-tea-tea-tea-tea-tea,* each repeated phrase in the "*teeeeee*" portion of the song now slow and deliberate. Every male we pass now alternates songs or song fragments, interjecting *chewink* call notes as well, the towhee passion for dawn expressed in such frenzied and unpredictable patterns.

"All towhees," I call up to David, sparing him the details, "eastern towhees, once called rufous-sided towhees for their good looks. They're big, handsome sparrows. Named 'towhee' for their call note. Lots more birds to follow. This is great!"

Eastern towhee

Whit whit whit. Wood thrushes awake, sounding agitated as usual, and soon they're singing, too. With towhees in the understory and wood thrushes in the canopy, the parkway is waking! Riding the crest of the Blue Ridge on this remote highway in the sky, up to half a mile above the hubbub of the valleys far to the east and west, we and the birds now own this narrow pathway threaded through the heavens.

Eastern towhee. *Chewink,* an awaking towhee, then songs, with woodcock "skydancing" overhead. (35, 3:14)

Peeent! That simple nasal call in the clearing to the right tells of a woodcock strutting in his ground display. *Peeent! . . . Peeent! . . . Peeent!* Calling every few seconds, this plump shorebird with an outsized bill turns this way and that, broadcasting to females in all directions, and then he's silent. Five seconds later I hear him overhead, with each wing beat the wind whistling through the special stiffened feathers in his wings. I lose him in the distance, but about 15 seconds later he's back, whistling with his wings *and* simultaneously chuckling with his voice high in the sky. "Hear the woodcock overhead, David? He's an odd shorebird, like a sandpiper who's moved to the forest and sings high in the air." And then the woodcock is silent. He is plummeting back to earth now, and . . . sure enough, behind us he resumes his *Peeent!* calls on the ground, no doubt having returned to the same special display arena from which he had departed.

Wood thrushes. Sharp *whit whit* notes of waking (and going-to-roost) birds. (36, 4:50)

WHIP-poor-WILL WHIP-poor-WILL WHIP-poor-WILL WHIP-poor-WILL. He's up ahead on the left bank, singing from the grassy shoulder. "Whip-poor-will," calls David with confidence in his voice.

He (the bird, not David) continues, *WHIP-poor-WILL WHIP-poor-WILL WHIP-poor-WILL WHIP.* . . . We startle him, stopping him mid-song. Biking past where he was, we next hear him behind us, undeterred, over on the other side of the road. Two more battle up ahead from opposite sides of the road, and more beyond them. We cycle on through, creating a small ripple in their routine as they pause for these strange two-wheeled beasts passing by.

Whip-poor-will

Eastern whip-poor-will. Jarring the darkness all along the parkway, delivering a song every second. (37, 1:33)

Chip-burr. Chip-burr. High in the trees to the right. "Listen," I call to David. "Hear him? *Chip-burr,* a scarlet tanager. That's the way he wakes up." It's as if he also awakens a little irritated, like the towhee with his *chewink,* the wood thrush with his *whit,* the robin with a *tut* and *piik,* all notes that are used in daytime situations when these birds seem in some way disturbed. I imagine him up in the canopy, his throat swelling first with the *chip* and then the *burr,* the tail seeming to pump each syllable from the opened bill. Soon he'll be singing his halting, disjointed dawn song, punctuated every few song phrases by this call, nothing like the snappy four-to-five-phrase song that he'll sing every 10–15 seconds when the dawn frenzy has passed.

"Too much!" calls out David. "I'm hearing them all, but can't keep up, can't remember which bird is which. Each new bird bumps the last one out of memory."

Scarlet tanager

Scarlet tanager. In dawn song, alternating *chip-burr* calls with a few slowly delivered song phrases. (38, 2:21)

My encouraging response is garbled in the wind as we descend, picking up speed, the wind whistling in my ears and masking the songs as well. I brake, to enjoy the songs rather than the speed, to 20 miles per hour, David flying up ahead.

A chipping sparrow sputters from the ground on the right, then another, and another, left and right now again as we race through a roadside parkland thick with these dawn ground-singers. What variety I hear in their voices; though each male has only one song, he chooses it from 30 or so different patterns that eastern males can sing, and many of these songs are here.

As we slow to climb the next hill, I pull up beside David: "Hey, Mr. Physics Man, if I bike 20 miles an hour past a bird singing at 3000 hertz, what is the apparent frequency as I approach and leave the bird?" I should get some return on the megabucks it took to put him through Stanford, shouldn't I? And physics *was* his academic major, though Ultimate Frisbee was the real reason he went to Stanford, and he *was* captain of the national champion SMUT just last year (that's Stanford Men's Ultimate Team). "You know all about the Doppler effect, I assume?" I get a nod from him, and in my mind I play with some numbers, trying to appreciate what I heard as I biked past the chipping sparrows.

Let's see. . . . To determine the distance to lightning, I count, knowing that the sound of thunder travels about a mile in five seconds. A mile in five seconds is 12 miles per minute for the speed of sound, or about 720 miles in 60 minutes, an hour. I'm traveling 20 miles per hour, or about 1/36th or 3% of the speed of sound. The numbers flow all too quickly in my head, too fast to trust, but if I'm right, as I approach the birds I'd hear 3000 + 3% and as I depart I'd hear 3000 − 3%, 3090 Hz dropping to 2910 Hz. Yes, I heard that with the chipping sparrows, the Doppler effect, roughly the difference between two adjacent piano keys far to the right on the keyboard. David bikes on, apparently oblivious to the grand principle of physics on display here.

A field sparrow? Yes, excellent! "David, over in the clearing to the left. Remember the 'bouncing ball' field sparrow songs on Malvern Hill a couple of days ago? This one is singing his dawn song." It's as if his energy has been pent up all night long and now at dawn he explodes with a four-parted *te te te te te te te tew tew tew tew tew tititititititi swee swee swee swee seee*, a bit like two daytime songs sung back-to-back, but backwards, each half decelerating rather than accelerating. "And listen! He chips continuously between songs. I didn't know field sparrows did that."

Inevitably I hear a tale of two sparrows, field and chipping, as they are close relatives, both in the genus *Spizella*. Each has a relatively simple daytime song, delivered with seemingly little haste or enthusiasm, but I have just felt their energy at dawn. A male field sparrow devotes a special song to dawn; it's twice the length of his daytime song, and is delivered in great haste with frantic chips between songs. While the chipping sparrow is equally energetic at dawn, he uses the same song at dawn as during the day, but at dawn sputters it nearly continuously in brief, machine-gun-like bursts. Both sparrows get the job done, to be sure, each in his own way.

Field sparrow. Frenetic chips between complex dawn songs. (39, 2:00)

It's uphill again. I shift down, smoothly, so happy that we bought bikes with the lowest gears possible for these steep eastern hills. I glance down at the three chain rings up front, remembering that they have 22, 32, and

44 teeth; the nine-speed cassette on the rear axle ranges from 11 to 34 teeth, for a total of 27 gear combinations. On this grade of perhaps 10%, I'm in the lowest gear, using the smallest chain ring in front and the largest sprocket in the back, my legs smoothly powering 270 pounds up the hill.

"Yes, I actually use all those gears" was my response to repeated questions from non-cycling friends as I trained early this spring. Being the inveterate scientist interested in numbers, I had figured it all out. When I shift from one sprocket to the next in the rear, I make a small change of 13% to 18% in gear ratios. Bigger changes are made up front with the three chain rings, from 37% to 45%. One never begins in the lowest possible gear and works up to the 27th—the gears aren't meant to be used that way. Instead, but without thinking much about it, I choose one of the three chain rings in front, and then have a feel for the range provided by the nine gears in the back. From a standing stop to 15 miles per hour on flat terrain, I use only three or four gear combinations. And when climbing, one learns quickly that the trick is to shift before it is too late, so that the legs keep moving at about the same pace, roughly 80 to 85 revolutions of the pedals per minute.

I laugh outright at the thought of my friend Sandy. He was so perplexed by all these gears that he fell for an ad on late-night TV, buying a bike that shifted automatically so that he wouldn't have to fuss with it all. The bike was a disaster, the shifting mechanism useless. He soon unloaded the bike on some unsuspecting stranger who also had dreams of biking without knowing how to shift.

Slowly now, down to three miles per hour, I listen: to the slight creak of my saddle, twice with each revolution of the pedals; to my breathing, heavy but steady, not gasping; to the smooth roll of tires on pavement, a little crunch now and then over some gravel; to the well-oiled chain that runs smoothly through the cogs on the drive train; to a son, as he powers up the hill in front of me; to the total silence of internal combustion engines, as we have the parkway all to ourselves; to the symphony of birds all around, as we're now moving slowly enough to hear *everything*; and to my own thoughts, knowing it doesn't get any better than this.

We glide to a stop at a lump beside the road. "A woodcock." Dismounting, I lean the bike against my legs and bend over, gently caressing this creature lying lifeless on the road like so many other roadkills we've passed. I think of woodcocks as being downright plump, but the cold, stiff body beneath the feathers now feels so slight, and the bill is even more outlandishly long than I had realized. I spread a wing, broad and rounded, running my fingers over the flight feathers, pointing out to David the three outermost primaries that are so much shorter and stiffer than the others. "They're the wing's voice box, whistling in the wind with each wing flap during his sky-dancing courtship flights. We heard that a few minutes back."

Lifting me from the woodcock is an oh-so-familiar song that explodes from the small bush just across the road. He sings again, the song building gradually over half a second, then bursting into a brilliantly emphatic ending. It's the daytime song of a chestnut-sided warbler, its familiar mnemonic *very very very very pleased-to-MEETCHA*. Again he sings, and another bird just up ahead, and another, all *MEETCHA* songs. How odd that I have heard none of their warbling, pre-sunrise songs along the parkway this morning. "Time and place," I think, "what are these birds telling me?" Yes, that's it—it's early May, high in the Appalachians, and for the first week after these warblers arrive from Central America they arise late and begin singing near sunrise, launching directly into their daytime mate-attraction songs, the *MEETCHA* songs that we hear all around now. Only after the females begin arriving and pairs settle into breeding mode do the males sing the aggressive warbling songs during the dawn chorus.

How satisfying to know these birds . . . to have spent countless mornings over several years learning their habits in the Berkshire Mountains of western Massachusetts . . . to have sought them out across their range, from Michigan to Minnesota and down the Appalachians to Virginia . . . to have learned that males everywhere sing one of four or five different learned versions of the *MEETCHA* song with no local dialects, as if all conforming to some fairly uniform message that females want to hear; that, in contrast, adjacent territorial males share aggressive dawn songs that are unique to their small neighborhood, in a pattern of countless small dialects across their breeding range . . . each new encounter and revelation with these warblers builds on others, each cumulatively richer than the one before . . . and now this, on the parkway, just a few songs triggering a cascade of memories that otherwise would have lain dormant.

The chestnut-sided warbler sparks one more memory, the very reason I was so eager to bike the Appalachians in

American woodcock

Chestnut-sided warbler

Chestnut-sided warblers. Daytime, mate-attraction *MEETCHA* songs. Bird 1 (40, 0:41) Bird 2 (41, 0:37) Bird 3 (42, 0:27) Frenetic dawn singing expected later in the season. (43, 3:18)

Virginia: This ancient mountain range is a wrinkle in space, with dozens of northern birds extending their breeding ranges far to the south at these higher elevations. I yearned to hear the familiar birds that I knew from back home in New England, to reacquaint myself with old friends in new places, these experiences for me more satisfying than finding entirely new birds in new places.

We're descending again, speeding, flying by another cyclist standing beside the road. It's Jimmy! "*Hey Jimmy*!" We exchange hearty though quick hellos, and cycle on. We first met Jimmy three days ago when he was walking his bike along the road, so we stopped to talk, learning that he was biking from Yorktown to someplace in California. A day later we passed him walking near Charlottesville, and yesterday he was sitting outside a small store in White Hall. And this morning *he's walking his bike yet again,* this time near a parkway visitor center where he probably spent the night.

We like Jimmy's style, so different from ours. He has no bright, high-tech biking helmet, but instead a soft, dark, wide-brimmed hat for sun protection. Rather than our low-sweeping handlebars for increased aerodynamic efficiency, his bars are high and straight, for comfort. Our preferred pedals have clips, our special biking shoes cleats, so that when cleat hooks to clip, our feet are locked to the pedals and we can cycle most efficiently, pushing down on one pedal while pulling up on the other; Jimmy's are simple platform pedals, his shoes regular walking shoes, the unsophisticated system on which any child learns to ride. We have wicking shirts and spandex shorts with padding; Jimmy bikes in a sweatshirt and blue jeans. We thought we were traveling fairly heavy, but Jimmy's four panniers are even bigger, and his front and rear racks are piled high with gear. We ride in silence, listening to the world around us; Jimmy's radio, perched high on the gear stacked in front of him, plays his favorite music. And whenever we see him, we are riding and he is walking, yet he is always ahead of us. With Jimmy safely behind us, we both burst out laughing, asking, "How does he do this, always walking but always ahead of us?"

Up ahead, from high in the tree on the right, that's almost certainly the new southern warbler I've been listening for, the yellow-throated warbler. I had memorized the field guide's description, "clear descending whistles, last one rising, *teeew teeew teeew teew tuwe*." The whistles are pure, so striking, more beautiful than I had anticipated, but this warbler is puzzling. Those who know it best declare that each male has only one song, but the close relatives of this warbler have a two-song system, with the specialized dawn songs for aggression and day songs for mating that we've heard from prairie

and chestnut-sided warblers. Perhaps the yellow-throated warbler once had two different songs, like its close relatives, but for some reason lost one of them along its evolutionary way. For the next few weeks, I'll be listening carefully to these warblers with the yellow throat, as we'll be with them all the way through the forests of Missouri.

Yellow-throated warbler. Leisurely daytime singing; clear, gently descending whistles, with a terminal rebound. (44, 2:00)

The sun now peeks over the ridge on the left, and we pause at an overlook to marvel at the sweeping views of ridges and valleys fading into the distance to the west, all of it sweetened by two songsters overhead. There, in the direct rays of the morning sun, among budding reddish oak leaves high in a tree at the edge of the clearing, two beacons sing, one indigo, the other yellow. The indigo bunting delivers four paired phrases, all sweetly and deliberately enunciated, trailing off in pitch and intensity, *fire fire where where there there run run phew*; immediately following are the nicely slurred whistles of the hooded warbler, ending emphatically, *tawee tawee tawee TEE-o*. After a somewhat lengthy pause, the bunting again takes his turn, the warbler following immediately. Over two minutes, through 13 pairs of songs, the warbler always defers to the bunting and awaits his turn to sing, except once, when the bunting pauses just long enough for the warbler to squeeze in an extra song.

Indigo bunting and hooded warbler. Taking turns, but the bunting rules. (45, 2:00)

In a flurry of feathers, the singing pattern is disrupted when a darkish bird intrudes and the hooded warbler hastily retreats. With songs so high and thin and sharp, it's an American redstart, and he alternates two different songs, as if excited about something, *tsita tsita tsita tsit . . . tset tset tset tset tset . . . tsita tsita tsita tsit . . . tset tset tset tset tset. . . .* Were this his routine daytime singing, he'd sing another of his songs over and over, and if this were his true dawn mode, he'd also be chipping strongly between these two songs. For now his mood is somewhere in between those two extremes.

We bike on, climbing, then descending, only to climb again. A parula. A northern parula—to be more specific, it's the daytime song of the eastern dialect of the northern parula. "David," I call to him, "listen to this." He pulls over beside me, semi-expectant, looking up into the tree overhead. "Look here in the field guide. Here's a good-looking bird with an intriguing story. He's bluish gray above, yellow throat, white wing bars, a rusty chest, pretty stunning. Listen to the description of the song: 'thin, sputtering buzz that rises and then snaps down at the end, *zz-zz-zzz-zzzeeee-wup*.' We've been hearing that since Yorktown, and we'll hear it all day long *until* some special place in Kentucky. There, for some mysterious reason, the song will change to the western dialect, and the description of the song in the field guide will be only half right. It's your job to hear when the song changes." He seems interested, listens some more. "There's another one across the road," he points out. I think he's got it. We'll see if it sticks.

Northern parula. Daytime songs of an "eastern" bird. (46, 1:55)

Croak . . . croak . . . croak. . . . Just three, over about half a minute's time, three unmistakable deep, throaty *croak*s. His black eminence soars high over the ridge to the left. "Thank you, Mr. Raven," I call out over my shoulder, with a tip of the helmet to a bird of rare intellect and an astonishing vocabulary. Increasingly I find myself talking to these birds, complimenting them on a song well done or a story well told, or here, thanking the raven near the end of our Blue Ridge experience for just making his presence known and triggering a flood of good raven memories. In the East, he's another one of those boreal birds who sneaks down the Appalachians, making a southern home here in Virginia, but we'll soon lose the ravens, not hearing them again until the Rockies.

I want to linger on this parkway, to soar down the sweeping descents, to climb more of its hills and hear more of its stories, but by late morning we have traveled the 27 miles from Rockfish Gap near Afton to route 56. We have crept along, stopping, listening, treasuring, but here we are to turn west, dropping off the parkway into the Shenandoah Valley. It seems ages ago that the towhees first *chewink*ed among the glowing dogwoods early this morning, and thousands of birds have chimed in since then. I take some comfort in knowing that we'll ascend to these heights one more time in the

East, on Mount Rogers, the highest point in Virginia, where more northern birds will have their say.

I smile as I think back to the sign out front of the United Methodist Church in Kent's Store, just two days ago. "BYRD CHAPEL. EXPECT PERSONAL TRANSFORMATION." Amen. Nothing beats a morning like this in the sanctuary of the birds.

5.
A VIRGINIA HIGH

DAY 9, MAY 12: WYTHEVILLE TO MOUNT ROGERS, VIRGINIA

Wind 20 miles per hour, gusting to 36, from the west. That was the forecast, and at 5:25 a.m., 50 minutes before sunrise, with headlights beaming and red taillights blinking, we are already feeling every blast of this headwind as we depart Wytheville, our destination for the day Mount Rogers.

Three days ago we descended the Blue Ridge and dropped into the Shenandoah Valley, arriving at Lexington. We Yankees puzzled over our guide at the visitor center there, as she was still clearly fighting the Civil War. In the emergency room at Stonewall Jackson Hospital, the growing redness on my left shin was diagnosed as cellulitis and I was told to "get some rest." Though we had hoped to sleep outside every night, we accepted defeat and took a motel room. It was good to "slay the motel dragon," we reasoned, so that we would then be able to sleep wherever it would be convenient over the coming weeks.

Ignoring the "get some rest" advice at the ER, we got an early start the following day, biking 92 miles to Christiansburg. We cheered at the city limits where a billboard proclaimed the city a Bird Sanctuary, and were amused at another billboard prescribing a code for horn honking: one beep for "thanks"; two for "please"; and three for, well, "go to hell."

And just yesterday, our eighth day on the road, we biked the 55 miles to Wytheville, noting that we feel stronger with each passing day.

Now, biking in the dark and into the teeth of the wind, I holler up to David, only partly in jest, "Why are we doing this?"

"I'm doing it because you are," he shouts back. "It was your idea . . . well, actually, I'm doing it for the bike."

I've heard that before, that he'd bike with me cross-country if I bought him a nice bike that would satisfy his needs for years to come. Exactly what those needs would prove to be, I don't think even he knows, but he's talked about biking from his home in the Bay Area to the tip of South America, all the way to Tierra del Fuego.

As for me, the trip is far more than about birds. I read a succinct answer to *Why?* in the prologue of *My Old Man and the Sea,* in which the father tells why he wanted to sail around Cape Horn. He needed to raise his "personal and professional threshold for bullshit," he said, to "become fresh again." Professionally, I am a professor mired in a biology department that worships cells, genes, molecules, and most of all money. "Show us enough grant money and you don't have to waste your time teaching," I am told. The director of my graduate program crowed, "I'm willing to be an asshole to get what I want." I despaired.

So I left town, up and disappeared, telling only a few close friends where I was headed. E-mail and office phone messages would simply accumulate unanswered, and in three months I'd know better who and why I was. If anyone *really* needs me, I reasoned, they'll call my home and the message will be relayed to me. In the meantime, I'll freshen up a bit.

And, quite simply, I yearned to celebrate life, in a simple, pure, and unadulterated fashion, shedding nearly all of life's responsibilities for a few weeks, to strip life to its essentials, just sleeping, eating, and moving from place to place as we crossed the continent. And all of this in "bird chapel," where personal transformation would be inevitable. And with my son. This is heaven, I remind myself: a good bicycle, the open road, a fine companion, singing birds, and an occasional headwind to provide some perspective.

Eventually, as if by magic, after 28 miles of considerable effort and just 17 miles to our destination, at a large wooden sign on the right that announces the Mount Rogers Recreation Area, the winds die, and we enter what sounds to be an enchanted forest, with *warblers singing everywhere.* Although we still climb, I feel as if I'm flying uphill, the warblers pulling me along, the wind no longer pushing me down. *Zee-zee-zee-zee-zoo-zree*—there's the black-throated green warbler, several *zee* notes on one pitch, a lower *zoo,* then the higher, longer, more wheezy *zree.* And his next of kin, a black-throated blue warbler: *zur zur zur zreee*—Peterson's *I am la-zy,* the drawled, wheezy song rising on the last note. In a small clearing beside the road, a chestnut-sided warbler proclaims *very very very very pleased-to-MEETCHA.* A high, thin, seesawing *weesee-weesee-weesee-weesee-weesee* tells of a black-and-white warbler, and there's an American redstart, *tsee tsee tsee tsway,* sharp and piercing, but not quite as high as the black-and-white, this particular redstart song dropping at the end. A Blackburnian—*teetsa teetsa teetsa teetsa tseeeeee,* more wiry and higher than a black-and-white, that last note so high-pitched I barely catch

Blackburnian warbler

Black-throated blue warbler. A lazy, rising *zur zur zur zreee* at so unhurried a pace. (47, 1:46)

Ovenbird. One male's version of the *te-CHER, te-CHER,* deemphasizing the first syllable. (48, 2:00)

it. A blue-winged warbler, sounding as if inhaling on the first note, exhaling on the second, *beee-bzzzzzzzzz*. And more, these singing males probably all just back from their tropical homes, all singing their daytime, female-attraction song here, none of them their dawn, aggressive song. "David, back in Yorktown I warned you about these warblers! Just listen to them all." Yes, this *is* Heaven.

I hear ovenbirds, too, *te-CHER te-CHER te-CHER te-CHER,* loud and emphatic, a two-syllable crescendo that builds to a remarkable volume for so small a bird. "No dawn song for that ovenbird," I call to David, "just this *te-CHER* song, though they have a mysterious, middle-of-the-night song they give from high in the air, sometimes well above the trees."

At noon, we cross the Appalachian Trail, where our TransAm bicycle journey and the journey of those who hike this 2200-mile trail intersect at the base of Mount Rogers. There, two hikers have paused to eat. Sassy and Kodiak are their trail names, they tell us, carefully chosen *before* they began hiking so that others on the trail wouldn't anoint them with something less desirable, like Snotrocket, or Mudbutt, or Phobia, or Swampass. David and I realize that it might be a good idea to choose a name before someone does it for us, but creativity and exhaustion don't mix. We'll take our chances. Besides, we're into our second week and have seen only one other biker, Jimmy, who's probably walking somewhere up ahead right now. Well down the road in Wyoming is my best guess as to when we'll start seeing eastbound cyclists.

By 1:00 p.m., with only 45 miles on our odometers for the day, we're setting up our tent in Grindstone Campground. Lunch is peanut butter and jelly sandwiches, bagels and cheese, a banana, and fig bars. Dessert is a Milky Way bar, then another. We rest in the afternoon and hike some of the trails, and over dinner reflect on our journey so far, listing some of the highlights as they come to mind (I agree to leave birds off the list, at least for now):

1. "The campground after our first day of cycling, on the James River"
2. "Heidi and Cycle Ed"—they rescued us on our third day, getting David's bike on the road again

3. "Jim"—his invitation to camp in his backyard, his fine breakfast, and the camaraderie among fellow cyclists, on our second day out
4. "Rural Virginia and the battlefields"
5. "The Blue Ridge"—solitude and unbounded vistas (and birds!) lifting the soul
6. "Billy Rays Smith"—the fellow walking his dog along the road, and how, after we stopped to talk to him, he was soon a longtime friend. He was not the "Smith" I know from New England, but rather "Sme-uth," stretched to two syllables. He lived about halfway between Lexington and Natural "Bre-udge," he explained, one of the natural wonders of the world. He had stories of chickadees, titmice, finches, turkeys, deer, bears, house wrens, crows, and other wildlife, how he listened to them in the early morning from bed with the windows open, and summed it all up by saying, "It's a shay-um more people don't listen to birds." Yes, Billy Rays, it is a shame.

Billy Rays Smith. In his fine local voice: "I love . . . all wildlife . . . watching 'em and listening to 'em . . ." (49, 4:10)

7. "My watch alarm"—it's set to go off every 15 minutes, reminding me to eat and drink, to keep the body going.
8. "The intense fragrances of flowers at dawn." What are they? Honeysuckle? Cherry? Wild rose? Perhaps all those and a host of others intermingle in one vast orgy, all these scents the business of plants unabashedly trading food for sex as they lure pollinators to their flowers. And the birds are doing it, too. Every morning we're immersed in a pre-sunrise chorus of males seeking sex and a furtive audience of attentive females who encourage all the singing to help them identify the worthy males. Yes, sex is in the air, rampant, promiscuous sex, and I almost expect the signs outside churches to declare a springtime state of emergency, an abstention from enjoying the odors and sounds of such wanton lust.

Song sparrow. Singing for half an hour, telling of nine different songs he knows. (50, 32:57)

9. "You," I say to David, "and how you heard the song sparrow's game at that midday rest stop in Catawba." I had asked him to listen to a song sparrow, whose current song began with two or three raspy notes and ended with a series of rather nice whistles. I had heard a weak grunt from David, who lay in the grass with eyes closed, so perhaps he was listening. Again and

again the sparrow sang this particular song, but when he switched to another song, David startled, saying "Whoaaa. What was that?" We then listened to three more series of distinctly different songs before the sparrow fell silent. "You're catching on to this listening stuff. I'm impressed."

10. "*Welcome to Wytheville*"—from the lawyer, his two dogs tugging at their leashes. We had just arrived, having collapsed on the grass outside the community center, when he walked up. He was an outsider, transplanted there by marriage. It was the tone of his voice that so intrigued us. In how many ways could one utter the same three words, and with such different meanings? At one extreme I imagine the mayor of the town bursting with pride and boasting of this terrific place, his *WELCOME TO WYTHEVILLE* beaming with a fervor that would make one want to stay forever. Our friend, however, had a different message. The words were the same, but *welcome to wytheville* was now more of a whimper, a longing for elsewhere, almost anywhere but here, reminding me of the village in Costa Rica whose literal translation is "worse than nothing."

We are attuned to these subtleties of human speech, and we'd miss so much in our daily conversations if we could identify only the words and not their intonation. It must be the same with the songbirds we hear, their voices modulating the messages in ways far too subtle for us to know. Maybe we can identify their vocabulary, such as the eight different songs of a song sparrow or the four preludes of a wood thrush, but surely there's great meaning beyond the mere "words."

David shows me what he's written in his journal:

> Everything is very serious in my father's world. Everything has a list with check boxes to be filled out. Every item on this trip has been prepared, deliberated over, put on lists, weighed, thought about more, wrapped in Velcro. I did no planning. . . . My father is always going over checklists in his head, thinking of the precautions that will need to be taken. This is fine, except that it takes the place of fun thoughts.

I reflect a bit on that. He's right. I would feel more free without so many umbrellas and parachutes in my life, more like he feels. But attention to detail has its rewards, too, I remind myself. We are very different, to be sure, and

I know I have much to learn from him. Two to three months on the road will be a good start.

A day that begins early ends early. Well before sundown I'm in the tent, salivating at the thought of all those northern birds in the two thousand feet of elevation between us at 3700 feet and the peak of Mount Rogers at 5729 feet. I have read about these special birds in *A Birder's Guide to Virginia*. How many of them I will hear at dawn is, at the moment, the Great Unknown.

DAY 10, MAY 13: MOUNT ROGERS TO ROSEDALE, VIRGINIA

4:30 a.m. I'm up, nearly two hours before sunrise. I will miss nothing at this high point east of the Rockies, but it's *cooolllllddddd,* in the low 30s, maybe even the 20s. I bundle up with all the clothing I have, including rain gear, then begin stalking the campground, walking the loops, listening for anything that stirs.

4:36. Veeries are already singing. *Already.* Or maybe they have been singing sporadically throughout the night by the almost-full moon. A veery. How satisfying, as the first bird of the morning is one of those northern birds found in Virginia only at this higher elevation. I walk, and walk, hearing *veer . . . veer,* the call of this thrush, but for the next hour I hear nothing else.

5:39. I check my watch, as the veeries are singing more now. I've been told that they have a few songs and repeat one of them several times before switching to another. How unthrushlike, I think, nothing like the way the wood thrush sings, nothing like the way the hermit thrushes are probably singing high up on Mount Rogers this very minute—with these two thrushes successive songs are *especially different,* as if a heightened contrast in back-to-back songs is important. I listen as the veery sings, concentrating on each song, memorizing its pattern, comparing it to the next. He uses three different songs, I eventually surmise, singing them in the pattern A B C A B C. Fascinating. These veeries perhaps aren't as different from their cousins as I had thought, though their song repertoire is rather small.

Veery. From the rooftop of Virginia, a series of calls and songs. (51, 2:03)

5:45. *Chip chip chip chip chip . . .* from the rhododendron thicket by the campsite he calls, whatever he is. Something is awaking, chipping two to three times each second for half a minute, breathing fire, and then just a brief snatch of a jumbled song. *Yes,* it's a *Canada warbler*. How much more northern a songbird could I ask for?

Canada warbler

Canada warbler. Hyper-energized singing well before sunrise. (52, 2:00)

American robin. Expressive robin on a tear, with low carols and high *hissellys* and call notes. (53, 15:12)

Least flycatchers. Neighboring males bombarding each other with rapid-fire *cheBEKs*. Male 1 (54, 0:56) Male 2 (55, 1:05)

Chip chip chip and another jumbled song, this one a little longer, *chip chip chip chip* and another song, not a second of silence anywhere in the stream of calls and songs exploding from his bill. The songs are exhilarating and rapid-fire, each a sputtering explosive series of brief notes that jump high and low and all over the frequency spectrum, no two notes sounding alike. The pace picks up, the songs lengthening, and now between every few songs he *churrs*, too, as if chipping were not enough. It's frenzy piled on frenzy from this 10-gram spitfire, as if all night he's been winding tighter and tighter into a sonic coil and he's now finally letting loose, unloading his passion and fury in a torrent to beat all torrents. *Chip chip chip chip churrrr chip chip chupety-swee-ditch-ery-sue-swi-chippery-chee-the-chee chip chip chip chip*, and then another song and more *chips* and *churrs* and another song and I'm out of breath just listening.

For five minutes I stand mesmerized, but then force myself to walk on, soon stopping beneath a robin. Not "just-a-robin," but *a robin*! How extraordinarily expressive: There are his usual low carols and high screechy *hisselly* notes, but he throws in some of his call notes as well, the *tut tut* and sharper *piik piik*, and even what sounds like his high, thin hawk alarm. Then it's a long string of only caroled notes, then muted carols as if he's highly conflicted, now a long string of only *hisselly* notes. Delivering all of these notes and songs at breakneck speed, more than two each second, he seems highly flustered. I love robins! I only wish he could translate for me what it's all about.

Over to my right I'm bombarded with *cheBEK cheBEK cheBEK* songs. Least flycatchers, *Empidonax minimus*! The simplest of songs, they're unvarying, the same as least flycatchers everywhere, perhaps the very flycatcher species the noted ornithologist Frank Chapman was listening to when he commented that, when music was being passed out, the flycatchers had backseats. Walking closer, I hear another, and another, four of them here, all clustered within earshot, each *cheBEKing* from his own small territory. In my mind I see their range map, with one long arm extending down the Appalachian chain from Pennsylvania to the northern tip of Georgia. Nice.

Just how fast do these birds sing? From the near male I count 70 songs in one minute, more than one each second, and almost perfectly sandwiched between each of his songs are those of his immediate neighbor. Back and forth they sing, first one, then the other, a perfectly timed seesawing exchange of *cheBEKs* every split second. Only occasionally do they get out of sync, one of the birds pausing a little too long or not long enough before the next song, but no matter, they quickly adjust their tempo so that they are back in time.

Empidonax. What a fine Latin name, with its literal translation of *Mosquito Prince*, but I don't know if that's because birds in this genus are tiny and dart around like mosquitoes or because they're supreme mosquito eaters. To me, "*Empidonax*" means "wonderful group of tiny look-alike flycatchers best known by their distinctive songs." I hope to hear ten *Empidonax* species across the country, and here is the first good listen.

Just down the road beside the stream is a second *Empidonax*, an Acadian flycatcher! He calls sharply, a continuous series of sharp *seet* notes followed by an explosive *PEET-sah*: *seet seet seet seet seet seet seet PEET-sah seet seet seet seet seet seet seet PEET-sah*. Gradually, his energy will fade, and by sunrise he'll offer just a few *PEET-sah* songs each minute. He's not a northern bird, but instead a southeastern forest flycatcher who rises high up on the flanks of Mount Rogers. Two Mosquito Princes now sing side by side.

By late May a third *Empidonax*, the alder flycatcher, will be singing atop Mount Rogers, almost 2000 feet above me. They'll migrate to the clearing near the summit and sing their simple *fee-BEE-o* songs among the low shrubs, thriving in the cool, thin air that makes them feel at home. They're northern birds all right, breeding primarily across Canada from Newfoundland to the Yukon and into Alaska, stretching down to this far southern outpost atop Mount Rogers. Very nice—three of the ten Mosquito Princes sing here on Mount Rogers.

Alder flycatcher. *Fee-BEE-o* songs and calls from atop Mount Rogers. (56, 2:28)

How I would love to be at the mountain's summit for a boreal dawn chorus, among the spruce and fir and oak and rhododendron, to hear so many northern birds who make their southernmost home here. Winter wrens sing their seven-second wonder of tiny whistled notes; *ka-wa-miti-go-shi-que-na-go-mooch* the Ojibway called them, for their never-ending song. Hermit thrushes leap from one fluty, ethereal song to the next, successive songs so different, the overall performance one of the finest in all of North America. Black-capped chickadees whistle their *hey-sweetie* on this island in the sky, shifting the pitch of their simple song through a range of frequencies. I can

Winter wren. Male with two different songs. (57, 4:28)

Black-capped chickadee. From Virginia highlands, the typical *hey-sweetie.* (58, 1:15)

Brown creeper. Rich, sliding whistles accelerating to a delicate jumble in this compact song. (59, 2:20)

Red-breasted nuthatch. Tinhorn *yenks*, with surprising variety. (60, 1:18)

almost hear the rich, mellow warbles of purple finches; the high almost-out-of-reach notes of golden-crowned kinglets; the *weety weety wee-tee-o* of magnolia warblers; the sweet, delicate, gliding whistles of brown creepers; the throaty, heaven-bound spirals of Swainson's thrushes; the rapid, nasal *ank ank ank* tiny tin horns of red-breasted nuthatches; maybe even a *hip, THREE CHEERS!* of an olive-sided flycatcher.

Black-throated green warbler. Dawn song with calls (61, 2:16) and a transition to day. song (62, 2:20)

Dark-eyed junco. Eventually revealing four different songs. (63, 2:22)

My imagined visit to the rooftop of Virginia is broken by a familiar song over the neighbor's tent, *zreee-zreee-zoo-zoo-zree,* a slow, lazy, wheezy song, the two *zoo* notes lower than the others, the dawn song of the black-throated green warbler. And yes, he chips some between songs, retaining a bit of the fire from earlier in the dawn chorus. I'm thinking that it's getting close to sunrise, and . . . sure enough, as I listen, he switches to his daytime song, *zee-zee-zee-zee-zoo-zree,* the same song we heard late yesterday morning when we entered the forest here. How intriguing that black-throated greens everywhere agree on the pattern of these two songs, and just by listening one can know the mood of the singer.

I pause to listen to the junco . . . from a rhododendron patch he delivers a series of one song type . . . then abruptly switches to another . . . and another . . . gradually revealing all the songs he knows.

Just beyond our neighbor's tent sings a red-eyed . . . no, wait, that's a blue-headed vireo, still another northern bird, its measured phrases a little higher pitched than those of a red-eye, and purer, sweeter. He offers a song about every two seconds, each different from the one before. I've read that each male has about 15 different phrases, not all that unlike the red-eyed vireo, so I listen, wondering if I can hear what he's up to. After a minute or so, I recognize a unique phrase that he sings every few songs, but then it disappears, occurring again a minute later; I monitor a second unique song, feeling the same pattern. That's intriguing. He sings not like his relative the red-eyed vireo, who tends to sing all of his songs before repeating them, but more like a robin, who favors one set of songs for a while, then another set, delivering his songs in "packages."

Blue-headed vireos. Two males in confrontation, showcasing the essence of blue-headed vireo. (64, 3:15)

The tent moves . . . the zipper opens . . . a head emerges, attached to a body that has had far more sleep than mine, and I know the difference will be evident come this evening. As I share with David some of the morning's adventures, we move through the morning routine, preparing oatmeal, packing up the gear, checking the map, preparing to depart.

The Canada warbler behind our campsite continues to unwind. Now, a good hour after sunrise, I hear a single *chip* and then immediately a one-second burst of jumbled song, and then silence, nothing, nothing at all for ten seconds! "David, over there in the rhododendrons , you wouldn't believe how energetically that bird was singing earlier this morning." He has certainly transformed since then, from the hyper-energetic to the downright placid, though the jumble of each brief song still conveys the image of a frenetic mind. "Too bad we can't see him—he's pretty, a Canada warbler," though in my mind's eye I see his brilliant yellow below with a necklace of fine black stripes; he's grayish above with sharply defined spectacles; his tail is cocked and his wings flick, as active in body as in song, so active that he often darts out and catches insects on the wing, a habit that has led to some calling him the "Canadian flycatcher."

Our cycling neighbor comes over to chat, our two loaded bikes a magnet for those who recognize us as "thru-bikers," for those who either have been or yearn to be on such a journey themselves. Tom is from Cleveland, and the talk soon turns to gear, as it often does in these conversations. "Great tires," says David in response to Tom's query; "haven't had a flat yet on this trip." I express my great outrage at this very thought, though mostly in jest, but deep down I know there can be consequences for such a boast.

Fifteen minutes later, as we're sailing down the hill to Damascus, just after the chestnut-sided warbler has sung his *very very very very pleased-to-MEETCHA* from a brushy clearing on the left, and just as the Louisiana waterthrush offers his three clear, down-slurred *tew tew tew* whistles and jumbled song beside Laurel Creek on the right, yes, just 15 minutes after boasting of no tire problems, I hear a holler behind me and catch a glimpse in my rearview mirror of David hopping off his bike. Our first flat tire. I turn

back, amusing myself at his misfortune, berating him for his brazen carelessness when talking to Tom. "Your job," he calmly informs me because of his superior experience in group touring, "is to patch the bad tube while I install a new one; that way we'll be on the road more quickly and have the repaired tube ready to use when we have our next flat." I fall into line.

We soon breeze into Damascus and pause for a second breakfast at the first restaurant we encounter. As I work over my two eggs, four pancakes, bacon, and orange juice, the owner of the restaurant stops by to say hello, introducing himself as Bert. Reflecting on the state of the world, he says, "People are in just too much of a hurry. They just get in their car and go." I ask what he knows about birds. "There's lots of colors to them, they fly, and they eat worms." That's all he knows, he claims, but with a little prodding he tells how he loves listening to the dawn chorus when he arrives at the restaurant early in the morning, how his treasured bobwhite that he knew from boyhood days here could no longer be heard in the area, and how much he enjoys the honking of geese as they pass in the fall.

As with Charles Haupt, June the Cookie Lady, and Billy Rays Smith over the past week, I love listening to this local voice, to the slight twist in each syllable that would be enough to tell an experienced linguist exactly where on this planet we are standing. It would be much the same for many of the songbirds we've heard on our journey, too. They also learn the local voices, and a person who has done his homework could listen to any cardinal sing, for example, and pinpoint where on this planet that singing bird perches.

Bert Snodgrass. A local man speaking the local dialect. (65, 1:06)

We've done only 18 miles on the day and I'm eager to move on, but we linger. At the post office we mail another box of unused items home. A stop at the bike store. Then the internet café . . . I sit outside in the shade, waiting, listening to robins, to Carolina wrens, to house sparrows, to chimney swifts twittering overhead. Patience, I tell myself, as David has friends, and keeping in touch with them is important. As for me, my friends know that I'm "gone," and that's sufficient.

Chimney swifts. Have chimney, have chittering swifts overhead. (66, 0:59)

"Thanks for your patience," he says as he finally emerges. "You're welcome," I smile. I flash back to one Friday night of his eighth-grade year when he was sitting in his bedroom. His mother was out, so it was up to me to deal with this miserable, lonely boy who had no friends. No one had asked him to do anything that night, so there he sat, looking at the telephone, bored to tears and wallowing in self-pity. "Call someone," I urged. No, yes, no, yes. "Here's how it works," I tried to explain. "Call someone, any of your friends, and even though it might be too late to do anything tonight, think how good this other person will feel that you've said you'd like to do something with him. Then who's that person going to call the next time he is looking for a friend to do something with? You! It'll work. Try it. . . . Let me dial for you. . . ." How agonizing and gut-wrenching to watch this little guy in such misery. Where was his mother? She'd be so much better at this than I. Finally, perhaps half an hour later that feels more like an eternity, he made a call. The beaming child who hung up the phone was out the door before I could stop laughing.

Yes, "you're welcome," I say again as we check the map and make sure of our route out of town. We see a few challenges ahead. It's only 36 miles to Rosedale, where we hope to spend the night at the Elk Garden Methodist Church Hostel, but just outside of Damascus, after crossing the South Fork of the Holston River, the fun begins. We'll climb the ridge out of that river valley and then drop down to the Middle Fork, climb again and then drop down to the North Fork, climb again and drop to the West Fork of Wolf Creek, and then there's the steep four-mile climb out of Hayter's Gap to the ridgetop before we drop down into Rosedale.

We'll be riding across wrinkles in the landscape that formed long ago when ancient land masses (now Africa and North America) collided, the forces ramming the billion-year-old granites of the current Blue Ridge to the northwest, the forces felt even farther west here as the land was folded into what is now called the "Valley and Ridge Province." Here is our last full day in Virginia, as sometime around midday tomorrow we will have crossed the roughly 600 miles from the Atlantic Ocean to the Kentucky border.

David is still fresh from his long night's sleep, and, buoyed by contact with friends over the internet, he charges ahead, while I labor to keep up.

6.
APPALACHIA

DAY 11, MAY 14: ROSEDALE, VIRGINIA, TO VIRGIE, KENTUCKY

"WELCOME TO KENTUCKY. WHERE EDUCATION PAYS." We stop to celebrate the sign beside the road and take our pictures. We've crossed the first of ten states, almost 600 miles on our odometers in ten and a half days, about one seventh of our estimated 4200 mile journey. At this rate, it seems reasonable that we'll arrive at the Pacific by early July, before the birds have stopped singing for the summer. *Hooray!*

And here, on the border between Virginia and Kentucky, in the heart of Appalachia, here like nowhere else, I expect to experience the paradox of our language. In each syllable that we speak, we tell where we're from and who we are, and although we speak to communicate, it is the very act of speaking that identifies our ethnicity and constructs our social boundaries. Pronounce the third syllable in "Ap-pa-la-chia" with a long "a," for example, and we're instantly branded outsiders, but pronounce it with a short "a" and we have a chance. Such is the power of a single vowel, the power to determine who belongs and who doesn't.

Terry Owens. Telling of his love for the dawn chorus ("It's just beautiful!"), wobblers, pileated woodpeckers, and more. (67, 7:55)

I seek out the park naturalist at Breaks Interstate Park, which straddles the state boundaries here, hoping he's a local with the local dialect. "My name is Terry Owens, and I'm from Haysi, Virginia." *Yes,* I say to myself, that's the little town that we climbed through just eight miles east of here, with a population of 186 according to our Adventure Cycling map.

What a treat to hear the local voice tell of his love for the local birds, the singing of the Swainson's warblers ("wob-blers"), the "wild cackling" of the pileated woodpeckers ("I get so excited . . . one of the most beautiful birds in the park . . . my imagination kicks in . . . a big jungle . . . it's such a unique call"), the hooting of the barred owls, the clucking of the cuckoo, the increasingly rare whip-poor-will calling its name. At dawn "you'll get to hear that morning chorus. Awesome. It's just beautiful. With our gorge here, you also get a continuous breeze in the morning . . . and all of a sudden you get that chorus of singing . . . it's just an awesome place to be."

It is the remoteness of these Appalachian Mountains that has helped preserve the unique local dialects of people here. Take a bit of English or Scottish or Irish or German or Polish or Italian or other ancestry, mix locally with lineages of Native American or African descent; then limit the movement of people from one dialect region to another (such as from one hated clan to the next, as between the feuding Hatfields and McCoys); to hasten the process, give special power to a few community leaders who disproportionately influence the local tongue: the perfect recipe for hundreds of unique micro-dialects among the hills and hollows of Appalachia.

As with birds, there's one more important ingredient that helps maintain these dialects. Based on their songs, birds actively discriminate against outsiders. A song sparrow who sings the same songs as his neighbors achieves greater success than do sparrows with non-local songs. A female cowbird would do well to choose a male with the local songs because he tells of his history, that he is a survivor who has had roots in the local community for at least two years and is therefore not a vagrant. Among songbirds, the urge to conform is strong, the resulting dialects telling of acceptance and discrimination in much the same way, no doubt, that these human dialects do. Local traditions rule, and one's success, however it is measured, depends on belonging and conforming. Here in the Appalachians, where outsiders see isolation, the locals see independence and an active commitment to a special way of life. Independence means freedom from the rule of foreigners, those who talk differently. If the songbirds could reflect on their own ways, they'd probably say much the same.

I'm sad to leave the park, but we need to move on. First, though, we refuel with bagels, cheese, water, peanut butter and jelly sandwiches, water, bananas, pretzels, and more water, all while perched at an overlook with a stunning view of the 1600-foot-deep "Grand Canyon of the East" (the "break" of Breaks Park) that the Russell Fork River has carved into Pine Mountain. Exposed in the rock cliffs of the canyon are raw strata of sandstone and coal, vast layers before us having been ripped this way and that along fault lines to create the landscape here.

Down the road, I ask David, "Hear that? . . . Listen." . . . "Same one we've been hearing, that warbler, the parula," he responds. Yes, he's got it, the daytime song of the eastern northern parula. It shouldn't be long until we encounter the western song. Will he hear it?

On the left two cars have been abandoned beside the road, vines now totally enveloping them. The cars are recent models and look intact, as if they could be driven off if only someone were to clear the vines away; we half expect to round the next bend and find a castle also smothered in vines, the occupants of some long ago waiting to be discovered and liberated.

It's all downhill from Breaks Park, and we race back through geologic time, cutting through successive layers of sandstone and shale, and of course *coal*, the layers all named by geologists. At the top is the Clintwood layer, then Glamorgan, Hagy, Splashdam, and now, at river level, the Banner layer lies perhaps 50 feet below us. We're riding through the Carboniferous, 300 million years ago when the continents were fused into a single Pangaea that straddled the equator, through vast tropical swamps of giant amphibians and tree ferns and horsetails and primitive conifers that will all be converted to a "fossil fuel" that will be prized far in the future by a creature called "human."

The roads are a labyrinth through narrow valleys, each valley with its stream and requisite roadside garbage, piled even higher, it seems, near the "Do not Litter" signs. Clusters of small houses are packed into available spaces, and we wave to people on porches until we tire, because so many porches have people sitting on them. We follow Poor Bottom Road—"rather aptly named," remarks David.

On a narrow winding road about 15 miles from Breaks Park, we first hear them up ahead on the right, a riot of *cock-a-doodle-doo*s. Approaching, we see them, too, rooster after rooster tethered to his own little coop; each is just out of reach of his neighbor, a whole hillside of them, each with bright red comb and resplendent tail feathers, the plumage of each unique. Their crowing extends up the hill along lanes through the woods, as far as my ears can hear.

"What can they be but fighting cocks? But isn't cockfighting illegal?" I have read about cockfights, how these roosters are fitted with razor-sharp, three-inch-long blades over their spurs, then drugged with some aggressive potion so that they are especially vicious. Two of them are held close together, letting them peck at each other to get the juices flowing, and then released into a small arena where they have at it, breaking legs and wings, slashing and puncturing each other with those outrageous blades. Any faltering fighter is revived and thrown back into the fray—there's no quitting at this game. Big money rides

Fighting cocks. Crowing vociferously beside the road. (68, 1:48)

on the outcome, and the boisterous, cheering crowd is fueled with alcohol. Inquiring of a passerby on the road, we learn, "Oh, sure, it's against the law to fight these cocks, but not to raise them."

In Virgie, with little daylight remaining, we inquire about a place to tent for the night. At the service station our request stirs up conversation among several men. "There's a park at the end of the road . . . cops won't bother you . . . you take this here road. . . ." A fellow then emerges from the group, introduces himself as Jeff, and says, "Come on out to my place." He quickly gives us directions, a few miles down the road but directly on our route out of town. Once outside, he levels with us: "You don't want to go to that park. Too much goes on there after dark." I'm reminded of what we were told by a camper back at Breaks Park, only partly joking, I sensed: "Everybody in eastern Kentucky's into something illegal—you just don't know what it is."

DAY 12, MAY 15: VIRGIE TO PIPPA PASSES, KENTUCKY

Morning comes quickly, but we sleep in, recovering from a long day yesterday; half dozing, I lie in my sleeping bag and sing with a robin, a cardinal, a song sparrow. A yellow warbler across the road chips vigorously and leaps among his dozen or so different dawn songs, all of the *sweet sweet sweet I'm so sweet* variety.

Yellow warbler. Before sunrise, chip calls and an invigorating variety of songs. (69, 2:38)

By eight o'clock we're packed and saying thank yous and good-byes to Jeff and his family. Out onto the main road, we immediately begin climbing, and steeply, as Jeff had warned. Slow is good this early, I tell myself, and I shift to my lowest gear, warming the body as I spin up the hill.

As we cycle uphill, the world passes in slow motion. Millipedes two to three inches long cross the road, their "thousand legs" rippling the body, descendants of meter-long millipedes that roamed here back in the Carboniferous. And centipedes. And caterpillars. All *crossing* the road, none traveling with us down the road, any genes encouraging such folly no doubt having been squashed from the gene pool long ago.

A pileated woodpecker, dead in the middle of the road. David's up ahead, but I stop. She's huge, big as a crow, lying on her belly, the left wing outstretched and showing her large white wing patch. I study her fire-red crest, the yellow-brown forehead, the contrasting white and black stripes below, the hefty chiseled bill, the stiffened tail feathers, her right eye staring up at me.

All along these lush mountain roads the pileated woodpeckers have been our escorts. Up they fly from beside the road where they've been dining on the ants in some fallen log, launching into the air, flying alongside us and then down the center of the road up ahead, often with a disjointed, laughing *kuk-kuk-kukkukkukkukkuk-kuk-kuk*. Almost daily we hear their slow, distinctive drum rolls announcing the sunrise. Mates then often call to each other, too, reestablishing contact by their wild cackling at the beginning of the day, her call a little higher than his. In quieter moments, we hear them hammering and dissecting trees as they forage. Not this one, though. She's done, forever silenced. Gently I lift her to the tall grass beside the road, giving her a more dignified return to ashes than would be her fate here on the pavement.

Pileated woodpecker

Pileated woodpecker. Foraging sounds—hear the wood chips fly! (70, 1:50)

Roadkills—we find so many along the road that we've started the game of "Name that Roadkill." A male and female northern cardinal resting peacefully side by side. Two black vulture corpses just ten yards apart, one more decomposed than the other, the cannibal apparently getting its just deserts. The woodcock back on the Blue Ridge Parkway. Mats of feathers ground into the pavement, offering few clues. Chipmunks. Possums and more possums, one with the white line at the road's edge painted right over it. "I know why chickens cross the road," said David, "to show the possum it can be done."

We bike on, the slow pace on the steep uphills here providing wonderful listening. Outside of Bevonsville I shift to first gear as we begin the steepest of climbs, and I listen . . . a red-eyed vireo approaches up to the right . . . an American redstart passes overhead . . . a small waterfall on the right . . . another vireo and redstart approach on the left, soon to fade behind . . . the spell is broken momentarily by a car meeting us, but it's soon swallowed by the valley below . . . two Carolina wrens, a meticulously enunciated *LIB-er LIB-er* from the right countered by a rapid-fire *wheedle wheedle wheedle wheedle wheedle wheedle wheedle* on the left . . . the creaking of my saddle . . . my heavy breathing taking in oxygen kindly provided by the green lushness all around . . . the roll of thunder in the distance, up ahead . . . another waterfall . . . *chick-a-dee-dee-dee-dee,* the rapid call of the Carolina chickadee directly overhead, then falling behind . . . the full raspy song of a scarlet tanager . . . *tawee tawee tawee TEE-o* from a hooded warbler . . . a tufted titmouse . . . an indigo bunting . . . a goldfinch singing . . . a red-eyed vireo, this one calling harshly, *myeah myeah myeah,* perhaps commenting on us . . . all this and more as we tunnel through this songbird symphony on our steep climb toward Dema.

More thunder, a flash of lightning, and daylight itself recedes; just before the skies empty, we pull our bikes onto the large covered porch of a house in Dema. I knock on the door, hoping to gain permission for what has already been done. "I was just thinking about Jesus when I heard this knock at the door," she bubbles. Sure, it's OK, she says, though she admits her disappointment that Jesus himself did not greet her. She tells of her five children, how she's just planted 150 pounds of potatoes to keep them fed. She leaves us, and it rains, and rains . . . a coal train lumbers by, carrying the blackened produce of three hundred million years ago. Water cascades from the roof, but we are dry, and soon stuffed, as we turn to our food to pass the time.

Listening from the bicycle. The birdsong-richness of eastern forests slowly passing by. (71, 1:35)

Our odometers read only 20 miles for the day. Another nine miles down the road is Pippa Passes, where we can spend the night at a biker hostel, though it'll be a rather short day. Yesterday it was 80 miles, today 30—each day, whether long or short, a modest bite from the continent. Dressed in full rain gear, we eventually depart, arriving at Alice Lloyd College in a little less than an hour.

On the campus, we ride on Purpose Road, crossing Integrity Lane, then Faith Avenue, then Action, Character, Conscience, Perseverance, Courage, Consecration, each intersection giving pause. The college radio station is 91.7 WWJD, as in "What Would Jesus Do?" All this was started by an Alice Lloyd who came here in 1915 to a county with one college graduate and where only one in 50 could read and write. Here's the cabin where she lived. Her motto: "Appalachia's future leaders are here, just waiting to be educated."

We bike past the college and up to the hostel, where we spread out to dry. I return to the college to use the laundry, and here I sit, depressed, dejected, wet and cold. Why? Why, I ask myself, am I feeling so low? Not enough miles on the day? Wet and cold and not eating and drinking enough? A bicycle that gives hints of failing me? A leg that needs healing?

I think back to the bit of wisdom written in the log book that I had glanced at in the hostel: "Enjoy every minute. It'll soon be over." That was written by an eastbounder with perhaps ten days left on the trip; we probably have two months left, but I quickly realize that the log was about something bigger, life itself. I'm here, with David, on the adventure of a lifetime, I remind myself. And I'm feeling down? I kick myself. Live. Live in the moment. Lighten up. Enjoy. It doesn't get any better than this, I soon convince myself. The all-you-can-eat cafeteria in the college's mess hall helps a lot, too.

DAY 13, MAY 16: PIPPA PASSES TO BOONEVILLE, KENTUCKY

We're both up at 4:45 a.m., eager for a bigger day than yesterday. The gear all loaded onto the bikes, we're off well before sunrise, and well before the other cyclists in the hostel have begun to stir. Through the college we ride down Purpose Road, a bubble of robins and phoebes and cardinals singing by every streetlight, these well-lit birds getting an early start on their day. Beyond the lights is darkness and silence. Out on the main road, we head west, our headlights beaming, taillights flashing. Roosters hurl their *cock-a-doodle-do*s into the coming dawn. Dogs howl from the left and then the right, each pack alerting the next down the road. A swarm of swallows swirls about us as we bike through a roadcut with rock faces rising steeply above on both sides of the road.

You-alllllll—a female barred owl remarks on these strange creatures moving by. A bluebird rapidly alternates two of his 50 or more songs, *cheeeer-e-o* falling in pitch, *tur-til-ee* rising, *cheeeer-e-o* falling, *tur-til-ee* rising, chattering continuously between songs. Three male cardinals hurl identical songs at each other. A phoebe alternates his songs, *FEE-bee FEE-b-bre-be FEE-bee FEE-b-bre-be*; "phoebe," David announces. I smile; he's listening. Our path is soon a dawn chorus of all that eastern Kentucky has to offer, and in passing, I try to acknowledge them all: gnatcatchers, chats, chipping sparrows, song sparrows, Carolina wrens, Carolina chickadees, tufted titmice, wood thrushes, indigo buntings, white-eyed vireos, each announcing dawn in its own way. Through all of this we are granted passage, privy to the outward signs of each singer's innermost thoughts, though for us the translation is

Eastern bluebird. At dawn, with fire in his belly, aggressively chattering between songs. (72, 3:49)

WARBLERS IN ENERGIZED DAWN SONG

Blue-winged warbler (73, 1:03)

Black-and-white warbler (74, 1:00)

American redstart (75, 1:01)

Hooded warbler (76, 1:01)

muddled. Dogs add their voices, too, one spooking me in a surprise attack from the left, bellowing at my ankle before I see him.

"That biker's log book back in Damascus," David calls up to me, "seemed to be mostly about how to deal with Kentucky's dogs. Advice included large projectiles, yelling loudly, or a Super Soaker gun filled with ammonia. Or just stop biking—go home. We'll see how we do."

All around us warblers announce the dawn, each species in its own special way. Males of three species repeat a single song over and over: The blue-winged warblers offer their stuttered *tsi tsi tsi tsi tsi tis tsi zweeeeeeeeeee ti ti ti*; the black-and-white warblers add their three-parted song, beginning with thin, rapid *wese wese* notes, then dropping to a few lower notes, then rising to the ending; the northern parula songs swirl upward with a buzzy quality, more complex than the simple rising buzz of the daytime. Males of three other species hasten through a larger repertoire of dawn songs: Yellow warblers work through a dozen or so, each with its typical "*sweet*" quality; each American redstart bounces among two to four songs, hardly pausing in between; the hooded warblers sing with the same clear quality of their one daytime song, but each song now feels more jumbled, and successive songs are always different. But on one matter these warblers all agree: Each male chips wildly between songs.

"David, I hope you're hearing all these warblers!" I try my best to point some of them out, but I can't expect much to sink in—hearing them for the first time like this is no match for the decades of relationships that I've had with these birds.

About sunrise I pause briefly beside the road to listen more carefully. Nearby a blue-winged warbler sings his lazy *beee-bzzzzzzzzz* daytime song, a black-and-white warbler in the forest just beyond his *weesee-weesee-weesee* song, a parula his daytime song as well. The yellow warblers, redstarts, and hooded warblers have also changed their songs, each of them now singing their one special song they've saved for non-dawn, non-aggressive situations, such as courting a female.

Blue-winged warbler

WARBLERS IN DAY SONG

Blue-winged warbler (77, 1:00)

Black-and-white warbler (78, 1:00)

American redstart (79, 0:54)

Hooded warbler (80, 1:02)

Two warblers continue as before, because they have no special dawn songs, though each sings an occasional jumbled flight song. Ovenbirds from the forest offer their *te-CHER te-CHER te-CHER te-CHER* crescendo, yellowthroats from the roadside ditch their *wich-i-ty wich-i-ty wich-i-ty wich-i-ty*.

We have biked slowly through this dawn chorus, but now we pick up speed through the metropolis of Hindman, with a population of 787 according to the welcoming sign, though a ghost town at this time of the morning. Next are Carrie, Emmalena, and Dwarf, each a post office plus a little more. I love the slow climbs up these back roads through heavy forests of oak and hickory and walnut and sycamore and tulip; how I wish I could add the American chestnut back into the mix. The trees and rock faces beside the roads drip with the night's rains; birdsong and vegetation are equally lush, the sun's rays barely penetrating to the ground.

"Hey David, there's a Kentucky warbler—should be the state bird, but Kentucky has chosen the cardinal instead." We pull off the road, listening more closely to this rolling chant of several repeated phrases: *chuuree chuuree chuuree chuuree,* the second syllable of the *chuuree* always higher than the first. The song reminds me of a mourning or a MacGillivray's warbler, the three closely related species perhaps all retaining from their ancestor this rich, rolling quality to their song. I try a bit more on David: "He's another warbler with 'just one song,' but he adds a trick to his singing that other warblers apparently don't: Like a black-capped chickadee singing *hey-sweetie,* he can raise or lower the frequency of his song. That way he creates a little variety and can play singing games with neighbors by either matching or not matching what they do."

Kentucky warbler, hooded warbler. Repetitive *chuuree* songs of Kentucky alternating with *weet-weet-weeto* songs of hooded. (81, 0:59)

Four-lane route 80 outside of Dwarf is all coal trucks. The sooty Monster Macs tower above us as we weave our way among the large chunks of coal strewn on the wide shoulder. I watch each truck approach in my mirror, making sure of sufficient space between us, and count wheels by axle as they pass, 4, 8, 12, 16, 20, plus two for the front of the cab,

22 tires in all. Another, and another, an endless train of them, they lumber uphill and roar down, hauling yesteryear's surfeit to market. Raw roadcuts tower above us, telling of this region's lifeblood in coal-black, sedimentary layers. From behind me, David works his camera, trying to silence a thousand words with a perfectly timed click of the shutter.

In Chavies we cross the muddy waters of the North Fork of the Kentucky River and pause on the bridge. *FITZ-bew*! One song is all it takes and memories of willow flycatchers rush through my mind. "David, down there, from that tree overhanging the water on the right, is our fourth Mosquito Prince! When you were just a few years old, you helped raise some of their babies in the basement of our house when we were trying to figure out how a male gets his songs." I encourage David to listen for the three distinctive songs, knowing that they'll be there, because we learned that willow flycatchers everywhere, thanks to their song genes, have these three songs. We confirm all three, the all-buzzy *FIZZ-bew*, the *FITZ-bew* with a sharper beginning, and the *creet* as a rising, stuttering series of brief notes. "I got 'em," David assures me. It's the "eastern" willow flycatcher, I tell David, as on the other side of the Great Plains we'll encounter the western form, two different "species" by some definitions, as the song genes and songs of eastern and western birds consistently differ.

Willow flycatcher. Three different songs, the same that willows sing everywhere. (82, 10:43)

Dogs! To the right now, outside of Chavies, beyond the gentle falling *teedle teedle teedle teedle teedle teedle teedle tew tew twee* of a yellow-throated warbler, a vigilant little yipper is first to spot us, its yipping followed quickly by a yowling hound, and it's hard to count how many others chime in, some high-pitched, some low, some in between. What a variety of voices, such passion and fury in them, all apparently restrained, sadly, as I can hear the clatter of chains, the little yipper choking on his yips as we pass. A bend in the road takes us around them, the ruckus lasting the better part of two minutes. A cardinal, then a Carolina wren, then a song sparrow add their refrains as the howls recede.

Dogs! Oh-so-familiar sounds of Kentucky. (83, 0:58)

After passing through an especially energized pack of dogs, I often opt for two more helpings, turning back to bicycle through them twice more. It's a scientific experiment, I tell myself, to see if their enthusiasm wanes on the second and third passes. The answer: Usually not, at least for those in the free-roaming packs, who seem to get an adrenaline rush with each encounter.

We've tried another experiment, too, this one best done on an open road with no traffic. I know that when a flock of birds is pursued by a predatory hawk, the flock bunches up and moves rapidly about; the hawk then seems uncertain about how to attack, perhaps fearing an out-of-control collision with its prey. To test if a dog reacts like a pursuing hawk, we initially ride far enough apart that any good dog will attack the lead bike. The following cyclist then accelerates, pulling up behind the dog; as the front cyclist weaves away from the dog, the rear cyclist is now on the dog's heels, as if attacking the dog. The rear cyclist next weaves away from the dog, the lead cyclist weaving toward it, so that the dog is again in position to attack the lead bicycle. I've seen the fear and confusion in the dog's eyes as he looks back over his shoulder to see a bicycle now in pursuit of him, only an instant later to find himself pursuing, then pursued, then pursuing. After only a few weaves, accompanied by a few of our blood-curdling screams, of course, even the best of dogs pulls away.

"My favorite dogs," David wrote in his journal yesterday, "are the little shitty ones who run up to my ankles and then I kick them into the ditch." He nailed a big German shepherd in the eyes yesterday with his Halt! dog repellent; it was a quick attitude adjustment, as the dog stopped in the middle of the road to rub his eyes while we made our escape.

We bike into Buckhorn Lake State Park, on the Middle Fork of the Kentucky River, and stop to talk to a camper. "My name is Mary Lou Napier and I'm from Hazard . . . we've got a fire going and we're just listening to the birds." Now there's my kind of camper. "Cardinal. State bird. Yep, it's got a real pretty sound . . . My favorite bird would be, I'd say, a bluebird. They're so pretty. . . ." I tell her I'm intrigued by how different her voice and mine are, fascinated by these human dialects. "We make fun of each other, but I like [our different voices] . . . a lot of times people call us hillbillies . . . but a hillbilly is a friendly person . . . we got it made down here . . . just listen . . . [a pileated woodpecker cackles nearby] . . . Listen! . . . It's beautiful . . . You won't hear that in the cities. You probably have a lot of pigeons pooping on things, but you won't have this . . . just listen!" In Mary Lou's fine voice I hear a hillbilly wisdom that eludes most folk.

Back at the intersection in Buckhorn, we stop at the Kentucky Food Store. "We Sell Tombstones" announces the big sign in the window. Another sign:

"Warning. This store is protected by an attack rabbit." Huge window signs tout Winston, Marlboro, and Red Seal. Four chairs line the sidewalk out front, four local gentlemen holding each firmly in place. Talk is all about the person driving the pickup that just went by, then about the fellow who just left and who had moved here from someplace else, and who spoke "rather odd." Two men emerge from the store, good-humored, smiling, chuckling with each other, one of them approaching and saying something I can't quite catch. I politely ask him to repeat it, and then again. Embarrassed, three times and I'm out. Luckily, David caught it, translating for me that we were invited in to sign the log book. Yes, the log book, of course, the log that other cyclists would have signed as they pass through town.

Mary Lou Napier. A self-proclaimed hillbilly happy in her skin, just listening to birds. (84, 14:58)

Soon we're on the road again. David points up to the left: "Parula, eastern," he says with confidence. The divide between east and west can't be far from here—he's ready.

From Buckhorn, we climb to the county line between Breathitt and Owsley Counties, then descend, soon following Cow Creek, which drains into the South Fork of the Kentucky River. At Booneville our dangerously hungry chief cook (our only cook, he would claim, as my fixing oatmeal doesn't qualify for cook status) works my Visa to load up on groceries before we head to the pavilion kindly provided behind the Presbyterian Church. Dinner is chicken teriyaki, with onions and green peppers and broccoli and real butter, milk, brown rice, and huge cookies for dessert—a true feast, David's finest of the trip so far.

Bodies fueled, cookies in mouth, the picnic table strewn with the stove and makings of dinner and dirty dishes, every available surface under the roof covered with our gear drying out from the rains of yesterday, all this and who should bike in but Kris and Kes, two cyclists we first met back on the Virginia-Kentucky border. We've seen them since, too, the four of us staying in Pippa Passes last night. We applaud and cheer for them and for us for the 80 miles we rode today. David and I quickly consolidate our gear, making room for them, sharing food and stories.

Near sunset David and I walk the fields and use the binoculars to study whatever we can find. We get good looks at a bluebird, a brown thrasher, and so much more. "There's the cigar on wings," says David as he points out a swift overhead. "Best time of day to bird watch."

From what we hear and see now, it'll be a good morning. Our tent is pitched out in the grass away from the pavilion, and we call it a night, a last check of the watch showing 9:23, with the sun already well below the horizon.

7.
BOONE COUNTRY

DAY 14, MAY 17: BOONEVILLE TO IRVINE, KENTUCKY

Exactly when the chat became the dream or the dream became the chat I don't know, but I first check my watch about 2 a.m. Perhaps he sang for the bright moon that came up sometime before midnight, or maybe he would have sung anyway. He's just off to the east, no doubt in that tangle of bushes I saw beside the road last night. I listen. The yellow-breasted chat. Such an odd bird. Is it a warbler or isn't it? The experts can't decide. He lives with the warblers in the field guides, but that's mostly because no one knows where else to put him. Linnaeus thought the chat was a robin in disguise. Someone else voted tanager. Can't be a warbler, I say. No warbler sings anything like this. But, then, no bird sings anything like this.

He honks and squeaks and buzzes and twitters and whistles and chuckles and barks and mews and cackles and caws and squeals, and his manner of delivery is even stranger. He's so unpredictable. What's coming next? Just a single whistle, or a string of them? One honk or a whole series? And does he pause a second, or two, or longer, or none at all before exploding into the next sound? Such a unique songster, all the more exquisite as a solo performance on an otherwise silent stage, all other birds hushed in the dead of night. From my sleeping bag I drift in and out of listening.

About 45 minutes before sunrise, I'm out of the tent and walking about. An hour ago, a robin was already singing by the streetlight. Purple martins and other swallows are airborne, and a killdeer flushes nearby, *dee dit dit dit dit dit deet deet deet.* Exploding from a small tree near me is an eastern kingbird; he launches into the air, soon circling high overhead, singing *t't'tzeer, t't'tzeer, T'TZEET-ZEETZEE, t't'tzeer, t't'tzeer, T'TZEETZEETZEE,* two of his neighbors doing the same. I listen, smiling; yes, Walter Faxton, you were right, it is as if they are "trying to pronounce the word 'explicit,' but . . . making a miserable, stuttering failure of it." Familiar birds are all about—indigo buntings, cardinals, catbirds, bluebirds; but it is the chat who draws me.

Eastern kingbird. Dawn singing, energized and continuous, sputtering and stuttering and bickering. (85, 2:06)

Walking closer to him, I'm astounded at all that I didn't hear from the tent last night. Between the loud honks and cackles and whistles that carried the hundred yards to the tent are all manner of soft pips and pops. What an eccentric!

I aim the microphone at him and begin recording. *Cha-re cha-re cha-re cha-re cha-re cha-re* he sings, each distinctive phrase beginning with a harsh *cha* followed by a high, pure-toned *re*; over the next minute I hear a burst of *cha-re*s five times among perhaps 30 other sounds, but then it's gone. I pick another odd song and hear it about ten times among 50 or so other songs over the next three to four minutes and then it's gone, the vocalist having moved on to other sounds in his repertoire. For 14 minutes he cooperates before flying off.

Footsteps? "Good morning" I say to David, who has emerged from the tent and walked up to me. His "moornning" is followed by a stifled yet monstrous yawn. It's six o'clock and here he is, joining me for a listen. Nice. "Cardinal?" he asks as he nods toward the *birdie birdie birdie birdie* song to the left. He's right. "Carolina wren down by the river," I point out. "Get enough sleep?" I ask. "Yeah." I enthuse with him about the chat, though try to limit myself to what I think he can handle at this time of day.

Yellow-breasted chat. Extraordinary variety and unpredictability, heard best right beside him. (86, 9:59)

"Mosquitoes," I point out, slapping one on my arm. I continue: "Hear the titmouse? Indigo bunting is singing up there on the wire . . . song sparrow right here, see his all-striped breast? . . ." I hear a yawn, but he seems to be taking it all in. We listen to crows cawing in the distance, the indigo bunting, the orchard oriole in the tree by the church, the kingbird now perched on the wire near the bunting.

"I'm hungry," I offer. "I'm tired. I'm going back to sleep," he counters. It's 6:10, 15 minutes to sunrise.

"Want to try to get ready to go?" I suggest. "I'd like to lie down for a fewteen minutes." What was that he said? A deer snorts off toward the river, the orchard oriole sings over the church, crows caw, titmice *peter peter*, a yellowthroat *wich-i-ty*s, and he wants to get a little more sleep? "OK."

As David shuffles back to the tent, I head over to the pavilion to rustle up some breakfast, listening all the while. Chimney swifts *chitter* overhead. Down by the river to the north a yellow-throated vireo sings, slowly offering his

Brown-headed cowbird. Announcing sunrise, in flight whistles and songs. (87, 1:00)

burry songs to a sun trying to peek through the clouds on the horizon. A minute later a red-bellied woodpecker calls nearby, the first woodpecker of the morning. Two more woodpeckers, a downy and a flicker, soon chime in, the downy with his descending whinny, the flicker with a loud *wik-wik-wik* song lasting a full five seconds; in the distance a pileated rolls his drum. Completing the chorus are flight whistles and songs of the late-rising cowbirds over the church.

Around seven o'clock David joins me for breakfast. "Best not to get up before the sun," I suggest, smiling, though what fine company he's been to join me on some early rides over the last two weeks. Some kid!

It's Saturday, beginning of the weekend in Booneville, and a lawn mower starts up just across the fence at the neighbor's, signaling the end of my dawn listening and the beginning of the day for others.

We're soon packed up and ready to go, but the overnight rain resumes, so we sit in the pavilion to wait it out, first with Kris, who is an English professor at the University of Georgia, and eventually with Kes, an environmental lawyer working for nonprofits. Kes enjoys his morning sleep, much like David.

I read a little of David's *Zen and the Art of Motorcycle Maintenance*, but I have difficulty concentrating. I scribble random thoughts into my small notebook purchased at the Kentucky Food Store in Buckhorn. I list all of the fine birds heard this morning. I work at being patient.

Yellow-breasted chat

Our biking friends have a radio tuned to NPR's *Weekend Edition*. We love these two liberal Southerners, but how painful to hear the news, to be reminded of the mess in the Middle East and how we got there, and of all the other messes in the world. I slink a little farther away in the small pavilion, trying to think my own faraway thoughts.

Yellow-breasted chat. Imitations and "song packages," so evident in this male's singing. (88, 10:25)

The rain continues, but I soon realize my escape. I get out my morning's recording of the chat, slip the headphones over my ears, punch "play," and listen, ready to scribble important thoughts in my notebook, but in the end just listen, taking mental notes instead. Once is not enough, so I listen again, and a third time, smiling at what I hear.

He sings in "packages," just like the robin and blue-headed vireo, each package consisting of four or five different songs that he works for several minutes, and then he moves on to another package, and eventually another. If he has about 60 different songs, as I've read, it would take a dozen or so such packages to present his entire repertoire.

With a lull in the rain, David and I choose to make a break for it, but Kris and Kes decide to wait. It is the customary "we might never see you again . . . have a great journey . . . hope we cross paths down the road . . . leave messages in the log books," and after taking some pictures of each other, we're off, fully decked out in our yellow rain gear.

Five minutes out of town we hear the sound of tires skidding on wet pavement behind us and turn to see a car sliding down an embankment, coming to rest against a tree. We head back, getting the unharmed driver's assessment: "Slippery road." And had we been a few seconds later? I remind myself to watch our tails in my mirror.

We're soon climbing into his forest, the *Daniel Boone* National Forest, though there's no sign to tell us where we are. Back in Virginia, "route 76" signs marked the 1976 Bikecentennial route across the state, but not here in Kentucky. The roads seem deserted, and we bike in silence, listening to the chorus of insects and frogs and toads and birds, half expecting Daniel Boone himself to step out of the forest. A black pickup truck emerges from a curve up ahead and races down the middle of the road toward us, swerving at us as it passes and brushing us into the ditch. We bike on, more wary; half an hour later, a black SUV approaches from behind and drives slowly by, giving us plenty of space, but a young girl yells from the open passenger window, "Stupid . . . stupid."

More dogs. Off to the left, they see us before we crest the hill and race from the house to the road. Big ones, little ones, six in all, yipping and howling, some then swinging around to our right, a half circle around us. We kick and flail with our feet, yelling, soon leaving the pack behind as we start down the hill, gaining speed. Looking back, we see a man sitting on the porch, as if quietly amusing himself at this game. Like train robbers in the westerns who swarm the slow-moving train on an upgrade, these dogs seem to know to have their fun when we cyclists are most vulnerable.

***More* dogs!** Catching the unsuspecting, vulnerable cyclist just before the hill's crest. (89, 3:05)

Minutes later, the forest of Boone again envelops us, singing birds abundant. And casually, as if nothing special is there, I point up to the left and ask David, "What's that?"

"It's the parula, the western song!" *He's got it!*

We pull over and I get out my recorder, and in mock radio voice, announce for all the world to know:

Here we stand, at this very place in our own national forest named after none other than Daniel Boone, here is the great parula divide. For 700 miles on our journey from the Atlantic to the Pacific, we have been listening to the eastern, daytime song of the northern parula. We've heard hundreds, if not thousands, each song a rising, sputtering buzz before bursting explosively over the top, *zeeeeeeeee-up*. To the west, beginning here, lies a bird of a different song; it begins with the same rising, sputtering buzz, but ends with a whimper on a high note, not an explosive lower note. And of what significance is this? We haven't a clue, nor does anyone else, except that birds to the east and west of here differ slightly in plumage and size, suggesting that this oral divide is of some special significance in parula history.

We ride on, and with 26 miles on our odometers for the day, I am enjoying all this forest has to offer, but on a delightful downhill in a remote section of County Road 1209, at 4:05 p.m., I hear trouble. With each revolution of the wheel there's a click, and now I feel it, too, the rear wheel slightly out of round, giving me a noticeable nudge with each turn of the wheel. I pull up, quickly working my hand over the spokes, grabbing pairs of them as I work my way around the wheel. Not one but two broken spokes! Aaarrrrghhhh! We have extra spokes, but I broke a spoke several days ago, too, and more broken spokes mean that the entire wheel is out of whack and the tension on the remaining spokes is seriously wrong.

Northern parula

Northern parula. Dawn songs, then at 1:45 is the wimpy western daytime dialect. (90, 1:51)

We walk the road until we find a small trail leading off into the woods, where David sets up shop. I pull the wheel off my bike, the tire off the wheel, and David masterfully goes to work replacing the two spokes. In *Zen and the Art of Motorcycle Maintenance*, he relishes the passages where he is reminded to calm down, to think, to become one with the bike. He has experimented with building wheels from scratch, given a hub, the spokes, and the rim, and his tinkering now pays off.

I watch and listen, the forest alive with song. Indigo buntings, tree frogs, warblers of all kinds, and scarlet tanagers everywhere, singing hoarsely from the treetops. If the older trees could talk, surely they'd tell of passenger pigeons, nicknamed the "blue meteor" for the male's slatey-blue head and grayish-blue back, for their long pointed wings and powerful flight, for their speed, grace, and maneuverability in the air. These trees would tell of passenger pigeons so abundant that the numbers are incomprehensible. It was almost 200 years ago when John James Audubon stood not far from here and watched a flock of these pigeons fly by, a flock so immense that it took three days to pass, and at times he estimated 300,000,000 pigeons passing in an hour.

I imagine standing here early evening in one of their night roosts, marveling at the devastation. Trees two feet in diameter have snapped, others are simply uprooted, and large limbs lie everywhere, as if a tornado has swept the area. Dung blankets the ground several inches thick. And just after sunset, the distant roar of the flock is heard, as is the cry of the gunners, "Here they come!" We hear them at first like a distant waterfall, and soon hundreds of thousands, most likely millions, pass by, their countless wing beats creating a gentle breeze. They pile on top of each other, large branches often crashing to the ground under their weight. Our voices, even shouts, are unheard in the din, as are the guns that fire again and again in the fading light.

Scarlet tanager. Treetops tell of the hoarse robin, the robin with a sore throat. (91, 1:22)

But that's all past. Like roadkill, the pigeons are forever silenced, our singing continent forever the less for it. Carolina parakeets would have been abundant here, too, perhaps even a few Bachman's warblers, or an occasional ivory-billed woodpecker. But no more.

While the pigeons and parakeets have come and gone from my imagination, I've been watching David replace the two spokes and then repeatedly spin the wheel on the overturned bike, marking where the rim is out of round, then adjusting the tension on the appropriate spokes to bring it back into shape. "It'll work out. It always does," says David. Yes, I nod; this kid has a few things to teach me.

It's a temporary repair, though, David informs me, and we agree to seek the wheel-building skills of a professional. From David's repair shop in the woods to Irvine is an easy 12 miles, and there we'll find a motel and determine how to fix the wheel more permanently. It's Saturday, and in the Bible Belt of mid-Kentucky, surely it'll be Monday before we can get help, at least two days of biking lost. It seems somehow fitting that rain ushers us into town.

8.
A RIDE IN HEAVEN

DAY 17, MAY 20: IRVINE TO LINCOLN HOMESTEAD STATE PARK, KENTUCKY

A female orchard oriole streaks to the left across the road just in front of me, followed by a male, his wing beats slow and exaggerated, the entire black and burnt-orange bird in a gravity-defying butterfly flight. The black head and tail drop on the wings' downstroke, rise on the upstroke, the black wings rising to expose the rusty underwings and undersides, then falling to highlight the brilliant rusty rump against all that is black above, rising and falling, black and orange, like a huge monarch butterfly. My mind slows him further and brings him closer, every feather in his outstretched wings now visible, every detail about him as brilliant as a courting male in spring dress can be. And he *sings*, a rich, lively, jumbled outburst of buzzes and piping whistles, they, too, slowed so that I can savor every mini-puff of air through his two voice boxes, the flight and song seeming to last forever in slow motion across the road.

Orchard oriole. A rollicking jumble of rapid, leaping notes, most musical, some harsh. (92, 2:02)

We are just a mile out of Irvine and I am high as can be, having just survived what felt like a near-death experience. Over the last two days, we had to reenter *the race*, renting a car, driving the interstates at ungodly speeds to Lexington and Cincinnati, spending three nights in a motel room, all in search of a mechanic who could rebuild our rear wheels—I felt frustration and despair. But with the repairs finished, we are on the road again, the euphoria I am feeling having sharpened all of my senses.

I feel the wind in my face and rushing past my ears. I feel my thighs warming to the task, settling into a happy rhythm as I pedal and cruise down the road. In one fluid motion, the left leg rises, flexing at the hip and the knee, pulling the ankle and foot and pedal upward, gliding over the top of the arc, the hip and knee then extending to push the pedal down, the pedal and crank arm powering the chain over the front rings, transferring the power to the cogs on the rear cassette and then to the rear wheel itself. The right leg balances the left, and I feel the rhythm, feel the shifting weight on the saddle with each stroke, the bike and I one as we glide westward, my mind smiling at every bit of the fluid machine that we are.

I see, in detail I never before appreciated. Trees are no longer just green, but all shades and hues and textures of green for which I have no words, and all shapes and sizes; I scan up their trunks along the road, admiring patterns in bark, unique branching patterns, the shapes of the leaves. I celebrate oaks and maples and tulips and hickories and ashes and willows and redbuds and others whose names I do not know. I see the road ahead, the fences, swallows darting over open fields, cows ruminating, farmhouses and barns, the entire landscape in focus all at once. A pileated woodpecker flushes from beside the road, flying on ahead, inviting us onward. And I see the dogs before they even think of chasing us. I see their fun, see the puppies in training as they follow the older ones to the road. To them it's all a big, delightful game, though at times I fail to see their fun, as three times already this morning I've had to physically defend myself, twice kicking a dog in the chops and once dismounting from the bike, always keeping the bike between the dog and me.

I smell the honeysuckle, the wild cherry, and more, a rich medley of spring aromas luring pollinators . . . the alfalfa . . . the manure recently spread in the pasture . . . the freshly mown grass at the farmhouse . . . a dead possum beside the road, the rich fragrance of rotting flesh as salivating to a turkey vulture as a baking apple pie is to me.

A butterfly, black and yellow, floats from the left across the road, kissing me on the nose before rising up to the right, soon disappearing behind me.

And best of all, I hear this continent sing! Chipping sparrows are everywhere, and I listen to each in turn, admiring the variety in their delivery and in the rhythm and tonality of their songs. Mockingbirds race through their repertoires; I pluck songs from midair and hang on, waiting for the next, and the next, spellbound by how versatile these birds are, relishing the mimicry I hear. From field sparrows I hear the classic bouncing ball, with a series of whistles

Field sparrow. A distinctly two-parted song, beginning slowly, an abrupt transition to a rapid trill. (93, 2:00)

that progressively shorten; other field sparrow songs are distinctly two-parted, with a couple of slow whistles followed by a rapid dribble; still others are three-parted, the best beginning with a few low whistles, followed by a few high ones, and then a rapid lower trill. Song sparrows—as I approach a singing male, I feel the rhythm of his song and then listen for him to repeat it, with perhaps one in ten birds switching to a new song as I pass. Baltimore orioles sing from the treetops, so rich and brilliant their clear, whistled notes, each bird with his unique song-print, though on occasion I hear matching songs, but know not whether they're from mates or neighboring males.

Warblers speak their minds. Each yellowthroat in the roadside ditches sharply enunciates his particular version of the *wich-i-ty* song . . . well after sunrise now, yellow warblers repeat their daytime *sweet sweet I'm so sweet* . . . prairie warblers spell their name, *p-r-a-i-r-i-e*, with their rising, buzzy daytime song . . . the parulas are unanimous—they all sing the western version of their daytime song, we having clearly crossed the parula divide just a few miles back in Daniel Boone's National Forest . . . and Kentucky warblers, named after the very state we bicycle, *chuuree chuuree chuuree chuuree.*

Yellow warbler. Sweetness in daytime mode, one special song repeated. (94, 0:50)

I glance over my shoulder, just to make sure David is still back there. Yep. He seems pretty miserable this morning, tough to get going. I think it's partly his breakfast, the fatty burrito at McDonald's, that holds him back; my six hotcakes and orange juice have supercharged me. I worry some that he's upset at me, tired of riding with a birdsong maniac who so relishes the dawn. "Just keep your distance" are the unspoken instructions. I know to carry on, that time and caffeine will restore him.

Indigo buntings line the road, telling their stories for all to hear. To each I listen in turn, and I hear songs as either sharp and crisply enunciated, with successive songs the same, those being the songs of a well-practiced adult two years or older, or sloppy and slurred, the notes inconsistently repeated, successive songs different, those the babbled songs of yearling males just learning to sing. Each of these yearlings is on a mission, and I catch him in the act of whittling and honing his song, perhaps adding an appropriate phrase or two, all in an attempt to match the song of the older bird he's settled next to, thus joining the local "mini-dialect." Were I into the looking, I could verify through the binoculars that the good singers are an immaculate indigo, the babbling singers mottled with brown-and-blue plumage, the mark of a yearling—but I bike on.

Whit whit whit chewink. Wood thrush and towhee up ahead, I say to myself, as I know those calls, but what are they so upset about? Slowly, up the hill I climb toward them. *Whit whit whit chewink.* And now a white-eyed vireo quickly sings after them. Wait . . . No . . . One more time. Yes, it is! It is *all* a white-eyed vireo—at the beginning of his song are imitations of both the wood thrush and the towhee calls, and then he finishes off the song with his usual snappy jumble of notes. Nice. Again, and again. As I ride by, he switches to another of his dozen or so songs, this one with an unmistakable *klee-yer!* call of the flicker, ending with the *jeeer* call of a male Carolina wren. What a master he is at pilfering the calls of other species, but never their songs. And how expressive he is when agitated, whether mobbing an owl or arguing with another male; he deconstructs his songs then, using separately all of the angry-sounding call notes of other species, sounding like a mob all by himself.

Indigo bunting

Lives of indigo buntings, as told in song.
Two brilliantly indigo adults with identical songs.
Bird 1 (95, 1:03)
Bird 2 (96, 0:58)
A yearling male rambles, searching for his voice.
(97, 2:00)

We bike on through the greenest of valleys, the rugged terrain of Appalachia having given way to smaller, isolated knobs. We pass a cemetery, reminding me of Robert Fulghum's cemetery plot. First he dug it up to make sure no one else occupied it, and after that he'd occasionally pull up a chair and sit over his plot, thinking into the future. It helped him focus, he said, on choosing wisely how to spend his remaining, limited time on this planet. As I bike, I also look to the future, to the Pacific and beyond, and I await news from my friend Bruce back at the university, as he's promised to alert me of any retirement incentives offered while I am away. Like Fulghum, I am eager to spend my remaining time wisely, as I am increasingly aware that I work to earn money, but the currency I use to buy stuff is actually life itself.

White-eyed vireo

After nineteen and a half breathtaking miles, I pull up at the Horseshoe Bend Grocery. David follows a minute later. After gathering some supplies inside, I wait outside, hoping that David will emerge energized after gulping down some mega-dose of caffeine.

White-eyed vireo. Mimicked calls of wood thrushes, chickadees, and more. (98, 1:58) In deconstructed song, sounding agitated. (99, 2:05)

"What a beautiful day!" he remarks, as if now seeing it for the first time. He's back to his amiable self, soon on his bike and happy, leading the way. Some kid!

Beyond the grocery we pass the Red Lick Baptist Church, then follow the map by turning left just after Pilot Knob, an isolated stub of a well-worn mountain rising perhaps 500 feet above us. In a mile and a half we hang a right at Bighill, climbing a few hundred feet and then dropping down to Berea. What different country we enter here. From Berea, we look back and see the knobby hills, the last hurrah of the Appalachians. Ahead lies rolling countryside, with cows grazing in green pastures, blackbirds and meadowlarks galore, and grasshopper sparrows, too, all birds of open country. Bicycling is far easier, and we float along, a gentle southeast wind at our backs.

Lining the road are friendly yellow-breasted, black-bibbed meadowlarks. Each sings a plaintive yet spirited medley of whistles that slur together and drop gently, trailing off: *Hey you, I'm heeere*, or *spring-o'-the-yeeaar*, each male with 50 to 100 different versions of this song. They flush from the roadside as we pass, their white outer tail feathers glowing as they fly on stiff wings with shallow wing beats, like little quail. Some voice their mild displeasure with an explosive, raspy *dzert*, or more excitedly with an extended *dzert-ert-ert-ert-ert-ert*. Kansas . . . I smile at the very thought, as I'll be listening eagerly for the first song or call that hints of the transition from eastern to western meadowlarks at the tipping point of the continent.

Eastern meadowlarks. Two males beside the road, countering each other in song. (100, 4:59)

Male red-winged blackbirds perch high in roadside bushes and up on the utility wires, and I listen to their familiar *konk-la-reeeeeee* songs, marking how repetitious each male is, noting how songs differ from one male to the next, but I listen especially for her. From down low in the grass or cattails or stubby bushes, a male's secretive partners (males are typically polygynous) often respond to his song with a harsh *ch-ch-ch-ch-ch-chit-chit*—and repeatedly I hear these social bonds as we bike by.

Red-winged blackbirds. Busyness of a blackbird marsh, female calls, male calls and songs. (101, 1:51)

Up ahead beside David a female blackbird flushes from the roadside, calling in flight, *ch-ch-ch-ch-ch-chit-chit*. "Hey David, did you see that? . . . It's a female blackbird leaving her nest. It has to be right there, at the base of that small bush. Good thing predators don't know what I do or there'd be no blackbird nests with any babies left in them." Odd, I reflect, how females so brazenly announce where their prize is. Doesn't make sense.

Beyond Berea it's Kirksville, then Bryantsville, nearly 70 miles of giddiness already, and then, just outside Burgin I hear an entirely new song from a bird perched on a fence post to the right. *What is that*? I slow and circle back, the lone singer cooperating and staying put. I take in his sparrow-sized features, the black bib, the striking yellow and gray and white pattern on the head, the smudgy yellow breast, the rusty shoulder—a dickcissel! Have I ever heard or seen one before? I don't think so. Here, with the Appalachians just behind us, how nice to hear a welcome reaching out to us from the prairie grasslands to the west, from the continent's heartland. And how he sings, as if he stutters his name, *dick dick ciss ciss ciss . . . dick dick ciss ciss ciss*, the only song he knows enunciated sharply over and over.

In Burgin, David's on ahead, but I'm startled by the raucous hooting and hollering of boys in a parking lot beside the high school. They are, what, throwing books into a dumpster? Curious, I pull up beside them. When their teacher asked for volunteers, I am told, they selflessly stepped forward, now throwing entire boxes into the dumpster, sometimes giving special attention to certain books and first ripping the covers off. Books of all kinds were flying, but when one of the students got to Strunk and White's *The Elements of Style*, the professor in me cringed; I tried to slow him down, tried to tell him how I had volunteered to teach writing in my university department when no one else wanted the task, and how learning to write can take you places. But I was soon laughing, as I realized that in them I saw myself, or high school freshmen almost anywhere, as I could have enjoyed the same kind of fun back then.

"We love Kentucky" I offer to them.

They pipe up, in turn: "Kentucky is a great state . . . yeah, Kentucky is the best state . . . YES! . . . [I love] the countryside . . . not as polluted as California . . . we don't like gay marriages . . . yeah, that's even a better thing. . . ." I tell them we're from Massachusetts, and they know what's happening there. "Now it's going to be legal to marry gay people . . . will you marry me, Ben? . . . yeah, I'll marry you, *not!* . . ."

High school students. Making merry as they trash old textbooks. (102, 6:42)

When I ask their names, their apparent leader speaks up: "I'm Jason, this is Ben [pronounced *"Bee-un"*], David, also known as Boner, Moose, Travis, and Rob . . . and if you want nicknames, it's Weasel, Brady, ah, Boner, Moose, Slob, and I'm just me, I can't say my nickname on air . . . la Bitch . . . with a capital B. . . ."

"Where you headed from here?" I ask. "Everybody stays in Burgin?"

"I'm going to California . . . he's going to get laid . . . I'm going to Texas. . . . *You cannot say 'get laid' on video camera* . . . I'm going to Michigan . . . with that accent?"

A teacher emerges and urges the students to get on with it, explaining, "We tried, but nobody wanted these old books." I leave the students to their fun and find David patiently waiting up ahead.

In just four more miles, about two o'clock already, we stop for lunch in Harrodsburg. While I tend to the bicycles outside, oiling the derailleurs, chains, and brakes, David shops inside the Safeway, eventually emerging with two large bags and a big smile on his face. Soon we are sitting at a picnic table at Old Fort Harrod State Park, which we learn commemorates an early permanent settlement just west of the Allegheny Mountains; one bike leans against each end of the table, and we sit across from each other. I start on my Starbucks Frappuccino with a lifetime supply of caffeine and calories, then tackle my share of the strawberry shortcakes, the parfait cupcakes, bananas, star chips, doughnuts, bagels, and orange juice, topping it off with a pint of Ben & Jerry's vanilla ice cream. With 76 miles on the day already, our chief cook and dietician reasoned that we could eat anything and everything, and we do. "The rain's held," I dare say, tempting fate. How fortunate we've been, as we've missed the heavy rains forecasted for the morning, and we just might be lucky enough to outrun the afternoon rains as well.

Over David's slurping on his strawberry shortcakes, I listen to the starlings overhead. Widely considered trash birds, these varmints were introduced from Europe into New York City's Central Park in the early 1890s. The millions upon millions of these creatures across the continent now threaten bluebirds and woodpeckers and other cavity nesters. Immense flocks numbering into the millions foul city streets and water supplies with their droppings. All this makes them one of only two species (the other the house sparrow) not protected by law, so we can shoot and trap and poison them, though with little effect. But just listen to them up there . . . they whistle and click and snarl and screech . . . and warble and chirp and creak and gurgle . . . and what extraordinary mimics, though in such understated tones.

European starling. Magic in the song of this virtuoso, one of the finest songsters on the planet. (103, 0:55)

One enthusiastic singer perches beside a nest hole about ten yards up in the maple overhead. His next song begins with a throaty, subdued whistle, clearly pilfered from the down-slurred *kee-eeee-arrr* scream of a red-tailed hawk; he pauses, then offers another. Now he accelerates into complex, paired phrases with imitated songs of meadowlarks and calls of blue jays. Next he flicks his wings in rhythm with the song, and from one voice box he clicks rapidly while from the other he continues with a variety of wonderful sounds, though with no clear mimicry. Now the finale! He waves his wings frantically, his body quivering from side to side, and he sings his loudest and highest. After a good 30 seconds, he rests, though not for long, as there's the red-tail scream again. How can anyone who has truly listened to these close relatives of mockingbirds still hate them? "Hear the starling up there?" I point out to David, sparing him the details.

After a full two-hour break, we roll on, five miles to Rose Hill, cycling through beautiful rolling hills, often riding ridgetops with vistas in all directions. Then it's another nine miles to Mackville and five to Fenwick, and we're increasingly amazed at the huge lawns and monster ride-on lawn mowers that we see, sights so foreign to eastern Kentucky. In Fenwick, we detour and bypass Springfield so that we can go directly to Lincoln Homestead State Park, where cyclists are welcome to camp overnight. It is on Miles Creek Road, 99 miles from Irvine and just four miles to our destination, when the rain begins. But no matter—we suit up and within 20 minutes take shelter under the pavilion at the park.

All alone, we spread out on the picnic tables, with David preparing one of his specialties: pasta, cheese, and carrots. Recognizing my limited skill set in the kitchen, David keeps it simple for me: "Just slice the carrots and cheese." We've even managed to save some Harrodsburg cookies for dessert. What a day, a gift of 103 miles, and in spite of the dire weather forecast, only the last four in rain. Our stomachs full, our bodies and bikes having served us well today, we are dry and warm and happy.

To the roar of rain pelting the metal roof over our heads, we each lay our pad and sleeping bag on a picnic table, well off the water that now puddles on the concrete floor. As I drift off to sleep, this must be heaven, I think, a carefree day of a good 100 miles on the bike; breathtaking sounds and sights and smells and thoughts all along the way; perfect weather, with clouds moderating the temperature and holding back on the rain; a gentle tailwind and terrain so fine that it all seemed downhill; good food; and a traveling companion beyond comparison, once he's caffeinated. Yes, heaven is an endless string of such fine spring days, though sprinkled now and then with just a pinch of adversity so that we can savor the whole all the more.

9.
LAID UP

DAY 19, MAY 22: BARDSTOWN, KENTUCKY

It wasn't as if there had been no warning. Twelve days ago in Lexington, Virginia, on the fifth day out, my left shin was red and sore. "It's cellulitis," concluded the nurses at the hospital's emergency room. Antibiotics would cure. To try to stem the growing inflammation, six days ago in Hazard, Kentucky, I picked up some stronger antibiotics. Whenever I could, especially when eating and sleeping, I've kept my left leg raised as I tried to encourage some healing.

House sparrow. An exuberant male, at Lincoln's homestead. (104, 2:41)

Yesterday began with promise, a dawn chorus of kingbirds and orioles and house sparrows and all manner of good songsters at Lincoln's homestead, where Abe's parents met. The overnight rains had yielded to a heavy mist and drizzle, making the 20 miles to Bardstown cool and wet, yet pleasant. It was there, at the corner store that advertised liquor and guns but no bathrooms, that my left shin began to complain loudly.

The staff at the hospital's emergency room worried first about a blood clot; a miniature ultrasound device soon traced veins from my groin to my ankle, the nurses gathering to fawn over the fine, clean vessels of a long-distance cyclist. With a normal white blood cell count, I had no infection, they concluded, not cellulitis, so I should throw out the antibiotics. X-rays showed no bone issues. The diagnosis: phlebitis, an inflammation of the veins. The cure: Rest with leg elevated, hot compresses, and 800 milligrams of Motrin three times a day until symptoms dissipated. *No biking!*

Meanwhile, David had found a motel room, and he soon led me there, by bike. After I settled in, he was off to the pharmacy and grocery and video store for supplies. My leg attended to, we ate, watched *A Scary Movie*, and went to bed. Lying awake at sunrise, I listened to a rousing chorus of house sparrows outside the tiny back window. A nighthawk *peent*ed somewhere in the sky, the occasional booming *VROOM* of his dive display bursting through the open window. Ate breakfast. Watched more movies, the likes of *Analyze This* and *Jackass, The Movie*. David headed for town, eager to explore the museums and enthusiastic about seeing the new *Matrix* movie in the local theater.

I lie in bed now, following the word "APEX" as it methodically traces a path about the darkened television screen. The last movie is over, the DVD player and TV long ago having entered sleep mode. I want to get up and turn them off, but I'm immobilized, lying on my back, a chair perched atop the bed, my left leg resting atop the chair; my shin is wrapped in a heating pad, and a fresh dose of Motrin courses through my veins. It's been over 24 hours since checking into room 102 of Bardstown's Old Kentucky Home Motel—a full day of inactivity and resting so that the leg can heal.

How high I felt two days ago with our 103-mile ride from Irvine to Lincoln's homestead, the ride that probably did me in, and how low I felt yesterday. If the leg doesn't heal, the trip could be over. If we do get moving again, these delays mean fewer birds will be singing by midsummer in the far West. And we were going to meet Tom, David's college friend, in Denver, so that we could bike the Rockies together. "How about Kansas, Tom?"

Increasingly, though, I feel at ease. "It'll work out," David said just a few days ago while replacing the spokes. In his *Zen* book, the boy is eager to get moving on a trip, but the father says there's nothing up ahead that is any better than right here. I work on that idea, the father-son roles reversed. I work on relaxing, too, replaying the advice in the Pippa Passes log book: "Enjoy every minute. It'll soon be over." I see the turtle next to the flower pot on the poster in my office back home: "There's more to life than increasing its speed." I recall the great mystery throughout the book *The Precious Present,* how "the present" remained a mystery until the reader finally realized that that "the present" is simply "now," living in the present, not reliving the past or worrying about the future.

I recall countless stories that I've heard in the songs of birds over the nearly three weeks that we have been on the road. I travel back in time 150 million years to the first dinosaur-birds, then follow lineages forward, imagining the magnificent life tree for all birds. I page through the field guide, studying the range map for each species, reliving encounters with eastern birds, anticipating the western ones.

In the range maps I see stories of how the Mosquito Prince *Empidonax* flycatchers and other groups are distributed about the continent, and I try to imagine the geological and evolutionary forces that have gifted us with such a variety of these birds today. The continent-wide stories are a joy to contemplate, but so are the smaller ones, such as told by the parulas, their daytime songs telling of a great division into East and West in fairly recent times, with a current-day boundary in Boone's forest; or as told by many songbirds who with their local dialects tell of isolation and influence; or as told by many of the warblers, who in their choice of song tell their innermost mood.

And I reflect on my traveling companion. He was born at home because we wanted something special for his birth. I can see this happy, curly-headed blond on his first tricycle, then his little bike with training wheels, and soon I am running beside him, holding him up and preventing a fall as he learns to balance without those trainers. His bikes grew as he did. When he was in junior high school, his fast-growing knees hurt, and while biking together I once, only partly in jest, pulled him up a hill with a tow rope that I brought along just for the occasion. He thrived, athletically and academically. In high school, he set swimming records, and was a national champion with his classmates in an engineering competition. As a high school junior, he practiced with the Stanford University Ultimate Frisbee team, and they recruited him, calling him "super frosh" when he arrived on campus; a few years later he would captain their national championship team.

Yes, some kid, and here he is, this twenty-four-year-old now pulling me uphill. It'll be good, whatever happens. We'll get on the road whenever we get on the road, or not. But I have this feeling that once we pass the city limits of Bardstown, the bike and body repairs will be behind us, and we'll be into Illinois and Missouri and Kansas and Colorado and at the Pacific before we know it.

With David out on the town, I reach for the phone and call home, talking to my daughter Kenda. She's a physician at the local hospital, an ex-Cornell gymnast who knows all about working through pain. She listens to my story, and in the end summons all of her well-trained doctor-speak to say, essentially, "Suck it up and get back on the road."

I like that. Tomorrow I have an appointment back at the emergency room in the afternoon, but what more could I learn there? I think I'd learn more on a 35-mile test ride over to the county park in Hodgenville, which just happens to be on our route west. And when that goes well, and it surely will, we'll be flying!

10.
ON THE ROAD AGAIN

DAY 21, MAY 24: HODGENVILLE TO UTICA, KENTUCKY

Sometime in the early morning hours, with all of Hodgenville asleep, to the hooting of a distant great horned owl and the relentless serenade of a nearby mockingbird, I turn over in my sleeping bag and pinch myself. Yes, I'm here, we're here, camping happily in Hodgenville, the birthplace of Abraham Lincoln, not just the homestead where his parents met. And I hardly dare even think it, but I do believe we might be flying. By yesterday morning my leg had improved some, so we left Bardstown a little after noon, a friendly tailwind and Motrin making light of the 35 miles. The weather was the best yet, with a high in the low seventies, low humidity, sunny with no threat of rain, and even singing birds seemed to celebrate the day.

We felt the landscape changing, with corn and hay fields increasingly lining the road, and we saw our first Catholic church, the thousands of others before this, on every street corner it seems, all Baptist. The Baptists took great pains to remind us of our failings with big signs out front declaring that "SIN DESTROYS," or imploring that we "Fight Truth Decay. Read the Bible," or inquiring "Eternity. Where? Heaven or Hell?" More emphatic was "TURN OR BURN." Also fading were the Ten Commandments posted in front yards and Confederate flags aloft.

An hour before sunrise, I slip out of the tent and over to the mockingbird who sings from a small tree beside the park's pavilion. He bares all, telling of the phoebes and swallows and bluebirds and cardinals and kingfishers and wrens and flickers in his life, but mostly, telling with his exuberance that he is alone and seeks a mate. Robins by every available electric light have already been singing for over an hour, and exactly when the tree swallows began over the open playing field I do not know. I walk out beneath them, gazing up into the stars, finding the Swan and the Eagle, now standing beneath a swallow who seems to hover in the darkness perhaps 20 yards overhead. How continuously he pours out his liquid twitters, about two phrases per second, the variety seeming endless; but gradually I feel a rhythm, and it seems he has just a few, maybe three or four simple phrases that he offers to the dawn.

The towhee has warmed up in the bushes beside the tent, and he alternates two different *drink-your-teeeee* songs, soon introducing a third, bouncing his way among all three, and there's a fourth song; I lose track of which song is which, but he knows, as each is a well-practiced routine that he's memorized and burned into the neurons of his singing brain.

Northern mockingbird. The night-singer, a bachelor seeking a mate. (105, 14:44)

Just above the towhee is a bluebird, also in full dawn mode. His harsh chatters alternate with the songs that bluebird lovers gush over, each song a rich, low, melodious and mellow and cheerful and pleasing and charming outburst of affection, so the lovers would say. I hear him alternating two songs now, as in *ayo alee,* chatter, *alee la-o,* chatter, *ayo alee,* chatter, *alee la-o,* as if he asks a question on the rising *ayo alee* and then answers it with the falling *alee la-o*. With 50 or perhaps even 100 different songs in his repertoire, he has a lot to say, and I leave him to his task.

Eastern bluebird

I walk the tree line bordering the eastern side of the park, marveling at the always surprising songs of a yellow-breasted chat . . . admiring the methodical performance of a great crested flycatcher in dawn song, playing out his infinitely variable *wheeee-up* calls, punctuated with that curiously faint, low-pitched buzzy note . . . enjoying the persistent, mournful *cooowaah, cooo, coo, coo* of a mourning dove, so owl-like, the rhythm so distinctive . . . indigo buntings . . . more towhees. . . .

Mourning doves. A chorus of mourners at dawn, with cardinals. (106, 10:22)

Back at our campsite, the brown thrasher who sang last evening over the tent has resumed his monologue. He continues all through our breakfast and into the packing, and still sings as we mount our bikes to leave about an hour after sunrise, on this our twenty-first day since leaving the Atlantic.

Yes, Willie Nelson, we are "On the Road Again," making music with our bikes, a band of two gypsies going down the highway, the best of friends, going places we've never been, turning the world our way, and I now dare imagine us on the banks of the Ohio River late tomorrow afternoon, some 180 miles to the west and boarding the ferry that will take us into Illinois.

Dickcissel

DICKCISSEL DIALECTS

Dialect 1, southern Illinois, Shawnee National Forest (107, 0:18)

Dialect 2, southern Illinois, Levee Road beside the Mississippi (108, 0:21; 109, 0:20; 110, 0:24)

Dialect 3, Perryville, Missouri, just across the Mississippi (111, 0:33; 112, 0:29; 113, 0:31; 114, 0:24)

Riding, I check in with singers all around, but am soon overwhelmed with dickcissels in an overgrown field that apparently has not been farmed in some years. *Dick dick cis cis cis cis,* they all seem to say. Beyond a farmhouse and after passing through a woodlot, we encounter another field of dickcissels, all of them now saying *diicckk diicckk diicckk diicckk*.

I pull up beside David: "Do you hear this? Listen to how all of these dickcissels have the same song. Concentrate on the quality and rhythm of the song, how each song consists of a slow repetition of a single longish phrase, and then compare these songs to those we'll hear in the next field, where they'll all be different."

"So what's the big deal? You've told me about these dialects in chickadees and cardinals and lots of other birds," counters David. Good, I think—he's taking this stuff in.

"Yeah, but it's tough to hear the dialects in those other birds," I explain. "Either you have to travel a long way before you cross into another dialect, or you need a great memory for sound to recognize how the dozen or so different songs of each cardinal might change over a short distance. It's easier to hear the dialects in these dickcissels because each bird has just one song, and two fields a minute apart can have entirely different songs."

And sure enough, just down the road David notes how the dickcissels are singing yet another song. "An astute listener would wonder why dickcissels sing all day long," I try to bait him. No reaction. Did he hear me?

We bike on, past wet meadows brimming with red-winged blackbirds and yellowthroats; through drier fields singing of eastern meadowlarks and grasshopper sparrows; beneath single mourning doves or pairs up on the utility wires, sometimes whole flocks beside the road flushing in a whir of whistling wings. On the wires up to the right, a flock of male cowbirds sing, as if in competition, then one whistles its flight call, announcing impending flight, and all depart in a great rush. Robins, orioles, mockingbirds, buntings, all singing along the way.

Woodlots near farmhouses swarm with starlings and grackles making a racket. Intimate past encounters with each species flash through my mind, but I settle on grackles, recalling how I would sit quietly in a colony, savoring their "rusty hinge" songs, the creaking, squawky, far-from-musical, third-of-a-second *readle-eaks* to my naked ears, knowing that they're full of exquisite detail when slowed down. I hear each bird with its slightly different song, how females sing, too, how pairs in a flock are identifiable by how rapidly they respond to each other's song, how intimate exchanges within the colony are mediated with all of those harsh *chack* and *chaa* and *jit* calls.

Brown-headed cowbirds. Males displaying and singing, a flight whistle announcing departure. (115, 2:05)

With ultra-low humidity again this morning, the sunlight is indescribably delicious, the light reflected to our eyes so sharp and pure. An indigo bunting briefly lands on the road in front of us, a flash of radiance as intense as I've ever seen. He's followed soon by a male cardinal, and then a bobwhite, who runs toward us from a truck in the other lane, flushing up and over our heads, every tiny detail of this cute little chicken on vivid display. Tall thistles line the road, with huge purplish-red heads all nodding to the east. With a nod to "smell the roses," we stop to sniff the thistles; we find little fragrance, but are impressed with the detail in the massive flower heads. American goldfinches call overhead, *per-chik'-o-ree, per-chik'-o-ree,* they, too, having an interest in the thistle, as they'll wait until July to breed when the thistles provide down for their nests and seeds for their gullet.

Common grackles. Unique squeaky-hinge song of one bird, among colony sounds. (116, 0:52)

Near Eastview, the yard of a white farmhouse on the right is decorated with dozens of birdhouses, from large purple martin condos to single apartments for house wrens and bluebirds, and attending them, we soon learn, is Reverend James Love. He's dressed in faded bib overalls with a patch over the right knee, in his late seventies, I'd guess. James has served in a number of local churches and been involved in several revival meetings, he says, but at the moment he's "just growing old and now semi-retired." In his fine local dialect, he tells how he loves to make feeders and nesting boxes, how he thinks birds have more joy than many people he knows, how he loves to awaken early and enjoy the morning birds. He tells of titmice and starlings and cowbirds, but enjoys most the mockingbird who sings all night long at his son's house next door, then comes to his house and sings all day. "Do you know why he's singing?" I ask. "Happy! I think he's enjoying life and he's singing to make people enjoy and take notice." I yield to his interpretation of the mockingbird's song, though perhaps James would have liked knowing that his night singer was unpaired and seeking love. But if I go there, somehow I'd have to reflect on how unhappy and joyless the female mockingbird must be, as she doesn't sing, and then on how he inevitably sings for sex, and it is far too nice a day to counter James's view of the world.

Reverend Love. Telling of birds in his yard: redbirds, jay birds, the mockingbird, and that problem bird the starling. (117, 11:13)

Beyond Eastview, we dip into the valley of the Rough River, riparian voices again abundant, Acadian flycatchers, Louisiana waterthrushes, and Kentucky warblers among them. Now a prothonotary warbler joins the chorus, a loud, emphatic, beautiful *weat weat weat weat weat*, each *weat* a pure tone sweeping up the scale. "David, here's a new bird!" As we pass, I point out the song and strain to catch a glimpse of a feathery patch of radiant sunshine in the tangles, but to no avail. Seeing is highly overrated, I remind myself.

By noon we've logged 50 miles. We soon fly through Fordsville, Reynolds Station, Whitesville, and, with 100 miles on the odometers, arrive at Utica, where all is quiet this Saturday night at our campsite beside the elementary school. We've crossed into the Central Time Zone, and for the first time on this trip I feel we're making a beeline for the Pacific. We sleep quickly, and depart early the next morning, eager to make Illinois by nightfall.

Prothonotary warbler. From a wooded swamp beside the road, a splash of sunshine in song. (118, 2:02)

DAY 22, MAY 25: UTICA, KENTUCKY, TO CAVE-IN-ROCK, ILLINOIS

We're just out of the schoolyard when a bobwhite calls from a fence post so near on the right that we hear and see it all, *oh BOB WHITE!* Immediately, from the TV antenna of a house across the road, a mockingbird outdoes him, *BOB WHITE! BOB WHITE! BOB WHITE! BOB WHITE! BOB WHITE!*, springing up into the air and flashing the white in his wings and tail for good measure. David hears it all, too. "It's going to be a good day," I predict.

Northern bobwhite. Distinctive, unmistakable, riveting: *BOB WHITE!* (119, 2:05)

Cows line the fence in the next field, all raising their heads and staring at us with that blank bovine look on their faces. "*Baaaaaa . . . baaaaaa,*" I hear from David. Responding to my questioning *Huh*?, he explains that he plans to test cow intelligence from here west. He wants to know if they respond differently to a *mooo*, a *baaaaaa*, or a *neigh*. I like that—pure and simple, it's research as fine as some of what has been published in my field of animal communication.

It's heavily overcast today, the light so different from yesterday, and soon the skies empty on us. Thoroughly soaked but in good humor, 25 miles down the road we arrive in Sebree, taking refuge there in a local restaurant. With our bikes outside in the rain, we sit at a window where we can keep an eye on them, draping our bright yellow rain gear over the four chairs at our table. As small puddles accumulate beneath us in the all-but-empty restaurant, we down a hearty breakfast, our second of the morning, then order lunch. When the Sunday crowd arrives, straight from church, the tables fill quickly, those at some distance from us the first pick. David stands to adjust the shade at the window, hoping to gain a better view of the weather, but the shade comes crashing down onto the table. "Thanks for the help," says the owner as he comes over. "I was going to replace them all next week."

We eat on, admiring the huge painted flag on the side of a building just down the street. The 13 red and white stripes are there, and the blue field is to the upper left, but in it are listed the Ten Commandments, framed by 50 white stars around the perimeter of the blue field. Above the flag are the words "NEW GLORY," and beneath it "RESTORE THE LAW. RESTORE THE GLORY."

The rain having subsided, we pay at the cashier, where an older gentleman with gray hair and moustache and fine Sunday attire approaches me, commenting, "I really admire what you're doing." Thank you, I think, wondering who he is and whether he has had his share of fulfilling his "someday I'd like to" yearnings.

House finches sing us through town with their rollicking, cheery warbling, each song so lively and rapid, with an occasional longer, husky, down-slurred *veeeeer* note thrown in. How remarkable to have them here in western Kentucky, how explosively these finches settled the eastern half of the continent once a few California birds were released on Long Island, New York, back in 1940.

House finch. Rollicking and cheery, two different songs from a Sebree bird. (120, 2:02)

Sedge wren

Sedge wren. Songbird who sings snowflakes. (121, 3:46)

Just outside of town in an extensive old field on the left, I hear what I had so hoped for somewhere on this journey: *cut-cut-cut-trrrrrrrrrrrrrrr*. "David, sedge wrens!" It was 29 years ago when I first met sedge wrens, and since then, in an attempt to understand them, I have traveled Canada and the United States, and south to Costa Rica, Venezuela, Brazil, and the Falkland Islands; to learn how they sing, we have raised babies in our house and followed their song development for a year.

"David, these are wonderful birds."

"Aren't they all?" he interjects.

I ignore him: "They're an exception that proves the rule. They're songbirds, and each male has 100 or more different songs, but he makes them all up based on a simple, standard pattern. As a result, no two birds have the same songs—each song is like a snowflake, different from all others yet of the same design. And the reason I think they do this? These wrens in North America move around a lot, being somewhat nomadic, and each bird might breed at two different places in the same year, maybe once early up in Saskatchewan, maybe a second time down here in Kentucky while on the way to the Gulf Coast for the winter. So instead of imitating the songs in one neighborhood and breeding there for life like many songbirds do, these sedge wrens improvise a generalized song that is instantly recognized as that of a sedge wren anywhere they might go. But, and get this, from Costa Rica south these wrens don't migrate; instead, they're resident year-round and sing in the same neighborhood throughout their lives, and, here's the best part, in those stable communities they *imitate* each other's songs."

"So what's the rule?" asks David, "that if you're going to be talking in the same small community of people throughout your life, learn to talk like them, like back in Appalachia? Otherwise, talk so you're understood anywhere you go?" Yeah, that's pretty much it, the rule being the same for humans and songbirds.

Another 20 or so miles down the road, just past the town of Clay, I hear the sharp *pit-i-tuck* call of a summer tanager. He seems to call from a dead tree beside the road, so we turn back, hoping to get a look at him. Yes, there he is, though he's only a silhouette against the bright sky. We move over so that he's in line with the trunk of the tree, then close our fist, making just a pinhole through which we can see only him against the dark trunk. With our eyes adjusted to the darker lighting, a rich, rosy red bird springs into view. Seeing is good sometimes, I admit, though the binoculars stay in the pannier.

Summer tanager. *Pit-i-tuck . . . pit-i-tuck* he calls repeatedly, as if perturbed. (122, 1:02)

With just under a dozen miles to the Ohio River and only mid-afternoon, we're making good time, averaging 13 miles per hour, the wind in our ears and the sun at our backs as we head north, when I hear a small something over to the right in an old field beyond a barbed wire fence. I slow, wondering, maybe just an insect, a branch breaking . . . no, there it is again, an unmistakable *tsi-lick*, the oh-so-brief song of a Henslow's sparrow. When I slow down these songs on my computer, the split-second song stretches into a rippling cascade of delightfully pure notes, perhaps the way the birds themselves hear their songs, but now, with the unaided ear, I hear only a hiccup. Add the Henslow's sparrow to the dickcissel and the sedge wren as gifts from Kentucky that foretell the prairies to come. "Sang 13 songs during the last minute," pipes up David. It's the scientist in him, I think, irresistibly collecting interesting numbers.

A few miles from the Ohio River, David asks, "Why do indigo buntings sing so much?" Ah, he had heard me yesterday when I commented about how an astute listener would wonder why dickcissels sing so much. Or maybe not; he gives no clue. "Some of the buntings," I explain, "are first-year birds just back from migration—they seem to practice nonstop as they learn the songs of an older neighbor. But the rest of the songs are from older males who seem to sing in an attempt to impress the females in the neighborhood. You see, a male and female will pair off to raise young together, because young birds apparently need two adults to care for them, but there's no guarantee that the babies in the nest are going to be his, because the females engage in what ornithologists call 'extra-pair copulations,' mating with males other than their partners. So a male who is especially impressive might have babies scattered in several nests around the neighborhood. And as for dickcissels, should one ask, males are more openly polygynous and can attract a 'harem' of females, and the more impressively a male sings the more females he might attract to pair with him and raise his young."

Henslow's sparrow. *Tsi-lick*: Brief, yes, but not pathetic. (123, 1:49)

We whoop and holler as we approach the Ohio River, *"Adios Kentucky, Aloha Illinoooiiis!"* Barn swallows dip down to the mud in the roadside ditch, pack a beak full, and fly off somewhere to build their nests. Kingbirds twitter incessantly, flying about the small trees beside the road. From high in the cottonwood on the riverbank just to the right of the ferry landing a Baltimore oriole sings his *here here, come here dear*. The river itself is brown and broad, almost half a mile across, the rapid current sweeping to the left. We stand in wonder at the overall scene, impressed by the progress we've made in the last few days, eager for Illinois, humbled by the image of Lewis and Clark passing by here on the river 200 years ago back in 1803.

We wait for the three vehicles in line to board the small ferry, then walk our bikes up the ramp, lashing them to the railing on the right just behind a red pickup truck. The ferry is soon underway, but as we float across the river, it seems that we're cheating again, now by not biking the entire distance to the Pacific—our only other choice, though, is to bike around the deck, and that seems silly. Having put on our rain gear for warmth on this cool spring day, we mill about, taking pictures of each other and the bikes and of Kentucky receding and Illinois advancing.

"WELCOME TO ILLINOIS NATIONAL SCENIC BYWAY" declares the large sign just off the ferry, adding a subtitle of "OHIO RIVER SCENIC BYWAY." Even more welcoming is the smaller sign just below, the likes of which we've not seen since Virginia: On a green background is the white outline of a bicycle, with the words "BIKE ROUTE." We feel instantly at home, more so than we have throughout Kentucky.

We head directly to Cave-in-Rock State Park, set up the tent, shower, and hurry to the restaurant, finding a large Sunday crowd inside. How puzzling that no one is outside on the deck overlooking the river, where half a dozen tables lie empty. Inside, we order our roast beef and mashed potato dinner with extra helpings, arranging to have it delivered to us outside. Pecan pie and ice cream follow, and we linger as the day gradually darkens and cools.

A pair of fearless chipping sparrows feed their nestlings in a small potted plant beside us. Down near the river a Carolina wren sings, and purple martins dip and frolic over the river itself. Two Baltimore orioles sing from high in the trees. Chimney swifts twitter higher overhead.

David writes in his journal: "We're up high, above the Ohio

River. We are finally moving with confidence, speed, and fun. In only two days of biking, we will be at the Mississippi River. The hardest climbs of the trip are behind us. Excitement and open roads ahead. It is good to be biking!"

There's a warbling vireo! It's the best we've heard yet, from the canopy of the nearby oak. Bursting with energy, he sings so rapidly he almost trips over his own notes as he races in his husky voice through the highs and lows of his song, emphasizing the highs, as in the mnemonic *If I SEES you, I will SEIZE you, and I'll SQUEEZE you till you SQUIRT*. He sings as if to remind us to listen carefully, because beyond the Great Plains we're promised an altogether different song from warbling vireos, so different that the western bird should probably be recognized by us humans as a different species.

The Great Plains! Just the thought of them . . . It took us ten days to stutter through eastern Kentucky, with four days for bike and body repairs alone, but in the last two days we've swept through almost 200 miles of western Kentucky. Looking north on our Illinois map, we see that we're already west of Chicago, and at the rate we are now flying, we'll be camping on the Mississippi River in two days, and in another five days be through Missouri and into Kansas. With our stomachs full and both bodies and bikes happy, that seems an entirely reasonable plan.

Warbling vireo. Just above us on the bicycle-friendly Ohio River Scenic Byway. (124, 2:00)

11.
DAWN SWEEPS THE SHAWNEE

DAY 24, MAY 27: FERNE CLYFFE STATE PARK TO CHESTER, ILLINOIS

Acadian flycatcher. *PEET-sah!* song, repeated in non-dawn mode. (125, 1:58)

All night long, it seems, I've heard the Acadian flycatcher and the eastern wood-pewee, about every 15 minutes or so as I've listened from my sleeping bag. The Acadian sings only his explosive *PEET-sah*!, never the repeated *seet* notes that he'll introduce during the dawn frenzy. And the pewee sings only the *wee-ooo*, never his *pee-a-wee* or *ah-di-dee* songs that he will add during the dawn chorus. They, like I, stir throughout the night, perhaps also in anticipation of the dawn.

A little after 4 a.m. I awake for good, dressing quickly and ambling down the path to the top of the cliff where we watched the sun set last night. Ferne Clyffe State Park, I smile. As we planned this trip, I'd read about this place and marked it on the map, knowing that when we reached the Shawnee National Forest in southern Illinois, camping here was a must. So yesterday we biked only 60 miles from Cave-in-Rock on the Ohio River to here, then parted from the four other cyclists with whom we rode much of yesterday—they'd stay in nearby Goreville or continue on to Carbondale. How sad, I thought, that they'd miss this place.

By flashlight, I lay my microphone at the edge of the hundred-foot cliff and aim it out into the darkness, then stretch the cable over to my camp chair. Settling in, getting comfortable, I connect the cable to the recorder and punch "record." "Ferne Clyffe. 4:22 a.m., May 27, sunrise about 75 minutes from now," I announce. For the next three hours I have nothing better to do than listen to the world awake.

The faint, whitish glow of the rocks at the cliff's edge drops off abruptly into the black abyss below. On the left, anchored well down the cliff, looms the shadow of a magnificent tulip tree; to the right is another large tree, perhaps a hickory, these two trees framing what will eventually become a remarkable view over the forest below. Soon dawn's first light will energize the hundreds of birds now all roosting in this scene, and I am here, perched among them, waiting eagerly.

I inhale deeply, as if a deep breath could capture within my chest all that is before me, when the purple martins again burst into my consciousness. I had heard them a good hour ago while I was still in the tent, and now they sing high overhead. They're off to the right, maybe a hundred yards or more up. How many? Two birds singing continuously, each with his own distinctive song, could sound like a large flock, but I'd guess at least five, maybe a dozen birds sing up there. In their bluish-black iridescence, with all the wisdom of experienced birds who have already migrated to Brazil and back at least twice, I wonder just how much they "know."

In their singing club up there, do these studs know that yearling males and females are still straggling in by night from Brazil, and that all this super-early dawn singing just might attract some of these young birds to join the colony? At some level, consciously or not, they know that having young pairs in the colony is good, especially because a young female is going to be impressionable and do what's best for her. With each older male already taken by another female, a yearling female will likely pair with a yearling male, and he will help raise her babies, but she'll likely hedge her bets by raising broods of mixed paternity, maybe two fathered by her own young and unproven partner, another two by whichever older male most impresses her.

Purple martins. High overhead at Ferne Clyffe, older birds singing hours before sunrise. (126, 2:54)

Humans can play much the same mating games, with as many as a third of women in one "monogamous" society giving birth to a baby fathered by a male other than her marriage partner. Why should I accept the "scientific view" that these purple martins are automatons driven by their genes, and that, unlike humans, they have no awareness of the mating games they play? I would wager that they know far more about their lives than we have given them credit for.

Lightning flashes on the distant horizon to the northwest, but miniature flashes are all about me, fireflies everywhere. I count out their rhythm, concluding that most flash three times over about two seconds and then pause briefly, but some seem to stay on continuously for several seconds, while others flash just once. Different species, perhaps. Just behind me a metronomic insect buzzes once a second, on and on.

In the darkness, I try to imagine being here some 500 million years ago, when an ancient landscape slowly sank and shallow seas invaded. Over roughly the next 300 million years, the seas would come and go, with ancient rivers emptying here and dumping so much sand that sandstone would form to a thickness of 20,000 feet, almost four miles. The cliff on which I now perch is merely an erosion-resistant wrinkle atop that vast sandstone layer all over southern Illinois and adjacent Indiana and Kentucky. More recently, during the two million years of the Pleistocene, I'd have felt the chilling winds off continental glaciers that advanced over most of Illinois but stopped just three miles to the north, leaving these Shawnee Hills unscathed.

From my weathered perch, I concentrate, listening. In the distance below, as a slight breeze briefly rustles the leaves of the trees around me, I hear the isolated *kow-kow-kowlp-kowlp-kowlp* ending of a yellow-billed cuckoo song. Beyond the cuckoo, a lone dog barks. The cloud of purple martins has drifted south toward me and is now almost directly overhead. Just below me a chuck-will's-widow repeatedly says his name, *chuck-WILL'S-WID-ow*, and two others answer, one to my left, one to my right, so distant that all I hear is *WILL'S-WID* from them; how appropriate the name "nightjar," because of their often endless singing that jars the night, probably unpaired males seeking a mate. And a distant whip-poor-will joins, four nightjars now rattling the night. Behind me, near our campsite, I hear the simple *you-all* of a barred owl, most likely the male.

Chuck-will's-widows. Sight unseen, two widows duel. (127, 4:02)

At 4:45 it's still very quiet and very dark, with stars still visible through the light cloud cover; best of all, it's perfectly calm. Just to my left an indigo bunting bursts into song, but just one, as if in anticipation of the frenzy to follow; in the canopy below, a great crested flycatcher sounds off with a raucous series of *weep* calls, as if unloading some nightmare, but then he's silent, too. I check in with the regulars, the martins, the fireflies, and the buzzing insect behind me; I listen for the occasional cuckoo and barred owl. I count two more *weep* outbursts from the great crested flycatcher before he settles into his regular dawn singing, at 5:01 a.m. according to my watch, 37 minutes before official sunrise here; now it's two versions of a rather gentle *wheeee-up* call alternated with his barely audible, low buzzy note. Nice. The dawn chorus has begun, and it is the great crested flycatcher who leads the way.

Great crested flycatcher. Methodically alternating two different *wheeee-up* calls with a curious low buzzy note. (128, 1:32)

Almost immediately, two indigo buntings, one on either side of me, fire in rapid succession. To the left a phoebe warms up with a string of *FEE-bee* songs, and another flycatcher, the peewee, begins his full dawn serenade from the tulip tree to the left: *pee-ah-wee, wee-ooo, ah-di-dee, pee-ah-wee, wee-ooo, ah-di-dee,* these three songs over and over, about a song every two seconds, 30 per minute. An Acadian flycatcher joins the chorus, *seet, seet, seet PEET-sah*! *seet, seet, seet PEET-sah*! Now two phoebes sing, left and right, the pace escalating, each rapidly alternating his two songs, *FEE-bee, FEE-b-bre-be, FEE-bee, FEE-b-bre-be*; a cardinal just behind bursts in, offering two different songs back to back, then settles into his routine of repeating one of his songs several times before switching to another. *Rain!* . . . but with stars still visible overhead, I realize that it will be only a light shower, and it quickly passes.

At 5:09 I hear, from the 20-foot oak to my right anchored partway down the cliff, the *pit-i-tuck, pit-i-tuck* calls of a summer tanager, but then emerging from the oak is a song I've never heard before. It sounds much like a robin's series of low-pitched carols, but unlike a robin this mystery bird sings continuously, never pausing, and the tempo is far too slow for a robin, about one phrase per second instead of the robin's two; nor do I hear any of the robin's high-pitched *hisselly* phrases. The singing program is also different from that of a robin; a robin favors one group of songs for a while and then switches to another group, but I hear a particular unique phrase from this bird about every six phrases. It has to be the songs from the summer tanager who just called there, I conclude, as he sings in the tanager way at dawn, taking the phrases from his discrete daytime song of about two seconds and playing them out more slowly and continuously. That's just what the scarlet and western tanagers do, too, though these other two tanagers always punctuate a brief series of their song phrases with their distinctive call note.

Acadian flycatcher. In dawn mode, each series of *seet* calls culminating in a single explosive *PEET-sah!* (129, 1:59)

Summer tanager. In halting dawn song, just below on the Clyffe. (130, 4:29)

From the canopy below me, a parula offers his dawn song—it's raspy and multi-parted, swirling upward, far more complex than the daytime song. Behind me in the campground two tufted titmice hurl the identical odd song back and forth at each other. The whip-poor-will and one chuck-will's-widow continue down below, the *you-all* behind me telling that the barred owls haven't yet gone to roost.

Northern parula. In dawn song, chipping madly between raspy, ascending songs. (131, 1:22)

At arm's reach to my left lands a smallish bird, soon belting out a beautiful song, so clear and sweet, a series of slurred

Louisiana waterthrush

Louisiana waterthrush. My companion, perching with me atop the Clyffe. (132, 0:36)

whistles followed by jumbled chips and chirps. Highly animated, he bobs his tail, seeming to circle it around, then sings again, and again. I'm momentarily confused by the sight and sound of a bird so close, but I concentrate on his looks, noting the dark streaking on his whitish undersides and the brownish back, the dark brown eyeline and cap, the striking white line above and behind the eye. "Hey, Mr. Louisiana waterthrush," I whisper to the feathered being who enjoys the same perch here that I do.

David arrives—what a surprise! Little is said, but he finds a comfortable place on the cliff edge and settles in next to my microphone, where he proceeds to open a pecan Danish and make far more noise than I can bear. I offer a few tentative thoughts about where he might better sit, but quickly realize it's easier to move the microphone, and the Danish will soon be devoured.

Below David, from a small tree about halfway down the cliff emerges a dry, rapid rattle, my subconscious declaring "chipping sparrow." But between songs, from the same little tree, I hear a persistent chipping. Ah, I know who that is! No chipping sparrow chips like that between songs, but I've heard that kind of chipping between dawn songs from so many warblers. Smiling broadly, I whisper confidently "worm-eating warbler," and although we've probably heard them on occasion before, this is the first time I know for sure. Very nice.

Worm-eating warbler. As the chorus fades, still chipping before some of his dry, rattling songs. (133, 2:00)

While I'm immersed in the warbler, David gets up and quietly departs.

At 5:30, eight minutes before sunrise, I take note of changes in the chorus around me. A family of crows caws behind me in the campground, and down below I hear the *chuuree chuuree chuuree chuuree* of a Kentucky warbler. Overhead I can now see the martins, but they're far quieter, with none of the continuous nighttime singing heard before. In the bushes behind me, a blue-gray gnatcatcher wheezes what sounds like the identical note over and over so continuously that he sounds like a young bird begging for food, though it seems too early in the season for that. "How different they all are," I think, "different beings, yet equal, all 'survivors' from the beginning of time," as I quietly celebrate the variety of minds who have come to the chorus this morning.

David returns, this time with his sleeping bag, laying it on the rock surface and crawling inside. He knows this is a special place. In his journal last night, he wrote how beautiful it was to watch the sunset from the cliff here, and something like "I felt as if I could jump off the cliff and run along the treetops. A brown thrasher sang, earning its place as my favorite bird so far."

But it's not so beautiful this morning. He wants to be here, but feels it can be better appreciated in a semiconscious or even unconscious state. How he has struggled with mornings since the very beginning, when as a child he often awoke crying. And how memorable that one morning when our family was backpacking in the Cascade Mountains of Washington, when David was last to awake in the tent, announced by his happy, ringing "It's Morning Time!" He's an outside person, whether hiking, biking, swimming, rock-climbing, playing Frisbee, or the like, though any such activity still remains all the sweeter when carried out well after sunup.

American crows

At 5:38 a.m., the very moment of sunrise, I once again sweep my ears over the soundscape. Two woodpeckers have just checked in, as they often do about sunrise, the red-bellied down below with its *chig chig* calls, the downy behind me with a few *piik* calls and a slow drum. Other than those two new birds, it's the usual ones, though each in its own way seems less frenzied, the dawn effort past. Only one bunting now sings, in the tulip tree to the left, and he pauses far longer between songs than he did earlier. The cardinal repeats each song many more times before switching. One titmouse continues singing to the right. The great crested flycatcher has abandoned his methodical delivery of dawn songs and now offers only an occasional outburst of throaty, raucous *weep* calls. The phoebes sing with less energy, no longer rapidly alternating their two songs but instead singing more slowly a string of three or four *FEE-bee* songs followed relatively quickly by a single stuttered *FEE-b-bre-be*. The wood-pewee continues, too, but he's abandoned the special *ah-di-dee* and rapid singing of dawn and now sings a leisurely string of *pee-ah-wee* songs followed by a single *wee-ooo*, only four or five songs per minute. The parula also marks the hour with his daytime song, the rapidly rising buzzy song with a high wimpy ending; the worm-eating warbler uses his same song, but chips less often. In his own way, the blue-gray gnatcatcher behind me seems to wheeze less emphatically. The nightjars and barred owl are silent, as are the purple martins overhead.

Eastern wood-pewee

Eastern wood-pewee. In daytime song mode, a leisurely six to seven songs a minute. (134, 2:10)

With David sleeping on the rocks nearby, I linger, looking, listening, smelling, thinking happy, contented thoughts. I note new birds: the late-rising pileated woodpecker, red-eyed vireo, and brown-headed cowbird. After a distant, thunder-like boom, a turkey gobbles emphatically in the forest below. Three blue jays now fly and *jay* over the canopy before us, and a ruby-throated hummingbird buzzes and chatters as he flies by my nose. Well past dawn now, a chipping sparrow offers a daytime song far sweeter than any worm-eating warbler could ever muster. And others.

The landscape has transformed in the gathering light. Shadows have found form and color. The tulip tree to my left is a rich, deep green, and the pines in the valley below have a reddish tinge. From the forest below to the gentle slope leading up to a ridge on the far side of the valley, the crowns of individual trees stand out as hues of yellow and gold and green and red, a medley of trees in spring dress.

Nearly two hours after sunrise, I reluctantly rouse David, and we head back to the campsite. As we prepare oatmeal and pack up, the red-eyed vireo I heard earlier sings from the tree above us. Inevitably, I listen for a few phrases that stand out from the others, and every 15 to 20 phrases I hear that he returns to them. Very nice.

Chipping sparrow. No great hurry in his daytime singing. (135, 1:52)

By 9:30 we're at the Goreville post office, mailing home items like the heating pad that I've been carrying since

Bardstown, Kentucky, and picking up some new items, as this is one of our mail stops. In just over 25 miles, we wind our way through picturesque Shawnee National Forest and Crab Orchard National Wildlife Refuge, arriving in Carbondale by noon. There we rendezvous with Kris and Kes, the two cyclists we met back on the Virginia-Kentucky border, as well as John and Mark, two British cyclists we met just yesterday. After some preventative maintenance on our bikes at a local bike shop and a whopping lunch at Mary Lou's fine eatery, we set out at 3 p.m. with Mark. It's 50 miles to Chester on the Mississippi River, with five hours to sunset. Should be plenty of time, we reason, to take the longer route down by the river. John, Kris, and Kes will follow later, taking a shorter inland route, and we'll all meet in Chester.

We're soon following the Big Muddy River, dropping down into the flood plain of the Mississippi River itself. The land is flatter than any we've seen since the coastal plain of the Atlantic, and we bike effortlessly along, heading north and paralleling the big river that we know lies just a mile or so away. To the right, forested hills and ridges rise up, bounding the flood plain about four miles to the east; similar forested hills lie to the west, well beyond the river. Red-winged blackbirds and dickcissels line the way, with killdeers scattering from the roadside. Though we know exactly where we are, it isn't until we climb up onto Levee Road and actually see *The* River just a hundred yards away that it truly hits us.

The Mississippi River. I say it out loud, *The Mississippi River*. It's a beautiful sound, the words rolling off the tongue. The *Misi-ziibi*, the "Great River," in Ojibwe. It's the river of Mark Twain and Huckleberry Finn, narrow enough to jump across at the headwaters near Lake Itasca, Minnesota, about half a mile across here, roughly 2500 miles long, draining most of 31 states and two Canadian provinces, a huge chunk of inland North America. Five tons of rich farmland flow by in sediment *every second*, I've read, an entire farm in a matter of hours in these brown waters. A few facts come easy, but words to capture our elation do not. In 24 days, we have biked from the Atlantic to the Mississippi, biting off three states of the ten we'll pass through, and Missouri is in sight, literally.

Killdeer. Calling their name, *ki-dee, ki-dee,* they scatter. (136, 0:34)

No Indian canoes or steamboats ply the river today, only a barge being pushed upstream by a tugboat, and soon more barges hug the shore. On the right is a huge coal facility, the Cora Coal Works, and a long train slowly works its way around a two-mile oval loop. As each of the hundred or so cars enters a small building, the entire car appears to be lifted and turned over, the coal emptying into a conveyor system that then transports the coal up and over the road to the barges waiting along the riverbank. Car by car the train advances, an endless supply of coal having made its way here by train from the extensive mines right here in central and southern Illinois (hence the name "*Carbon*dale"). I've seen the numbers, about 20 different mines, 30 million tons of coal produced each year, trucks and trains running 24/7 to deliver the goods. We watch, mesmerized by the automation with no obvious human oversight, though a few vehicles are in the parking lot nearby.

It's getting late and we press on, eager to make Chester by nightfall. We take turns pulling, drafting the two bikers behind, first David in front, then Mark, then me, then David again; David plays drill sergeant, announcing every 60 seconds to change positions, with the lead biker then pulling over and falling to the back of the line. We push hard, working the hardest for our 60 seconds in the lead, then resting somewhat for 120 in the slipstream of the others. Mark pulls in front of me now: He's tall and lanky with wild red hair, in his mid-thirties perhaps; his aging bike shorts have lost a bit of their elasticity, and all of his gear seems tied haphazardly onto his rear racks. His handlebars are straight and absurdly high on his Dawes Galaxy English touring bike, so he rides far more upright than seems efficient, but how he can fly! He's good fun, a bit goofy, reminding me that all of us who bike cross-country are just a bit insane, if healthily so.

I'm exhausted, my early morning vigil having taken its toll. We make Chester by sundown, but traveling in a paceline is no fun for me, as I hear or see little except the road or someone's rear tire in front of me. Not so with David. He loves a good paceline, and the more cyclists the better. He

speaks of "efficiency . . . speed . . . intensity . . . sense of accomplishment . . . riding high on each other's spirits" and the like.

I collapse in the camping area at the city park, but David has energy to spare. He first bikes down to the river to get a closer look, snapping his sunset picture there, then returns from town with our Subway dinner. He loved the late afternoon ride and the fast paceline, and seems energized by the evening light.

Tomorrow, I think from my sleeping bag, tomorrow we cross the Mississippi River. I like the turn of the words, their implication, and say it a few more times in my mind. Then it's the Ozarks of Missouri, and for only a few more days will we get to enjoy all of the eastern forest birds that have accompanied us since the Atlantic. I hear their chorus as I drift off to sleep.

12.
THE OZARKS

DAY 25, MAY 28: CHESTER, ILLINOIS, TO PILOT KNOB, MISSOURI

We're up by 7 a.m. in Chester's city park, packing and listening to a red-bellied woodpecker drum around the pavilion. First he drums low on the trunk of a large dead tree, the resonance deep and hollow. Next he flies up to a smooth dead branch, its bark long peeled away, the resonance there sharp and far higher pitched. Then this master percussionist is on to other surfaces nearby, as if experimenting, as if testing everything he owns to determine what sounds best to his ear this morning.

Red-bellied woodpecker. A master percussionist searching for just the right resonance. (137, 4:42)

Our four cycling friends are stirring, and after warm good-byes and a "see you down the road," we depart, heading to a downtown Hardy's for some monster pancakes. Just outside of town, David warns me: "The kids I met last night when I went for dinner told me that the most exciting thing in town is just ahead." Soon we are flexing our muscles beside the statue of Popeye the Sailor Man—what memories surface, of that first black-and-white TV in my childhood home during the '50s, of the Mickey Mouse Club, the Lone Ranger, and of course Popeye and his spinach, Olive Oyl and Swee' Pea and Bluto and Wimpy, all based on actual residents here in Chester, we're told, the birthplace of Popeye (or at least that of his creator, Elzie Segar).

"*It's the Mississippi! Oh yeeeaaaaah! Whoooohooooo!*" Such is the eloquence of my traveling partner as he rides ahead, approaching the bridge, with me right behind him. Traffic fits tightly into the narrow lanes, so we can't safely linger, but what a feeling to be here, now, above the Mighty Mississippi and crossing to Missouri. I mark the moment: "8:42 a.m., May 28, our 25th day on the road."

We pause at the "Welcome to Missouri" sign, reflecting, taking pictures, when David asks "*What's that?*" I heard it, too, and was unsure what it was until I saw the bird on the wires above us. "Oh, that's nice," I reply, "a male red-winged blackbird with a very different call from any we've heard since Virginia. Must be the wide river, the blackbirds on the Missouri side not able to hear those on the Illinois side, so the birds here are free to develop their own dialect of call notes."

Red-winged blackbirds. The local call dialect beside the Mississippi River. (138, 0:37)

I have been listening to blackbirds since the Atlantic, enjoying the subtle variations on *check, chick, chuck, chunk, chink* and the like, and some distinctive calls, like a piercing, downslurred *tee-yeer*. I know that each male has about ten different calls and that birds in a given field or marsh tend to share the same repertoire of learned calls, much the way dickcissels have song dialects from field to field, but what's so special about the blackbird calls is how they're used. It's as if the males monitor all that goes on about them, and all neighbors use the same call until one of them spots a disturbance; he then switches to another call, as if to alert others, and then they all switch to the new call. The call repertoires have probably changed gradually all along our travels, but the river here isolates birds on opposite banks, thereby allowing for such an abrupt change from one dialect to another that even we hear it.

I try to convey much of this excitement to David, but soon reflect on what is revealed by his seemingly simple question "*What's that?*" He knows enough about all the birds that we have been hearing to know that this sound is different from anything he has heard before. *That's impressive!* He is absorbing more about these birds along the way than he has let on. "I love birds . . . birds are the greatest," he says with not a trace of sincerity in his voice.

By 9 a.m. it's sunblock time, and we liberally splatter SPF 50 cream on our face and calves, as we have about this time of day during the entire trip, but it's only in the fifties, so the arms are still well protected beneath our fluorescent yellow windbreakers. We're soon out of the Mississippi's floodplain, climbing on lettered roads up into forested hills to the west, first on County Road H, then Z, N; and up ahead, according to our map, lie P, B, F, W, V, then somehow N again. We've been warned about how steep the climbs are in the Ozarks, far worse than the Rockies, and I'm soon in my lowest gear, crawling uphill at four miles per hour, then in my highest gear as I scream down the hill, then my lowest again. "Oh, my, we have hills, we have hills," reports David, as if I hadn't noticed.

Just past a farmhouse with all of the world's house sparrows making an enormous racket, a kestrel flies across the road in front of us. In acrobatic flight, its all-red tail is fanned, the black tip conspicuous, the bluish wings contrasting with the red-and-black-striped back, the falcon facial pattern distinctive, and in hot pursuit is an eastern kingbird. Yesterday we watched a kingbird chasing a red-tailed hawk high in the sky, the kingbird just a speck bouncing about the far larger hawk. "Tyrannical tyrants" these kingbirds are, according to their Latin name *Tyrannus tyrannus,* so aggressive that they chase crows and hawks much larger than they are, often in defense of a nest, but sometimes I would wager just for the pure joy of the chase, because they can.

House sparrows. Typical farmyard din. (139, 4:19)

My mind wanders, to the latest world news, to what might or might not lie ahead, to the last four weeks on the road, but I remind myself to focus on *now,* the present, and I consciously take note of all around me. Trees of all kinds line the road, maples, oaks, wild cherry, basswood, and my favorite, the tulip tree, as many have been in bloom since Virginia. The weather, cool, in the low sixties, with a gentle headwind. A good starling, some would say (but not me), lying on the road; my roadkill list grows by one. My labored breathing, the continuous whir of the drive chain as we tackle an 8% grade. Wildflowers—yellows and whites and purples. The wind in my face, the sheer thrill of 40 miles per hour downhill. Constantly adjusting to the best gear, trying to maintain 80–90 revolutions of the crank arms per minute. A brown thrasher singing from the right, a glance his way revealing his typical hunched-back, treetop silhouette, with tail down and slightly forward.

"David," I call up to him, "better get a good listen to all of these birds. Most of them are eastern birds and we'll soon lose them when we hit the prairies of western Missouri." I'll miss the warblers especially, the *beeeee-bzzzzzzz* of the blue-winged warbler; the high-pitched, seesawing *wesee-wesee-wesee-wesee* of the black-and-white; the high, sharp songs of the redstart; the buzzy songs of the parula; the rising *p-r-a-i-r-i-e* of the prairie; the chipping sparrow–like songs of the pine and worm-eating; the *turree-turree-turree-turree* of the Kentucky; the jumbled ecstasy in the flight song of the Louisiana waterthrush. I listen for the last of the wood thrushes, white-eyed and yellow-throated vireos, tufted titmice, Carolina chickadees, Carolina wrens, eastern bluebirds, summer and scarlet tanagers, cardinals, eastern towhees, field sparrows, eastern meadowlarks, orchard and Baltimore orioles, and more, all to be replaced by so many new birds in just a few days.

Whoooaaaa, County Road W out of Farmington is tough. There's no shoulder, and mad, speeding drivers seem intent on pushing us off the road. The door of a passing car opens, a man shouting something unintelligible from the passenger seat; angry horns blast; an oncoming car pulls into our lane in front of us, sending a clear message. It feels like eastern Kentucky all over again, perhaps not surprisingly, as we're told that Missouri was settled by people from Kentucky's Appalachia.

Pine warbler. An Easterner, in dawn song with chips and song variety. (140, 0:46)

Two hours later we arrive at Pilot Knob, where we leap at the chance to take a $40 motel room, reasoning that it'll be much easier to get an early start in the morning than if we camp. Mark and John arrive later, also taking a room and telling their own tales of County Road W. We seem to have lost Kris and Kes somewhere.

David decides on his bike's name and painstakingly spells it out on the top tube with stickers he bought at the last stop: "'QDR 10000'—that's 'Quality Dickin' 'Round,' which is all bike touring really is," he explains to me. "Just a great way to waste some time. Ten thousand is just an appropriately large number that sounds cool."

I name my bike, too: "'YIS.' Short for *Your Inheritance Spent*. Big number is implied."

DAY 26, MAY 29: PILOT KNOB TO ALLEY SPRINGS, MISSOURI

Just a few hours later, it seems, we're eating the cold oatmeal concoction prepared the night before, and at 5:55 a.m., ten minutes after sunrise, we're on the bikes again, the temperature maybe in the high forties. It's tough to tell the tight, tired legs they have to start pumping again, tough to get the butt to find a comfortable place on the saddle, but within a few hundred yards I'm comfortable and happy. Not so for David. His entire body rebels in pain over the early start, and he soon lags behind—it doesn't help, he has written in his journal, that I'm often "beyond giddy" at this time of the day.

A house wren, there, just to the right above the fence post. I've heard them since Virginia, though always at some distance, but this one bubbles and gurgles right beside the road in plain view. Just past him, I slow to a crawl, hoping he thinks I'm history, and he continues; I glance back at this bold, saucy creature as he bounces about on his perch, so full of energy, tail cocked over his back, and he sings again, beginning with a series of brief, stuttering chatters, and then he's higher and louder still, the song next cascading and

House wren

House wren. Effervescing beside the road throughout the East. (141, 1:38)

bubbling down the scale. His energy is inspiring, contagious, and I spring ahead, hoping that David hears him.

On the left, is that a new song for the trip? My mind races among the possibilities, but seconds later, on the next song, blurts out "Bewick's wren!" I had hoped that we'd hear one in Kentucky, but although they once ranged widely over the East, it seems they're now largely extinct east of the Mississippi River. I make note: "6:27 a.m., May 29, at 4.32 miles out of Pilot Knob on County Road N, a Bewick's wren!" I wait for David to catch up and try to explain, in an appropriately subdued tone, the enormity of this occasion. In graduate school, from 1969 to 1972, I studied Bewick's wrens for my doctoral thesis; I worked mainly at the Finley National Wildlife Refuge in Oregon, where we'll stop just before hitting the Pacific, but I also surveyed these wrens throughout the West and east to Maryland, when they still existed there. The songs from Colorado east are the best, this male now singing from the cluttered front porch of a house one of those long, complex, six-parted songs, much like a glorified song sparrow. These wrens taught me so much during my first four years as an ornithologist, and to them I will forever be grateful. He now flies from the porch to the top of a nearby tree, singing there again, then across the road and into the distance, where he switches to another of the ten or so different songs he can sing, and then he's silent.

Bewick's wren. A remarkable virtuoso, delivering five different song masterpieces. (142, 12:30)

Continuing, David soon asks, "Why is it so quiet?"

"Yes," I agree, "it's quiet. There's always a lull for about an hour around sunrise, after the dawn chorus, but I think it's especially quiet today because it's so cold. Insects are less active in these temperatures. That makes it more difficult for birds to find the energy they need, so there's less time for singing."

With 30 miles on the odometers, we arrive in Centerville, where David stokes his engine with some coffee. "Best not to wake up before it gets light," he reminds me. For me, it is the best of mornings: relatively little traffic, no wind, cool, the birds at their best, though just a little quiet.

Only three miles down the road, on a long uphill and quite possibly the steepest of the entire trip, a smiling Mr. Caffeine passes me. He's back! My watch alarm beeps again, every 15 minutes today, reminding me to eat and drink some more; two fig bars and some Gatorade do it for now, together with two ibuprofen to keep the swelling in my left leg down ("Could take six weeks to heal," my doctor daughter had said). David stops to help yet another turtle across the road, lest it join its battered and squashed buddies, and I crest the hill first, then roll down what looks like an endless slope disappearing into the distant trees. At 40 miles per hour, now 44, I approach a turkey vulture dining on a carcass beside the road; by the time he spots me, he has barely enough time to get airborne, and for a brief moment we glide down the hill together, he barely lifting above me as I overtake him, I almost airborne with him. This close, he's huge, the wingspan enormous, the outlines of each feather in his massive wings just overhead forever burned into memory. What a fine roller-coaster day this is!

At Owls Bend State Park on the Current River, part of the Ozark National Scenic Waterways, we break for lunch. David is giddy. He walks upstream, then floats down. He relaxes, even naps in the shade. If that weren't enough, he's thoroughly rejuvenated after talking to the college-age women who are floating down the river in their canoes.

I work on my patience, as I'm eager to finish the day's ride to Alley Spring, just 18 miles down the road. I wade into the water, trying to prove I'm not as grumpy as I am, but the rocks hurt my feet; relaxing is tough; sleep is out of the question. But I breathe deeply, taking in the beautiful river, the cleanest water we've seen on the trip, the serene setting, listening to all that is about. Many old friends are here, but I enjoy the belted kingfishers most. A male and female parade up and down the stream, their dry rattles announcing the birds well before I see them. Best of all, it's wonderful to see David happy, and I am envious that I can't enjoy the stop as much as he does.

Belted kingfisher

Belted kingfishers. Typical banter of male and female heard from Virginia to Oregon. (143, 1:51)

Departing Owls Bend, David rides circles around me. "Best break of the whole trip," he beams. I grumble a little internally, but only a little. He's doing the trip largely my way, arising before sunrise and biking the early morning hours. That's quite a gift to me.

Alley Spring arrives quickly. We set up camp beside the Jacks Fork River and I scout the campground, planning my good-byes for the morning. Here is the last fully forested stop before the prairies. Here will be the last stronghold of so many eastern birds, and tomorrow morning I will walk among them and listen one last time.

DAY 27, MAY 30: ALLEY SPRINGS TO HARTVILLE, MISSOURI

Throughout the night, a barred owl calls, and well before sunrise I'm walking the campground. Swallows call over the river; a yellow-breasted chat and a chuck-will's-widow sing just beyond. Overhead an eastern wood-pewee awakens and begins his dawn prelude; how intriguing that he sings first for a minute or two at a leisurely daytime pace, with a string of four or so *pee-ah-wee*s followed by a single *wee-ooo,* as if warming to the dawn's task. I linger here, listening to the cardinal's song, a simple repeated *whoit whoit whoit whoit whoit,* each whistled *whoit* beginning low and sweeping high, each beginning with the left voice box and ending with the right. Robins are everywhere. An eastern kingbird circles over the clearing by the road, energetically stuttering his dawn song. And there, the peewee abruptly shifts into his dawn mode, introducing his third song, the *ah-di-dee,* now singing at five or six times the pace of his slow prelude. Nice—all is in order.

Eastern wood-pewee. Hurried but exquisite dawn song, a staple of eastern forests. (144, 7:05)

On the pavement just up ahead four chipping sparrows hop about each other, each sputtering its songs at the others. In another half hour, still before sunrise, each will have returned to his own territory, but for now they gather in this small arena and have at it, matching wits and songs, perhaps settling scores, each finding its place in the local male hierarchy, the females no doubt listening and taking notes.

I walk into one of the campground loops and marvel at four species who mark the dawn in a special way. The Acadian flycatcher continuously calls *seet, seet, seet* between his explosive *PEET-sah*! songs. Two yellow warblers work quickly through their dawn repertoires of perhaps a dozen songs, three redstarts through their three or four, all five males chipping madly between songs; and all five birds hold in reserve one special song that they'll use later during more relaxed singing. The northern parula overhead also chips frantically between his songs, but he has only one dawn song that he repeats over and over; he'll save his one daytime song for later.

Four chipping sparrows. On the pavement at dawn, sputtering brief songs at each other. (145, 5:01)

And there's a fifth species, an orchard oriole. We've heard them daily from high in the trees, but now he sings from a small bush beside the road. Wow, I've never heard an orchard oriole sing like this before. He sings so rapidly now, about a three-second song every five seconds, filling most airtime with a wonderful jumble of buzzes and piping whistles. I edge closer until I'm standing just a few feet from him, each note in his songs now sharp and penetrating. Picking out a unique phrase, I listen for it to recur: there, at the end of the song, next in the middle, then absent altogether, no two songs alike. And poof, he's gone, this phantom resuming his dawn effort in another small bush on the other side of the road.

Orchard oriole. Dawn song, intense and exhilarating, from the bush within arm's reach. (146, 4:54)

More eastern birds oblige as I focus on a few songs from each: a Carolina wren . . . an indigo bunting . . . no, two . . . a distant wood thrush . . . a Carolina chickadee . . . a Kentucky warbler . . . an ovenbird . . . a towhee . . . two titmice . . . a phoebe by the restrooms . . . the pewee still in dawn song. Surprisingly, the chuck-will's-widow continues. They're all males, though it is the discriminating females who are in charge, because their choice of a mate dictates which songs and singing behaviors are passed to future generations. "Thank you, lady bird," I find myself thinking, "you are the silent composer of all that I now hear."

By the time the sun has risen somewhere beyond the heavy cloud cover, the singing has slowed and it begins to rain. I take shelter beneath a tree, noting still other birds: a red-bellied woodpecker, crows, a warbling and red-eyed vireo. Flying by, a brown-headed cowbird whistles in flight, right on schedule. Just what do cowbirds do before sunrise? And why are vireos such late risers?

Cedar waxwing

A small flock of cedar waxwings land in the tree above me. "A songbird without a song," I smile to myself. I hear their high, thin *seeee* calls and some of the raspy *bzeeee*s, but know how expressive they can be by varying features of these two calls. How curious that never does a male waxwing rise to the treetops and deliver anything that one might think of as a song. Never. It's just not in their lineage, even though they're "true songbirds." And why not? The female waxwing must know why they've lost their song, as waxwings have almost certainly evolved from singing ancestors, but the answer remains a mystery, as we have no way to ask.

Cedar waxwings. A songless song-bird, a puzzle. (147, 1:18)

Yet another vireo chimes in, a yellow-throated vireo, sounding much like a slow, hoarse red-eyed vireo. I walk over and seek shelter under his tree, listening to him shower his songs on me, about a song every two seconds. I play the same game I do with the red-eye, picking a unique song, soon realizing that he's alternating two different songs, A B A B A B. Perhaps ten times he does so, and then he plays with three different songs, C D E C D E, regularly repeating the pattern. After a minute I hear what sounds like a return to those first two songs, and in another minute he works the three songs again, suggesting he has a repertoire of only five different songs. How fascinating, not just the small repertoire but especially the packaged style of delivery, so different from what I'm accustomed to based on years of listening to red-eyed vireos.

Yellow-throated vireo. Missouri male singing in two packages. (148, 2:20)

The rain continues, so I return to the tent, crawling into my sleeping bag about 6:30 and listening to the patter of rain on the fly, to the daytime song of the western parula overhead, to the current song package of the chat across the river, drifting off . . . By 8:00 the rain has stopped, and I'm out of the tent again, David soon following. "Phoebe," he points out. "Notice how he's alternating his two songs again? You should pay more attention." He's baiting me. I give him a simple "Yeah," but beam inwardly, knowing that he's listening more than he often lets on.

By 10:00 we're on the road, soon stopping to put on our sunblock. David also puts on his white legionnaire's headdress, which flows in the wind behind him; this costume prompted a cyclist back in Kentucky to dub him Osama bin Ridin'. She had already christened me with TransAm Birdman.

Late morning, a little over seven miles out of the campground and just after the friendly *honk-honk* and wave from a white pickup passing by, we come to the intersection of route 106 and State Highway D. There, in the middle of nowhere and reaching high above the canopy, is a wooden observation tower, the stairs inviting us skyward. We accept, and are soon looking out over the rich forested landscape of the Ozarks.

I look to the east, mentally tracing our route from Yorktown to Jamestown to Mechanicsville to Mineral, through the length of Virginia and Kentucky, clipping off the southern tip of Illinois, now halfway through Missouri, acknowledging each leg of the trip. Images and sounds flash through my mind, all eastern.

I then turn and look to the west, toward the Pacific. The Great Plains lie just ahead; then the Rockies, the Tetons, and Yellowstone; Montana, Idaho, and Oregon beckon. By this time of day, the plains and far western birds are singing all the way to the Pacific, and I can almost hear them. I'm ready!

13.
A PRAIRIE GEM

DAY 29, JUNE 1: ASH GROVE TO PRAIRIE STATE PARK, MISSOURI

Because I love chickadees, I had learned all about Prairie State Park in far western Missouri. "Yes, the hybrid zone between Carolinas and black-caps runs right through there" confirmed my friend Mark Robbins at the University of Kansas; birds to the east are pure Carolina chickadees, to the west pure black-capped chickadees, and in this narrow zone are odd birds with mixed-up songs. That was enough for me—I spotted the park on the map, just off the TransAmerica bicycle route, and learned that we could camp right there among the hybrids. They would be in the wooded campground area, but perhaps the real treat would be a glimpse of the prehistoric Great Plains in the untamed tallgrass prairie.

It's now late afternoon on June 1, and we're "off route," having jogged six miles north on highway 43, two miles west on route 160, and one mile north on highway NN. From here, gravel Central Road leads west, a route to another time. We pull up to take a look. "What do you think? Deflate the tires some to protect the rims on the gravel?" I ask David. "Yeah." With a whoosh of air, each tire is deflated from 80 psi down to about 45. We'll get better traction on the loose gravel and, with the softer ride, reduce the risk of damaging the wheels.

Heading slowly west over the gravel, I reflect on the exciting changes in the landscape over the last few days. Two days ago, in the 75 miles from Alley Spring to Hartville, the forests were heavily fragmented, the open land more of man's design than nature's. Yesterday, in the 68 miles to Ash Grove, the forests and plains battled, neither having the upper hand. But today the plains have clearly won, tentacles of eastern forest reaching to the west only in the river bottoms. Here, after three miles on Central Road, we take a right onto NW 150th Lane, entering the heart of the park; a vast blue sky now

extends to the horizon in all directions, and prairie flowers abound. "Stay in your vehicle," David advises me, reading the warning on the bison sign. We bike on, a bit nervous, as a soft breeze ripples the sea of grass up the gentle slopes to the left and right. In the rich late-afternoon light, prairie songbirds line the way as we ride through pristine tallgrass prairie that has never seen a plow. Yes, on the hillside to the right are bison. Easily imagining that the tallgrass prairie and its buffalo extend endlessly in all directions, I imagine how this could well be 200 years ago, in 1803, when this land was bought from France as part of the Louisiana Purchase and when Lewis and Clark were beginning their epic journey to the Pacific.

Two miles north we turn left into a small woodland, fording the stream into the campground area and taking note of the sign that warns of flash floods. It is as if the chickadees were waiting for us. As we set up camp, I listen, disappointed that the chickadee nearby sings a typical black-cap song. It's a richly whistled *hey-sweetie,* the *hey* higher in frequency than the *sweetie,* the *sweetie* clearly two syllables with a noticeable break in the middle. Over and over he repeats it, and he soon transposes that song to a lower frequency, now singing a series of songs there—that's just what a good black-capped chickadee should do. But the next series is strange, a *hi-lo-hi-lo-hi-lo-hi* song, each *hi* whistle so much higher and more piercing than the highest *hey* of the black-cap—now he's mostly Carolina chickadee, but the song is twice as long as it should be! This is good—he *is* confused.

Chickadee hybrids! In Prairie State Park, what mixed-up singers. (149, 3:08)

With the camp secure, David heads out to the prairie, but I linger, eager to hear more from the camp chickadee. He obliges, soon singing *hey-sweetie-hi,* a typical black-cap song with the *hi* note of a Carolina on the end; eventually he offers a *hey-sweet-sweetie,* a beautiful cascade of descending whistles, multi-parted like a Carolina's song but with black-capped components.

A neighboring chickadee is just as innovative, also with at least five different songs. He has the same *hey-sweet-sweetie* and the typical black-cap *hey-sweetie,* but he has two typical Carolina songs of the *hi-lo-hi-lo* format as well. Then, instead of singing *hey-sweetie-hi,* he sings a *hi-hey-sweetie.* Mixed-up hybrids indeed!

How special to be here, straddling the boundary between these two chickadees. I have studied this narrow hybrid zone on maps, recalling that it extends from here to the northeast, cutting through Illinois, Indiana, Ohio, and Pennsylvania, with a long arm of the hybrid zone extending to the south along mid-elevations of the Appalachians, where black-caps own the high ground, as on Mount Rogers in Virginia, Carolinas the low ground. All along this line one senses a chickadee tug-of-war over territory, with compromises made at the border by chickadees who chose love over war and then mated with the opposition.

I join David on the prairie, where I find him lounging on the crest of a small rise with a magnificent view in all directions. I take in the scene, the buffalo, the gently undulating terrain, the soft breeze from the west rippling the grass, the distant horizons, the endless sky, a profusion of prairie birdsong, the entire landscape alive. What is it about this place that feels so magical? I feel energized, as if the prairie awakens a genetic memory of when my ancestors came down out of the trees to explore a whole new world. I want to run out into the prairie, whirling and twirling about, taking flight.

But I also notice those thunderheads to the west, so I head back to camp, making sure we're ready for anything. And, yes, anything happens, as the sky soon opens, just after David returns. He merrily cooks a meal in the shelter of the tent's fly, a little disgruntled at my lack of participation, and we're soon eating yet another sample of the "chef's choice," though in cramped quarters. Then we call it a day and drift off as the rains continue.

DAY 30, JUNE 2: PRAIRIE STATE PARK, MISSOURI, TO CHANUTE, KANSAS

I awaken to the sound of running water! Grabbing a flashlight, I unzip the netting and look out, seeing nothing but water, the tent an island in a river flowing by. In a flash I have my sandals on and I'm out of the tent, wading toward the stream, wondering if it is now a raging torrent that could sweep us away. I'm soon beyond the ankle-deep water and walking the road, and find that the stream is only a little higher than before. Back at the tent, I see that the rain falling on the entire campground drains through our campsite. There's no danger, and we're dry, thanks to trained eyes that chose a site for the tent that was just a few inches higher than the surroundings. Careful not to awaken my I-can-sleep-through-anything son, I crawl back into the sleeping bag, soon dreaming of a prairie dawn.

Well before sunrise, I'm back on the lane that bisects the park, eager to return to where the stream crosses about a mile to the south. The birdlife will be richest there, I reason, with birds typical of eastern forests in the streamside trees and birds of the prairie on the drier slopes above.

The overnight rains have saturated the prairie sod. Gravity has taken what the sod cannot, feeding the stream, which now babbles and gurgles with far more authority than it did yesterday afternoon. It's no surprise that an eastern phoebe has taken a liking to this place, and he's soon in full dawn song. And, yes, he's alternating his two songs, just as an excited phoebe should at dawn. I feel his rhythm, a song every two seconds, but his metronome consistently misfires just a bit, as he hustles from *FEE-bee* to *FEE-b-bre-be* a little faster than he does to the next *FEE-bee*. All innate, I muse, and because the song genes are the same everywhere, all phoebes know to sing like this. I'll give David the full phoebe report later.

Eastern phoebe. Streamside, bison nearby, alternating two songs in dawn mode. (150, 4:54)

I'm eager to walk off the road and out into the prairie, yet I'm wary, as it's still dark and I don't know where the bison are. The gentle rising slope on the east side of the road and just north of the stream looks most inviting and I'm soon wading through the grass, but in just a few yards stop short. Off to the left I hear them snorting, and in the beam of my flashlight at least a dozen pairs of eyes shine back at me. More snorting. Must be the normal sounds of grazing, I assure myself, nothing to be concerned about, but I backtrack, choosing instead to walk the other side of the stream, reasoning that they'd be unlikely to crash through the thick streamside vegetation to get to me. It's surely as safe as "staying in my vehicle."

Up ahead from a small isolated bush explode the songs of a Bell's vireo, the first of the trip. What extraordinary energy in his singing, each second-and-a-half song a jumble of rising and falling notes, every three seconds another song bursting from this unseen creature hidden deep among the densely packed leaves. With baby steps, I work slowly toward him, soon standing just a few yards away. He's unfazed.

I listen, recalling the few facts I know of his kind. He has maybe half a dozen different songs. He typically alternates two different songs, sometimes three, clearly heard by how distinctively some song endings rise and some fall. A given song may occur 15–20 times among 50 or so songs over several minutes, then disappear, returning several minutes later. He's a "package singer," like chats and robins, like yellow-throated and blue-headed vireos.

Bell's vireo

Bell's vireo. From the bush beside me, singing a question and then answering it. (151, 4:55)

Fascinated with this hidden spirit just beyond arm's length, I pick out a song that rises uniquely on the end, and then listen for him to sing it again. It clearly occurs every two or three songs, with different songs in between. Then it disappears, and over about three minutes, he sings 88 other songs, playing out two or three other songs in his repertoire before returning to my special song. "Very nice. Thank you, Mr. Vireo."

Just beyond the vireo and beside the stream sings a chat. What an odd assortment of sounds he proclaims, and in so unpredictable a rhythm. The next chat just 30 yards upstream provides an equally improbable performance, as do others in the distance who line this lush stream bottom. Their singing seems to have been put together by a large committee, no two stubborn members having agreed on how this bird should sing.

The willow flycatcher who sings from an exposed dead branch over the stream is just as odd in his own delivery. There's a *FIZZ-bew FITZ-bew,* two of his three songs given within a second of each other. He pauses seven seconds before another *FITZ-bew,* then five to the next *FIZZ-bew,* two to a *FITZ-bew,* a full ten seconds to a *creet,* the simple, rising, stuttered third song of his kind. He bumps his way through the performance in fits and starts as if he can't quite decide what to do next, as if he had taken lessons from the chorus of chats nearby.

Yellow-breasted chat. Concentrating on just one package of songs. (152, 3:20)

I check in briefly with other familiar birds in the streamside bushes and trees. An eastern towhee still alternates two of his *drink-your-tea* songs. A cardinal whistles the dawn away. A catbird squeaks and rattles and meows; a brown thrasher smacks in annoyance as I approach, then settles into a distant tree, *prairie dawn prairie dawn, sing it sing it, whoopee whoopee, glory be glory be. . . .* A yellow-billed cuckoo knocks, his *ka-ka-ka-ka-kow-kow-kow-kow-kowlp-kowlp-kowlp* lasting a good ten seconds. Yellowthroats—each proclaims his own version of

Willow flycatcher. With three distinctive songs, as his kind sing everywhere. (153, 4:00)

the basic *wich-i-ty wich-i-ty* theme. A red-winged blackbird persistently adds his *konk-la-reeeeeee,* as does a bobwhite his emphatic *oh BOB WHITE!* In flight overhead, a goldfinch calls *per-CHIK'o'ree*; landing nearby, he offers only a partial song, telling that it's still too early in the season to get excited about nesting. Was that a crow in the distance, or a chat's imitation of one?

Dickcissels. Six birds with local dialect. Bird 1 (154, 0:28) Bird 2 (155, 0:38) Bird 3 (156, 0:28) Bird 4 (157, 0:29) Bird 5 (158, 0:35) Bird 6 (159, 0:23) Birds 1–6, just one song from each. (160, 0:24)

Just up from the stream bottom is the prairie, and *prairie birds*. Dickcissels are everywhere! I approach one, listening to his *tup tup tup see-up sup sup,* three or four sharp introductory notes that then rise to the *see* and drop to the lowest *sup sup* phrases; that very song echoes from five other dickcissels within earshot, each bird on message, not one dissenting. How distinctive these songs, yet how different from the dickcissel dialect just a quarter mile up the road toward the campground.

Meadowlarks, all eastern. *Spring-o'-the-yeeaar* sings one from a bush nearby. I whistle the words of the mnemonic, rising at first, then dropping in pitch and drawing out the *yeeaar,* making it sound as one continuous effort. It's just a mnemonic, I remind myself—listen for the details in each song, the tempo and rhythm. He continues, but then seems upset at my approach and responds with a raspy, explosive *dzert,* then a rapid *dzert-ert-ert-ert-ert-ert-ert-ert.* He flies off and lands a good 100 yards away, where he offers a markedly different version of his *spring-o'-the-yeeaar* song. His territory is huge compared to that of the dickcissels.

And there's one of the most remarkable songs of all, the liquid gurgling *bublocomseeeee* of the brown-headed cowbird, though the mnemonic is more of a sad commentary on our human ears than on the song itself. Having captured these songs with my best parabolic microphone and then slowed them on my computer to one-tenth normal speed, I know the details that a nearby female would hear when he sings in her face. He begins with an immaculate series of tiny, pure-toned notes that bounce quickly from one to the next, uncannily low-pitched and so softly delivered that only she would hear them. With an acrobatic leap from his left to his right voice box, in a split second he rises to loud, pure, slurred tones so high that they're a challenge to human hearing. And he has not just one but several of these masterpieces at his command.

In a low bush just up the slope I can see two cowbirds in silhouette. He fluffs his feathers, then spreads his tail and wings, simultaneously bowing and almost falling forward off the perch as he sings. The *ch'ch'ch'ch'ch'ch'ch'ch'* rattle that follows his song gives her away. He gurgles his song again and she responds with another rattle, the pair of them then flying off as she rattles once more and he appropriately gives a flight whistle, a *whssss-pseeeee*, sounding almost as if he inhales on a squeaky *whssss* and exhales on a whistled *pseeeee*.

Eastern meadowlark. Soon to be replaced by western birds. Calls (161, 2:56) Songs (162, 2:49)

Brown-headed cowbird

These were the original "buffalo birds," I muse, once restricted to these prairies, where they followed the bison and ate the insects stirred up by the grazing herd. We humans created more than a few problems by clearing the forests and opening the land, the "buffalo birds" transforming to "cowbirds" and quickly taking advantage of their new habitat from coast to coast. And take advantage they do, as a female cowbird (villainously, some would say) lays her eggs in the nests of other species, duping other songbirds into raising her offspring for her. Suddenly, on an evolutionary time scale, bird species such as the Kirtland's warbler are threatened with extinction, because they have evolved no protection from the ways of cowbirds. Not the cowbirds' fault, though, I would suggest, and I toast the evolutionary marvel that these cowbirds are whenever I hear them.

Brown-headed cowbirds. Courting among bison, in their ancestral home. (163, 1:07)

Up higher still, where the prairie sod is shallower and the grass finer, two sparrows now sing. I've heard them before on this trip, but always isolated, a few birds here and there, never together and *never* in these numbers. This is the prairie, their home! In a flight of fancy, I am a Henslow's sparrow, delivering my own unique *tsi-lick* song, beginning high, dipping and then climbing again, feinting, jogging, warbling briefly, pausing, resuming precipitously lower, on and on, each dip and jog and feint lasting no more than 1/100th of a second, the entire cascade of notes in my high-speed roller coaster song beginning near 10,000 Hz and ending at 3000, lasting a glorious half second.

And grasshopper sparrows. A field guide might describe their songs as "insect-like, beginning with a few weak notes followed by a high, thin, hissing, unmusical buzz: *tip-tup-a-zeeeeeeeeeee*," but slow them down ten times and hear the magic there: In the first half second, four sharp introductory notes

Henslow's sparrows. Neighbors with different "hiccups," stellar when slowed for our (pathetic, human) ears. Bird 1 (164, 2:09) Bird 2 (165, 0:28) Slowed 4x (166, 0:07) Slowed 8x (167, 0:12)

Grasshopper sparrow

Grasshopper sparrow. Male at dusk, dazzling with his two songs. (168, 0:31)

are now the sweetest of musical tones to anyone's ears as they leap about in pitch and time; the *zeeeeeeeeeee* bounds from high to low five times each second instead of 50, revealing a vocal athleticism that my pathetic human ears would never be able to detect in real time.

Two of the grasshopper sparrows now seem to argue near their territorial boundary, each of them singing not only his relatively simple *tip-tup-a-zeeeeeeeeeee* but also a more extended song. In real time it sounds like a jumbled sequence of haphazard notes, but when the song is slowed down, I have heard the bits and pieces take shape, with tiny little sections over one hundredth of a second now dancing through time.

Time slips by, and I figure I best head back to the campground, to move on with our day. There, above where David works on his breakfast, are two remarkable songsters who take turns singing from the exposed branches of a dead tree. "Do you hear the special treat for you up there? One is a rose-breasted grosbeak, Peterson's 'robin with voice lessons', so gorgeous that Peterson put this bird on the front cover of one of his earlier guides; the other is a Baltimore oriole, just as fine."

We listen to the grosbeak's rich, melodious phrases, far brighter and more fluid than those of the robin, a three- to five-second spirited ramble of slurred whistles that hustle to an all-too-soon ending, no two songs alike. And then, in turn, the oriole—his whistled notes are rich and clear, the rhythm and tempo so distinctive. I gawk with ears and eyes, feasting, soon acknowledging the songs of a third bird who joins in. "There's the pewee again. Just his two daytime songs. Wasn't his serenade at dusk last night with all three songs beautiful?"

We talk about what lies ahead. These three songsters, as well as the chickadees, foretell the great sea change that is about to occur. Singing here are a rose-breasted grosbeak, a Baltimore oriole, and an eastern wood-pewee, but they will soon be replaced by their western cousins, the black-headed grosbeak, the Bullock's oriole, and the western wood-pewee. It's the same with this morning's eastern phoebe, willow flycatcher, eastern kingbird, eastern towhee, and indigo bunting—they'll be replaced by the Say's phoebe, the western form of the willow flycatcher, the western kingbird, the spotted towhee, and the lazuli bunting. "NEW TERRAIN, NEW BIRDS—HELP ME CONTAIN MY EXCITEMENT!" is the appeal David coins for his next mass e-mail to his friends.

Rose-breasted grosbeak, Baltimore oriole. Jewels in plumage and song, side by side, back-to-back. (169, 4:00)

"Blame the glaciers," I suggest, "and climate change, your specialty." Current range maps for these birds tell the history of glaciers that once advanced southward well into the heart of North America. Ancestral lineages that once occurred more widely were isolated in pockets to the east and west as mid-continent habitat became inhospitable. Over time, by chance and by choice, changes in plumage and song accumulated in these isolated populations, yielding unique eastern and western descendants that once were one.

As the glaciers receded, these formerly isolated groups expanded their ranges and met again along contact zones throughout the Great Plains. The song-learning songbirds, including the chickadees, grosbeaks, orioles, towhees, and buntings, seem to have been a bit confused upon meeting their long-lost cousins, leading to much cross-mating and hybridizing wherever they met, all of which led to the narrow hybrid zones now seen throughout the Great Plains. Not so with the non-learning flycatchers, including the phoebes, kingbirds, and wood-pewees; they weren't at all confused, as we surmise from how few hybrids are found among them. It seems that the inborn songs of flycatcher lineages guide mating with far more certainty than do the learned songs of songbirds.

Half past eight we're out of the campground, heading north on 150th Lane, then turning left onto road P, soon finding ourselves in a small town. Stopping for David's coffee, we check our tires, finding some significant gashes in them. To prevent a possible blowout, we remove two of our tires and insert folded dollar bills between the tubes and the worst of the gashes in the tires, then reinflate to 80 psi for the paved roads ahead. That'll get us to the next bike store, where we'll buy four new tires.

A police car stops, a policeman stepping out to inquire if we need help. David, noticing "Mulberry, Kansas" on the door, blurts out "Hey, we're in Kansas!" Though the days add up one by one, each special in its own way, we always celebrate entering a new state. We are in Kansas! And our friendly Kansan in uniform, initially with a "like, duh" expression on his face, as if we had just stepped out of Oz, celebrates the moment with us: "*Welcome to Kansas!*"

"One hundred forty miles to Rosalia," estimates David; "Tom told me he'd be there by noon tomorrow." We should be able to make that. Although

David's college friend had planned to ride the Rockies with us, we're behind schedule, so he'll join us for the rest of Kansas instead. It's no small change of scenery for Tom, but he's game, eager to make some music on his touring bike wherever the road may be.

A steady rain eventually forces us inside a restaurant in the town of Walnut. We wait it out by eating a second breakfast and listening in on conversations at neighboring tables. Four women in painting clothes are to our right, discussing color patterns and decorating in whatever building it is from which they're taking a break. Half a dozen farmers are to our left, all men; they're repairing farm equipment, discussing corn and dairy prices and the seemingly endless rains. One in his late twenties is taking a class remotely from some university in Idaho. The huge fellow in bib overalls is the comedian, fibbing and giggling and smiling with cavernous dimples, his entire body rippling in response to his own understated question "Getting any rain at your place?"

Finally back on the road, we face a fierce headwind that slows us to four miles per hour, five at most. We've been warned about these Kansas headwinds. We take turns up front, pulling each other along, but it's too much. Mid-afternoon, we pull into Chanute, only 60 miles on the day, and take a motel room. We'll get an early start in the morning, we agree. The wind can't be any worse then.

That evening in the motel room, with tent and gear strewn about and drying from the rains, we check a weather forecast: "For tomorrow, wind peaking at ten miles an hour, gusting at times, then fairly steady for the next five days, pleasantly cool from the northeast." We ask each other if we heard that right: *from the northeast*. Yes, we did. We'll be blown across Kansas!

14. KANSAS OCEANS

DAY 31, JUNE 3: CHANUTE TO HARVEY COUNTY EAST PARK, KANSAS

"Got our tailwind! 80 miles by noon or bust," says David about 4 a.m. as he sticks his head out the door to inspect matters firsthand. If we depart by five o'clock from Chanute, we reason that we will have seven hours to average about 12 miles per hour, all stops included, to Rosalia. We just might make it. David's friend Tom has about 50 miles from where he's spending the night in Wichita.

We roll out of the motel room to a Chanute that lies eerily asleep, a ghost town of quiet streetlights standing sentinel, their nighttime shift nearing an end. About an hour before sunrise, robins rule the planet and airborne purple martins chortle as second in command. David loops through the city park to see if any of our cycling friends camped there. "Nobody there," he reports. So we chart our course due west, on West 21st Street.

In the darkness, the wind in our sails, I imagine sailing through the spectacular water world that was once Kansas. It is just a few tens of millions of years ago, during the Cretaceous, when all of modern-day Kansas is inundated with a shallow, tropical ocean. On dry land back in Missouri, *Tyrannosaurus rex* is king of carnivores, *Triceratops* his dinner. Here in the waters beneath us are fantastic nautiloids and mollusks and echinoderms; giant marine turtles swim with ichthyosaurs and sharks and outlandish fish of all kinds. Reptilian "sea monsters" rule the seas: Plesiosaurs, with their turtle-like bodies and long snakelike necks will much later inspire the Loch Ness Monster; powerful, predatory mosasaurs up to 50 feet in length rule the oceans. Winged pterosaurs cruise overhead.

And best of all, birds are everywhere! There's *Hesperornis*! He zips through the water like a penguin with wings as flippers, five feet long head to toe, but how odd to see those sharp teeth in his prehistoric bill. Overhead flies the tern-like *Ichthyornis,* also with a beak-load of teeth, this ancient bird also making a living on the abundant fish here. What I wouldn't give to *hear* this water world and the world of *Tyrannosaurus* back on shore. But there would be no songbirds, as they're yet to come.

In time, the mega-dinosaurs disappear and the Pacific tectonic plate undercuts North America, hurling the Rockies three miles skyward. The mid-continent rises more modestly to perhaps half a mile above sea level, the great inland sea then retreating to what is now the Gulf of Mexico and the Arctic Ocean, leaving Kansas high and dry. What's left is an ocean of tall grass rippling in ancient breezes, all the drier for being in the rain shadow of the Rockies to the west. Mega-mammals then thrive here, lions, mastodons, ground sloths, tapirs, giant short-faced bears, mammoths, camels, saber-tooth cats, and so many more, until about 10,000 years ago, roughly when early humans arrived.

In the gathering light, the reality that is now Kansas emerges from the dawn. I've heard some silly debates about whether Kansas is flat as a pancake, but having just left the Ozarks, we have no doubt. With hardly a wrinkle in the roads and a gentle tailwind, we fly. "Roads are laid out at precisely one-mile intervals," remarks David after checking a few with his odometer.

Gracing the fence posts along the road are scissor-tailed flycatchers. Such elegant creatures with their ink-black beak on a pearly white head, a bright immaculate pink that shows beneath the wings when they fly, a graceful tail that is deeply forked, scissor-shaped no less, almost twice as long as the body. One would never guess that they are as tyrannical as other *Tyrannus* kingbirds, yet two on the left now attack a red-tailed hawk that screams and flies off a utility pole; one scissor-tail lands and rides momentarily on the hawk's neck, hammering away there. A few minutes ago another scissor-tail chased an eastern kingbird, one tyrant after another. Double or triple the size of any of these kingbirds and beasts as ferocious as *T. rex* itself would be in for a thrashing.

Nighthawks fly about wherever we look, offering an occasional *peent.* In the dark this morning they were more energized, their nearly continuous *peent*s raining down on us, punctuated every dozen or so with the loud *VROOM* of the wings feathers during a display dive.

Common nighthawk. With uncommon *peent*ing calls and explosive, diving *VROOM*s. (170, 4:04)

Whoooooleeeee, wheeeloooo-ooooo . . . "David, to the right, an upland sandpiper! He's a shorebird without a shore, taken to the prairie, nowhere near water." As he flies, his long neck extends far out front, his tail is spread, and he glides on extended wings with down-turned tips. I coast, entranced by a song as memorable as the best cry of wolf or loon. The "Long Mellow Whistle," ornithologists call it, pleasantly gurgling at first as it rises to the purest of whistles and then trailing off and down for the last of its three seconds, an enchanting *whoooooleeeee, wheeelooooo-ooooo*. Compelled to eloquence to mark the occasion on my voice recorder, I manage "Upland sandpiper. Unforgettable. 7:38 a.m., 29.72 miles outside Chanute, somewhere in the middle of nowhere." Soon every utility pole seems to sport an upland sandpiper, and as we pass, they often fly to another perch, landing there with wings up and outstretched over the back, then gracefully folding them in place.

Upland sandpiper. Shorebird without shore, graceful in flight and call, *whoooooleeeee, wheeelooooo-ooooo.* (171, 0:14)

The occasional chickadee puzzles me. Why, in wooded stream bottoms and occasional woodlots, do I hear what sound like typical high-low-high-low Carolina chickadee songs? I hear *chick-a-dee-dee-dee* calls, too, rapidly given like those of the Carolina. Wasn't there supposed to be a *narrow* hybrid zone back in Prairie State Park in western Missouri, with the chickadees to the west pure black-caps? The world seems a little messier than what I had thought.

House sparrows! Every farmyard has its own universe of them, a mob scene with a gazillion of them *chirrup*ing and *chatter*ing, but the ruckus fades as we pass, with none in the wide-open farmland.

It's the land of little oil, ancient pumps littering the landscape and squeezing but a few barrels out of the ground each day. And of dickcissels. And of armadillos taking a beating in their futile attempts to cross the road. And, surprisingly, of a few straggling eastern birds, with eastern phoebes at many roadside bridges, a few eastern wood-pewees and great crested flycatchers in wooded stream bottoms.

With three minutes to spare, at 11:57 a.m., 81 miles on our odometers and big satisfied smiles on our faces, we roll into Rosalia and find Tom at the local restaurant, where he had biked in just five minutes before. In our tight spandex biking shorts and bright jerseys and celebratory moods, the three of us are as out of place in the subdued lunch crowd of local farmers as David and I were in the Sunday dinner gathering back in Sebree, Kentucky. Beef and potatoes is the special, chased with pecan or coconut cream pie—the routine here for those around us in working jeans and bib overalls.

The 17 miles north to Cassoday is big sky country, not a tree to be found, cows grazing from nearby to distant horizons, nary a car in sight. Wild mustangs are reported to roam the area. "Hey, this is pretty fine," sums up Tom, somewhat surprised to be this happy touring Kansas. At the outskirts of Cassoday, David and Tom dance and whoop about the welcoming sign that celebrates the town as "The Prairie Chicken Capital of the

Oil pump. Deafening, increasingly the scourge of the Plains. (172, 2:02)

World"—being the professional that I am, I contain my excitement, more than a bit amused with the fun they're having at my expense, their playful mockery as big as the strutting chicken on the sign above.

You know you're in Kansas when . . . a pickup pulls beside us just outside of town, the driver asking "Did you leave a billfold at the store back there?" Sure enough, realizes David. On his fleet road bike Tom races back to the grocery, soon returning with the billfold, its contents all accounted for, of course.

Energized by the tailwind and by Tom, we bike another 30 miles to Harvey County East Park, 12 hours and 130 miles since David and I left Chanute. Tom and David soon fix a fabulous pasta dinner, interrupted by the friendly neighbors who are "camping" in the house that they pull behind their truck. "Come on over for dinner," is the invitation from the fourteen-year-old, mom and dad waving in the distance. Minutes later, impatient with our progress, they start delivering dinner to us, so we move on over to their table, feasting on an unlimited supply of pork chops, potatoes, corn, rolls, and ice cream, the more-than-ample servings clearly a routine meal for our more-than-ample-bodied, generous hosts.

"People from Kansas are way too nice," David concludes.

DAY 32, JUNE 4: HARVEY COUNTY EAST PARK TO QUIVIRA NATIONAL WILDLIFE REFUGE, KANSAS

By my usual three quarters of an hour before sunrise, I'm prowling the campground. Tom and David sleep on, but I'm hoping for a *Tyrannus* dawn fest, as I believe I saw all three in the campground last night. A great crested flycatcher is already at work, but I pass him by.

Eastern kingbirds are the first *Tyrannus* to catch my ear; four of them within earshot fly above their territories, sputtering and stuttering sharply, *t't'tzeer, t't'tzeer, T'TZEET-ZEETZEE, t't'tzeer, t't'tzeer, t't'tzeer, T'TZEETZEETZEE,* over and over again. Just what do the females listen for in all this? They must be comparing each male with the others, their listening somehow helping them decide who will be the father of their offspring. I linger a few minutes, entertaining memories of other eastern kingbirds at dawn, then move on. One down.

Great crested flycatcher. At dawn, doing his thing, the last of the great cresteds in Kansas. (173, 5:04)

From a small tree at the far edge of the campground sings a *western* kingbird. I sneak closer. The overall rhythm is the same as that of the eastern kingbird, with the stuttering and false starts, then finally exploding in a climax before he begins the sputtering bicker all over again. He stutters a *kip kip kip ki-PIP,* several low *kip* notes before one is eventually coupled with a higher and sharper *PIP* note, perhaps the entire sequence repeated once or twice before he explodes into the *ki-PEEP-PEEP-PEEP-PEEP,* the longer and less harsh *PEEP* notes trailing off in both loudness and frequency. How intriguing that the two kingbirds are about the same size, about 40 grams, yet the overall pitch for the western kingbird is so much lower. Two down.

KANSAS KINGBIRDS: EAST MEETS WEST

Eastern kingbird. Sputtering, bickering dawn song. (174, 3:33)

Western kingbird. Ditto, in his own voice. (175, 2:23)

Passing up all the other birds now in song, I home in on the pavilion, as I hear there the third "kingbird," the scissor-tailed flycatcher. I circle around, working my way closer from the side of the pavilion opposite where he sings, eventually standing beneath the shelter and next to his small tree. Ever so slowly, I lean out to peek up beyond the roof overhang, and *there he is,* just two yards up! He perches facing me on a leafless, horizontal branch in the middle of a small locust, his white chest glowing in the early light, his magnificent tail pumping noticeably with the exertion on each note. I listen intently. His stutter is low-pitched like that of the western kingbird, though sharper, and through the climax he becomes increasingly animated, ending on the loudest and highest note of all, surging at the end instead of fading like the western kingbird.

Scissor-tailed flycatcher. Third *Tyrannus*, just inches above me, in his own tyrannical voice. (176, 2:00)

"Kingbird" Notes

Species	Stuttered, repeated portion of dawn song	The climax
Eastern kingbird	*t't'tzeer*	*T'TZEETZEETZEE*
Western kingbird	*kip kip kip ki-PIP*	*ki-PEEP-PEEP-PEEP-PEEP*
Scissor-tailed flycatcher	*pik pik pik pik pik pi-DIK*	*PI-DIK PI-DAH PI-DAH-LEEK*

Scissor-tailed flycatcher

The inheritance of singing genes from their kingbird ancestor seems clear. Males of all three species stutter and sputter for a bit, then deliver their emphatic climax. Although the details of the song differ from one species to the next, the overall pattern is the same . . . but I wonder why only the eastern kingbird has taken to the air to deliver his dawn song.

This *Tyrannus* dawn fest is precisely what excited me so about biking across Kansas. It is the grand meeting of East and West, not just of kingbirds but of a host of other species as well, and here, as a special Kansas treat, a southern *Tyrannus* rises to overlap the ranges of the other two, the three tyrants all occurring here in this very campground.

The sun now rising over the pond by our campsite, I take note of the non-*Tyrannus* world around me. A male red-winged blackbird sings *konk-la-reeeeee,* a female vigorously responding *ch-ch-ch-ch-ch-chit-chit* from where she builds a nest in the cattails. Two Baltimore orioles, their ordinarily brilliant orange now stunningly fluorescent in the rays of the rising sun, fly slowly in tandem about our campsite, both singing—male and female in courtship, I wonder, or two males contesting our site? In the grassland beyond the campground are the usual dickcissels, eastern meadowlarks, and grasshopper sparrows; a ring-necked pheasant crows with a harsh double squawk, *koork-kok,* but I'm too distant to hear the whir of his wings that follows. And, of course, robins still *cherrily cheer-up cheerio,* as they will every day of our trip to the Pacific. In this man-made wooded oasis I still hear eastern birds who will soon disappear: a great crested flycatcher, cardinals, an eastern bluebird, the eastern version of the warbling vireo, as well as those Baltimore orioles and eastern meadowlarks.

My riding mates now awake, we down our oatmeal and break camp; our destination is a mere 80 miles distant, Quivira National Wildlife Refuge, a wildlife haven just a little over halfway through Kansas. "Today is meadowlark day," I alert Tom and David, informing them how to listen to the eastern meadowlark's slurred *spring-o'-the yeeaar* song and its *dzert* or longer *dzert-ert-ert-ert-ert-ert* calls.

Baltimore oriole. Far from its namesake city, in a dialect almost unrecognizable. (177, 4:01)

Newton is just down the road, and we soon jog north to Hesston, then head due west again, arriving in Buhler, where Tom whoops up the sign that proclaims Buhler as the "Glider-Rocker Capital of the World." So many towns claim some form of distinction, as if a competition; I think a couple of them have claimed to be the "hay capital of the world."

Western meadowlark

Beyond Buhler we cross a welcome ripple in the pancake, the Little Arkansas River, then pass through Medora, and just beyond Nickerson and over the Arkansas River itself, with sixty-some miles already on the day, the promised meadowlark sings from the roadside fence post up ahead! "Tom, David, there he is, a western meadowlark!" The song is unmistakable, so much richer than that of the eastern bird, beginning with clear whistles and then tumbling into a rapid, gurgling jumble, starting a bit like the whistled melody of a Baltimore oriole and ending with the jumbled notes of the wood thrush's magical *e-oo-lay*. And he flies, calling *chupp . . . chupp-vicicicicicicicic*! "Hear the call? That's nice. The calls are innate, so they assure us that he is a true western bird, not an eastern imposter who has merely learned the western songs." But songs of the next meadowlark are eastern, as are the next. We have not yet reached the true tipping point of the meadowlark's continent.

Western (!) meadowlarks. Males dueling at territory boundary, in songs and calls. (178, 2:56)

At Quivira's education center, we settle in for the night, and I later stroll beside the nearby pond to take in the evening. A pied-billed grebe sings out, *kuk kuk kuk kuk kaow kaow kaow kaow kaooo-gow-gow-gow,* though my silly mnemonic slaughters his effort; he *kuk*ed rapidly for perhaps two seconds, then more slowly *kaow*ed over six or seven, then added his donkey-bray on the end, *kaooo-gow-gow-gow,* ten or more seconds in all.

Pied-billed grebes. *Kuk*s and *kaow*s and *kaooo*s and *gow*s; "intelligent." (179, 3:32)

Admiring this grebe with me is a young woman with binoculars. "He's intelligent. God gave him that design," she comments. Dumbfounded, I quietly play with the words; I know them all, of course, but never before have I heard them in this order. "Intelligent design," yes, and Kansas is the hotbed of creationism, but it is the creator who was supposed to be intelligent, wasn't it, not the created? Kansas, I quietly accept, yes, you know you're in Kansas. . . .

DAY 33, JUNE 5: QUIVIRA NATIONAL WILDLIFE REFUGE TO NESS CITY, KANSAS

David and Tom are out with me early the next morning well before sunup. From the nearby marsh, the pied-billed grebe declares his intelligence. I increasingly like the woman's perspective from last night. In the grebe's voice is an accumulated intelligence that dates back to the beginning of life on earth, the *kuks* and *kaows* and *kaooo-gows* in his song the culmination of eons of evolutionary changes that now perfectly express what is on the mind of this grebe, at least to other grebes.

An American bittern speaks his mind, in his own way, from the far side of the marsh. Oh, to be close enough to see him in the act. He starts by clicking his bill, I know, then gulping air. His head jerks violently, and once his breast appears inflated, he pumps convulsively, head and neck thrown up and forward, *pump-er-LUNK, pump-er-LUNK, pump-er-LUNK, pump-er-LUNK*. It's an eerie, ventriloquial trio of low, resounding notes, repeated up to ten times, though at this distance I hear only the loudest note, the *LUNK,* sounding like a mallet whacking a stake, hence the bittern's alias of "stake-driver." In the distance is the din of a yellow-headed blackbird colony—we'll get closer to them some other time.

On the road, we magically float among the clouds, through patches of fog that hang over the fields and the road, the water-saturated air amplifying birdsong as if we ride through a great symphony hall. From small bushes beside the road rise the sharp, emphatic songs of Bell's vireos, accompanied by the *wich-i-ty wich-i-ty* of yellowthroats everywhere. A brown thrasher sings from a treetop to the right, pure and smooth, *Kansas Kansas, so sweet so sweet, nice day nice day, ride on ride on, Oregon Oregon, the Pacific the Pacific . . . .*

Dickcissels and grasshopper sparrows are always within earshot, but it is the clarion songs of meadowlarks that define the dawn. Three or four sing from territories nearby, but songs from the next ring of a dozen territories beyond them seem hardly diminished. The eastern's *spring-o'-the-yeeaar* still predominates, but that'll soon change. Though all the species this morning are the same as before, no two listenings are alike, as each individual speaks his own mind.

Beyond the oasis of the Refuge, we emerge into farmland again, much of it irrigated in circular patches a half mile in diameter. The 34 miles to Larned are a breeze, and by now it is only western meadowlarks who sing, their telling calls unequivocally declaring their identity. All too abruptly it feels as if we're in the West, the meadowlarks having made the call. The Pacific in California is now closer than the Atlantic back in Virginia, though we'll take a rather devious route to the Pacific by heading north in Colorado up to Montana and then west. We celebrate by buying a dozen cookies from the woman and her granddaughter who constitute the farmer's market in Larned this fine Thursday morning.

Western meadowlark. Innate calls unequivocally declaring *The West.* (180, 3:14)

The fog has long given way to big-sky sunshine, with only a few clouds on the horizon, though there was something in the forecast about rain later today. At every small creek that runs beneath the road, cliff swallows swarm about, sometimes tens of birds, sometimes hundreds. They're good lookers, their buffy rumps at the base of short, square tails standing out against the dark back, with each bird wearing a small miner's lamp on his forehead. And what a racket! As we pass, birds swirl about us, repeatedly calling *churrr*, a quarter-second, down-slurred, raspy note. Amid their vigorous protests, we stop on a bridge to admire these architects, their gourd-shaped mud nests attached firmly to the vertical surfaces of the bridge supports. It is bridges like this that provide nesting sites and entice these birds out of their native homes in the western mountains and into the plains.

Cliff swallow

At Fort Larned, we step back in time to when the flag had only 37 stars, learning about military life on the Santa Fe Trail just after the Civil War. We emerge from the visitor center to a different world, too, the sunshine having given way to a heavy cloud cover. Two hours later, about 30 miles from our destination of Ness City, the rains begin.

Cliff swallows. Busyness of a large colony, with intimate nest songs and flying *churrr* calls. (181, 2:15)

About ten miles out, a man pulls up beside us in his pickup truck, rolls down the window, and shouts "You'll want the first motel in town." An hour later, there he is, outside the Derrick Inn Motel and waving us in. "Here, go on in and get some warm drinks," and he puts a few dollars into David's hand.

We agree, as David put it two days ago, that "Kansans are way too nice."

We drink, eat, dry out, do the laundry, warm up, and rest after another 100-mile day, the last 30 in the rain.

DAY 34, JUNE 6: NESS CITY TO TRIBUNE, KANSAS

With every 100 miles we leap west, the sun rises five to six minutes later, so I wait until 5:30 before biking the streets of Ness City, soon stopping beneath one western kingbird after another, all of them in excited dawn song, their song genes all the same. Not an eastern kingbird is to be heard.

A black-capped chickadee sings unmistakable *hey-sweetie* songs. I hang with him, hoping . . . yes, there, after just a minute he shifts his song to a higher

Black-capped chickadee. Typical North American *hey-sweetie*, from The West. (182, 0:38)

Northern flickers. A spirited threesome, calling and drumming, last of the yellow-shafted variety as we cross the Plains. (183, 0:34)

pitch; three minutes later he shifts down. That's special, the first pure black-capped chickadee I've heard since the highlands of Virginia. There is no doubt now: I'm in black-cap country.

Other birds include a few mourning doves, blue jays, house sparrows, a house wren and house finch, and a northern flicker as known by its *klee-yer!* and *flick-a* calls, still the eastern yellow-shafted form, as the transition to the western red-shafted form lies just to the west in eastern Colorado.

After sunrise, with the birds now far quieter, I return to the motel. We pack up, though ever so slowly, as there seems to be so much gear to repack after a drenching rain.

We are in wheat country. The game of "Name that Roadkill" has lost its luster here, not for the lack of kill, but because of the low variety of birds and other animals in these prairies. The new game is "How far is it to those grain elevators up ahead?"

"Three miles," guesses Tom. "Four," from David. "Naw, exactly three and a half," I offer. We bike on, passing three miles, the grain elevators barely larger on the horizon, then five miles, then arriving ten miles later. The monstrous silo at Amy, on the railroad tracks, of course, and in the middle of nowhere, has 20 massive silos beside the road, and perhaps they are two or even three deep, for all we know. Despite our practice, our estimates never improve.

Late afternoon, after 70 miles, much of it into a surprising headwind, the wind shifts to the east. The last 30 miles of the day into Tribune are among the finest one can imagine on a bicycle. We float along, the winds at our backs and pushing us uphill, the hill just barely perceptible, though real it is, maybe 1/3 of a percent grade. We have climbed the better part of a mile since the Missouri border, and we rise ever so gradually to meet the Rockies.

In the waning light, we follow the sign to Tribune and turn right onto Broadway Avenue, stopping to admire the tracks of the original Missouri-Pacific Railroad that parallel our route. A side track here merges with the main track that disappears straight away into the distance to the east, and grain elevators loom large four miles away. During the late 1800s competition among railroad companies was fierce, and the famed Union Pacific lay just 30 miles north of here, the Atchison, Topeka and Santa Fe thirty-some miles to the south. Federal grants to railroad companies provided land along the route that would be sold to settlers; the land sales and subsequent commerce in turn funded the railroads, which had the power to build or dwarf towns along their route. Back in 1887, we soon learn, Tribune actually lay a mile north of here, but the town relocated to be here, on the railroad.

At the high school, we discover that hundreds of bicyclists have gathered, as tonight is the kickoff for the annual Biking-Across-Kansas event. The assault on Kansas begins in three waves, one leaving here from Tribune, a second from Syracuse 34 miles to the south, a third from Johnson City 28 miles beyond that, all converging about 500 miles later at La Cygne on the Missouri River. We crash the mob scene here, acting as if we belong, scavenging sandwiches and cake for dinner.

Amid the snoring and laughter and farting and occasional shrieks and assorted sounds of hundreds of others, David and I sleep on the school's gym floor overnight. On the grass beside the gym outside, Tom sleeps cocooned in his bivy sack, reveling in the fierce winds and torrential rains of overnight thunderstorms, learning as cheerily as a good Kansan that his bivy sack is not waterproof.

15.
SHORTGRASS PRAIRIE

DAY 35, JUNE 7: TRIBUNE, KANSAS, TO EADS, COLORADO

In the far-too-brilliant lights of the high school's gym, hundreds of fluorescent red and orange and green and yellow and lime and all manner of wildly designed biking jerseys mill about us. Though it's still well before sunrise, the eager Biking-Across-Kansas cyclists have already swept the kitchen clean. In desperation, the cooks dump a box of yesterday's leftover sandwiches and cake onto the table, and we grab our breakfast.

We're soon caught up in a 16-mile-long party moving west to the Colorado border, the official starting point for the eastbound Kansas riders. The mood is festive, with party hats and stuffed animals riding atop helmets, and all sorts of outlandish costumes. The carefree BAK cyclists have thrown everything but their bikes into a large truck for transport to Scott City, their next stop to the east, and streaming past us are speedy, lightweight racing bikes, and some tandems, too. With our loaded touring bikes, we keep pace with the recumbents, and almost run down a unicyclist. At the border, the queue grows for photos beneath the "Welcome to Colorful Colorado" sign, and a

"bugler" offers his rendition of what sounds like, to give him some credit, a dying cow gasping her last. David, accomplished trombonist that he is, asks, "Do you mind if I try?" Sure, and he quickly whips the crowd into a frenzy with his rousing version of the bugler's Cavalry Charge.

Leaving the warmth of the pack and heading into Colorado, we now feel the headwind, with the temperature in the low fifties thanks to the cold front that the thunderstorms delivered last night. But we're far happier in these conditions than in the 100 degrees it was here just a few days ago.

And how the landscape has changed! The giant grain elevators and endless wheat fields of the former "tallgrass prairie" in Kansas have given way to what is called, not surprisingly, the "shortgrass prairie." The nearby Rockies suck most moisture from the westerlies that rise over them, so perhaps only a foot of rain falls here each year, and most of that evaporates before doing much good. Instead of rain, the Rockies dump their waste here, the layers of weathered mountain debris below us ever thickening; we will rise another 1000 feet above sea level to the Rockies themselves. Towns are no longer ten miles apart, the area that even the biggest Kansas grain elevators can service, but instead 40 to 50 miles apart—it feels lonely out here. In the sandy, heavily grazed pasture off to the left, next to the "WE STAND UNITED! VOTE GOP" sign, a lone cactus blooms.

The soundscape has changed, too, with an entirely new song, the source unknown. *Oh meeeeeeeee, my my*. A brief pure note introduces a wavering *meeeeeeeee*, followed by two low whistled notes. These songsters are all along the road, all unseen, all with the same beautiful song pattern.

"New roadkill," I call out; "Game on!" Tom and David are up ahead, laughing and having a merry time biking side by side, but I circle back to pull up beside a sparrow of some sort. I pick it up and roll it over in my hands; the words "plain" and "nondescript" first come to mind, "subtle elegance" soon following. I now see the fine streaks on the breast, the grayish throat with dark lateral lines, the gray forehead and crown with fine streaks; he's a study in subtle shades of gray and brown and cinnamon and buff and white and black, each feather a masterpiece, the patterns among the feathers together showing fine spots and bars and streaks.

One voice urges me to open the bird book and identify this sparrow, to give it a name, but another voice prevails. *Seeing Is Forgetting the Name of the Thing One Sees* wrote Lawrence Weschler in his book title. It's the same for *Hearing*. When we truly see or hear, a name is irrelevant; give him a name, too, and I'll feel all too smugly that I know him, but leave him nameless and he invites further looking and listening. So we'll travel together for a bit, the

Lark bunting

Lark bunting. Colorado's state bird in perched and flight song. (184, 13:32)

mystery bird and I, and his lifeless body is soon nestled into the outer pocket of my left rear pannier.

The state bird of Colorado also stakes his claim here, with lark buntings launching into song flight from roadside fence posts. I watch the next one. His wings beat rapidly at first as he rises to about 15 feet, but he soon rows through the air more slowly, exaggerating his buoyancy in butterfly flight, all feathers now spread for maximum lift. His large snow-white wing patches flash against his overall blackness; the tail is spread widely, showing the white tips there. And how he sings! The slow, sweet notes of the opening phrase are followed by a high buzz, and he then alternates slow, pure notes with more rapid, harsher phrases for maybe ten seconds overall before gliding to a landing in a small shrub about 20 yards from where he began.

In this simple semiarid, shortgrass habitat, there's little support for a complex community of birds. Besides the meadowlarks and lark buntings and the mystery sparrow, I hear so few other birds. The ubiquitous mourning doves still explode from beside the road and from utility wires overhead, and I hear snippets from grasshopper sparrows and horned larks. But that's about it. Even the dickcissels are long gone, all behind us.

"Rock wren! Rock wren! He's all new, another westerner! To the right, on the metal post beside the railroad tracks." He sings, *kjree kjree kjree kjree kjree*, bouncing up and down in shallow knee bends, looking quickly this way and that. He sings again, a song different from the first, *ching ching ching ching ching*, then the *kjree* song and then the *ching* again, in an alternating pattern I came to know so well after weeks and weeks of studying their songs thirty-some years ago. In a flash he's gone. I mark the moment: "First rock wren, with far more to come. 3:31 p.m., 48.61 miles on the day's trip odometer, just a little over 30 miles into Colorado."

Rock wren

Rock wren. On the Kansas-Colorado border, alternating songs in a brief western welcome. (185, 0:21)

Mid-afternoon, after a relatively short, 60-mile day, we arrive in Eads, our first stop in Colorado. It has been a tough day, cold and with a considerable headwind, but Tom is fresh and strong, and he pulls us along with his high spirits and speed. The soreness in my right ankle is an unhappy Achilles tendon, as confirmed by a visit to the local emergency room. "Take it a little easier," I'm told there. Since mid-Kentucky, when we left Bardstown 16 days ago, David and I have been pushing hard with no rest days, averaging over 100 miles per day for the 500 miles across Kansas.

"You need more Vitamin I," prescribes Tom. "Ibuprofen. Kill the pain, reduce the swelling, and keep on biking." To ease the strain on the Achilles, we dig from David's panniers the platform pedals that he used back in Virginia; I'll use those instead of my more efficient clipless pedals, moving my foot position forward on the pedal and pushing down more on my heel than on the ball of my foot. I lower the saddle slightly, too. A test ride feels odd, the power now coming solely from my leg muscles, none from the fluid motion of the foot itself. I feel peg-legged, but it'll do.

In the park across from the motel room that I've requested, Tom and David prepare a festive barbecue cooked over an open fire: sweet corn, baked potatoes, and locally grown beef. They laugh their way through a watermelon-spitting contest. Joining us is Keith, a British cyclist who retired at age 50 and took up bicycling as a hobby. "I prefer motels," he admits. "Haven't yet used the tent and sleeping gear I've carried since I left Yorktown late April."

I'm tired. I plan to sleep in tomorrow, though I will regret missing our first dawn in Colorado.

DAY 36, JUNE 8: EADS TO BOB CREEK, COLORADO

Morning comes all too quickly, and we depart Eads late morning, as our pace seems to have taken its toll on me. By noon we're in Haswell, just 21 miles down the road, eating lunch at what might be the "Nation's Smallest Park" just around the corner from the self proclaimed "Nation's Smallest Jail," apparently the town's claim to fame.

"Mountains! I see the Rockies!" David shouts just outside of town. We pull over, get out the binoculars, and sure enough, there they are. Pike's Peak, we guess, the mountain with a road to the top. With the Rockies looming, no rock wren is needed to say "WEST" now. Our big emotions match the big terrain we are about to enter.

In 34 miles we arrive in Sugar City, where we celebrate with a whopping serving of pie and ice cream at the Sugar City Café, followed by a whopping beef dinner. "How about some dessert?" asks David, and half a dozen cinnamon rolls are soon delivered. The owner and waitress is friendly, yet also seems sad and maybe even a little bitter. It seems that National Sugar's beet-processing plant closed here 36 years ago, and the place has never been the same. We surmise that she stays open as a kind of community service, giving locals some place to go but hardly making an income.

Riding west on full stomachs in the softening light of the early evening, we search for a place to camp; it's our last night in the prairies and our last with Tom. Experienced touring cyclist that he is, Tom spots a good "bandit" site to the right along Bob Creek, less than ten miles out of Sugar City. We haul our bicycles across the ditch, over the raised railroad tracks, and down behind them into a perfect grassy strip between the tracks and a fence. In the field beyond the fence and along the creek itself, there'll be good birds in the morning, I assure myself, and the train tracks appear unused. For good measure, to cap off the caloric intake for the day, Tom and David cook up a carbo-loading pasta dinner.

Watching the sun set from the railroad tracks, I also see the lights of a large building emerging from the fading daylight half a mile to the west, and simultaneously a police car pulls to a stop along the road. The officer guardedly strolls closer, inquiring, but relaxing when he learns we're cyclists. "Just checking to see if I had a dead body up there," he remarks a bit puzzlingly. "That's a prison just down the road. You're fine there, but if the railroad complains, I'll have to ask you to move along." Given the mild caution, we hunker down, staying below the tracks and out of sight the rest of the evening.

DAY 37, JUNE 9: BOB CREEK TO PUEBLO, COLORADO

With sunrise about 5:30 a.m., I set my watch alarm for 4:00, as I'm eager to hear who will be about in the morning. But I sleep lightly all night long, mesmerized by a mockingbird who sings to the north along the creek. He must be unpaired, crooning through the night in hopes of attracting a mate, and what a fine sampler he offers of local bird sounds. He's a western mocker, all right, as I hear none of the mockingbird's favorite targets I've enjoyed from Virginia through Missouri: no cardinals, red-bellied woodpeckers, blue jays, wood thrushes, great crested flycatchers, eastern bluebirds, eastern phoebes, Carolina wrens, or tufted titmice. But who is he mimicking? They must be common western birds, but I am rusty on them. I concentrate . . . rock wren? . . . jay? . . .

Northern mockingbird. Mimicking all *western* songs! (186, 10:22)

Yes, rock wren! I know well the wren's buzzy, ringing call, an isolated *pdzeee*, and the mocker repeats it rapidly, six times over two seconds, pauses briefly, then eight more times over three seconds . . . and a robin's *qui-qui-qui* calls, then *tut-tut-tut-tut-tut-tut* . . . western wood-pewee . . . western meadowlark . . . western kingbird, not the dawn song but the calls that mates use when greeting each other . . . flicker *klee-yer!* call . . . blue-gray gnatcatcher? . . . rock wren song, slow, rhythmic . . .

killdeer . . . hawk? . . . flicker song, *wik-wik-wik,* carrying on for a full three seconds, a rendition any flicker would be proud of . . . western scrub-jay? . . . rock wren, yet a different song. He loves rock wren songs, the cadence and length perfect for his style, just as Carolina wrens were back East, and with each male rock wren having one hundred or more different songs, what wonderful material for this mimic to draw on. "A perfect primer on western birds," I smile.

Four a.m. comes quickly and I punch the alarm off, listening to the mockingbird some more, but within 15 minutes I'm standing beneath the stately cottonwood tree at the creek, a western kingbird singing above me. He sputters and bickers his way through the *kip kip kip ki-PIP* notes before exploding into his *ki-PEEP-PEEP-PEEP-PEEP,* completing this routine every five seconds. In the distance a great horned owl hoots, but nearby, in seemingly every sizable tree, are other western kingbirds hailing the dawn.

Though I walk among kingbirds, I am inevitably drawn to the mockingbird, the only other bird singing at this hour. He sings at almost double his night speed now. I hear all of the now-familiar pilfered sounds of western birds, but I especially love his flicker *klee-yer!* call and the *wik-wik* song, to which he's now added the *flicka-flicka* call. The mocker and I stand in a special place, in the dry rain shadow of the Rockies and right smack in the middle of the narrow hybrid zone between red-shafted flickers to the west and yellow-shafted flickers to the east.

Western kingbird. Ecstatic dawn singing, always enthralling. (187, 3:11)

Here in this dry shortgrass prairie the flickers are consistently intermediate, as they're all hybrids. Perhaps the flicker from which this mocker stole these sounds had the gray throat and brown crown of the western subspecies, the black malar stripe (sometimes called a moustache) of the eastern subspecies, and intermediate pumpkin-gold feather linings in the wings and tail—but almost any combination of eastern and western traits would be possible.

I've had fun with these flickers and other prairie birds in my ornithology classes over the years. We'd play with the words "species" and "subspecies," trying to make sense of a messy world, usually concluding that we humans often try to impose order where there is none. For the flickers, it seems that the birds most suited to this shortgrass prairie are the hybrids that contain genes of both yellow- and red-shafted forms. Outside this dry prairie, the hybrids do less well, so that pure yellow-shafted birds own the East, pure red-shafted birds the West. Because hybrids thrive only here, the hybrid zone is stable and seemingly permanent; but because the yellow- and red-shafted forms hybridize so freely here, these two forms are considered "subspecies" of a single species, the "northern flicker," even though pure yellow and red forms predominate over the vast majority of the continent.

The meadowlark case is different. If somewhere in mid-Kansas an eastern and western bird pair and raise a family, the offspring appear healthy, but they're infertile as adults, and their eggs never hatch. So the lineage of a meadowlark who pairs with a bird of the other kind comes to a screeching halt. That's strong selection against birds who make mistakes in finding a partner, as the genes of the mistake-prone birds are quickly cut from the population. Hence, the two meadowlarks are considered "good species," with none of the hybrid mess of the flicker case.

It's almost 5 a.m. now, and a western meadowlark draws me into the pasture to the east. But as I approach, I hear once again my mystery sparrow, the song of the bird who has been riding with me in my pannier the last two days: *Oh meeeeeeeee oh my oh my*. I'm close enough now to hear two high-pitched *oh* notes that I hadn't heard from my bicycle.

Cassin's sparrow. From the shortgrass prairie, *Oh meeeeeeeee oh my oh my*, airborne or perched, so fine. (188, 5:14)

Against the orange glow of the eastern horizon, I see him flutter up into the air, maybe 15 feet or so, singing as he glides back to a perch. Up again, steeply, singing on the gliding descent. His song flights carry him to the left, then to the right, then toward me, and I watch him through my binoculars as he crisscrosses his territory.

As the sun peeks over the eastern horizon, I pull my Kaufman Field Guide out of my pack and quickly scan the sparrow pages: there, Cassin's sparrow! He's "gray-brown, confusingly plain. . . . Best known by habit of fluttering up in air and gliding back down while singing." Cassin's sparrow. He now has a label, but do I know him any better? The Latin name helps, as I see he's in the same genus with other sparrows that I don't know in the arid Southwest, and with the equally fine singer, the Bachman's sparrow, that I know from Florida. Knowing his Latin name reveals to me a group of related sparrows and provides an evolutionary context in which to appreciate the Cassin sparrow's song.

I head back to camp to check on Tom and David. Still sleeping. I'm soon across the road, listening to a small cottonwood tree sing what sounds almost like the rambling song of a canary; each song is three or four seconds long, each a melodious jumble of rich trills and whistles sharply contrasting with churrs and buzzes, the pace fairly slow as he bounces about in volume and pitch. On the other side of the tree, about head-high, is the source, a lark sparrow. I listen, smiling; no two songs are alike, and he clearly ranks among the finest of singers we've heard since the Atlantic. The binoculars on him now, I see a striking bird: chestnut, black, and white patches and stripes on his head; below, his white throat gives way to a grayish breast with a central black stickpin; his back is a grayish brown with black stripes, the tail a rich brown with bright white corners at the tip. A lark sparrow. Very nice.

The field guide says they've been with us since Missouri, but how did I miss their remarkable songs for the last thousand miles? I know of others I've missed, too, as visions of their corpses beside the road flash before my eyes. The red-headed woodpecker, a brilliant red-and-white-on-black body lying on the pavement. A loggerhead shrike, easily taken for a mockingbird, until one looks more closely and sees the extra dose of black in the face mask, wings, and tail, and such a large head with that stout, hooked bill.

Lark sparrow. Stunning in looks and song. (189, 1:43)

Just as silent was the barn owl I picked up yesterday from the roadside weeds. The outstretched wings spanned four feet; long legs dangled below, the mouse-catching talons sharp but useless. The back was mottled grandly with buff, gray, white, and brown; fine dark specks adorned the light undersides, including the tawny breast. The head hung limply, but I held it up, imagined the eyes open, the bird alive. How friendly the face, all white except for the dark eyes, entirely bounded by a tawny, heart-shaped outline. I ruffled the white facial feathers, finding the magic beneath. Hidden there are the two ears, the left ear high with overlying feathers oriented down, the right ear low with overlying feathers oriented up, so that the left ear is especially sensitive to sounds from below, the right from above. When the owl hears just the faint scratchings from a mouse, the sounds arrive at different times and intensity in the left and right ears, thanks to the asymmetry in feathers and ear placement, and neurons then fire in appropriately precise locations within the brain so that the owl knows instantly the exact location of its prey. Intelligent, yes, with abilities honed over eons of mouse-catching.

I'm torn, as the lark sparrow continues to sing, the cliff swallow colony on the railroad bridge over Bob Creek hums with activity, a meadowlark sings to the south, kingbirds are all a-twitter, and the mockingbird continues to the north, but I see movement in camp, now a good hour *after* sunrise.

"What a morning!" Over breakfast, I share my listening adventures with Tom and David. They're happy for me, but clearly happier for themselves that they got their sleep. It's

Barn owl

no surprise that I'm usually dragging at the end of the day when they're still going strong.

Seven miles down the road in the small town of Olney Springs, I'm reminded of one of the reasons why we are biking from east to west: The morning sun is at our backs, not in our eyes, casting the most beautiful of soft morning light on, this time, a male Bullock's oriole! He flies with me, just off my handlebars to the right, every detail immaculate. His large white wing patches flash with each flap of the wings; on the upstroke he reveals his orangish undersides and underwing linings, on the downstroke his black back and orangish rump and tail. The black tip to the tail, the black throat, black crown, and orange face are pure Bullock's oriole. Western. Western! "*It's a Bullock's oriole! We're in the West!*" I call out, as if Tom and David needed a reminder.

I have studied research papers on orioles, and know that by now we're well past the western Kansas hybrid zone between the eastern Baltimore and western Bullock's orioles. With so many oriole hybrids in the transition zone between the tallgrass and shortgrass prairie, ornithologists have agonized for years over whether these two orioles should be considered one species or two. Before 1973, they were considered two; then the lumpers ruled, and they became one, the "northern oriole" (the name acknowledging other orioles to the south); then, in the 1990s, the splitters ruled again. Why the orioles are considered "two good species" and not the flickers isn't entirely clear to me, or perhaps to anyone.

Another 15 miles down the road, in Boone's city park, where we stop for a lunch, a Bullock's oriole sings, and sings, and sings. It is the rhythm that strikes me so, beginning with three gruff notes, *cha cha-cha,* followed by a wonderful medley of piping whistles mixed in with a few harsh chatters and rattles. Over and over he repeats the same song, so different from the mellow tones of the Baltimore oriole.

Bullock's oriole. Harsh chatters and rhythmic song declaring "This is now *The West*." (190, 2:17)

On toward Pueblo, we continue into a headwind and the heat of the day, with Tom and David taking turns in the lead and pulling me and my ailing ankle along. By noon, we're on Highway 50 within the Pueblo city limits, the traffic and noise unfamiliar to us, the highway busier than any we've seen since Virginia.

Though we're welcome to camp free at the city park, we choose a motel for the night. David and Tom check out the nearby bicycle stores, returning with various supplies and, most exciting, new touring tires for our bikes. David had limped into Pueblo, his rear tire so frayed back in Kansas that, at Tom's suggestion, he rode the last two hundred miles with a tube and two tires on his back wheel, one failing tire inside the other.

With more energy than I can muster, the two college roommates are soon out on the town together, leaving me to rest and to reflect on the day that began at Bob Creek in the wee hours of the morning . . . and to listen to house finches in the motel courtyard. Western house finches, part of the grand western tradition of house finches from which the eastern stock was derived in the early 1940s when a few were released on Long Island. I listen, words like "rollicking, cheery, bright, sunny, lively" all coming to mind, the occasional husky, down-slurred *veeeeer* note their signature. And "western," as no two songs sound alike, so different from the eastern birds that tend to repeat one song several times before switching to another.

House finch. Spirited, lively, cheery; western stock from which easterners were derived. (191, 4:57)

Tonight, Tom will leave by bus to rock-climb in Wyoming. Tomorrow, David and I will leave the prairie and begin our climb into the Rockies. Oh happy day!

16.
WESTERN BIRDS

DAY 38, JUNE 10: PUEBLO TO CAÑON CITY, COLORADO

On far less sleep than either of us would have wanted, we are in the motel courtyard with our loaded bicycles a good hour before sunrise. After a rousing send-off by the local house finches singing by the lights of the motel, we roll through the empty streets of Pueblo. I breathe deeply, happily, taking in the western air, listening for western birds . . . mainly western kingbirds at this hour . . . but it's still early, still quiet.

On Main Street, we cross the Arkansas River, but instead of following Adventure Cycling's recommended route 96 out of town, we drop down to the local bike path that follows the river. It is along the river where I'm convinced we'll hear the best of dawn singing.

As I look over at David, I recall the entry he showed me in his log book some days ago. It was something like "I do not know why I can barely stand the early dawn. It is almost always painful and groggy. Everything is cold and unalive. I prefer to be gently awakened by the rising sun. For Dad, it is yet another glorious morning."

Almost immediately we are greeted by an unmistakable trail-side *hey-sweetie* from a black-capped chickadee. I stop, and David pulls over beside me. "Vintage black-cap," I remark. "Hear the neighbor down the path singing a slightly higher version? . . . There, the neighbor just shifted to a much lower pitch. Count: One . . . two . . . *yes,* on the third song, the chickadee right here also shifts. They're matching each other now with identical, low-pitched songs." As we pull away, the neighbor shifts to yet another pitch of the *hey-sweetie* song. "They'll sing like this wherever we encounter them the rest of the trip, *except,*" I smile as I remind myself, "beyond the Cascade Mountains in western Oregon. Something for you to look forward to."

Black-capped chickadee

In the wet tangles that line the riverbank is one energized yellow warbler after another, each chipping between songs, each hustling through multiple variations of his *sweet sweet sweet I'm so sweet* dawn songs.

"Hey, there's the new phoebe for you, a Say's phoebe. Listen for his signature song, a down-slurred *pit-TSEEeeuur*. But he's got two other songs at dawn, too, both of them rising in pitch, one sort of raspy, the other with pure tones like the *pit-TSEEeeuur*. A little more complex than the eastern phoebe's two songs." I follow along, listening to his dawn program, breathing in *The West* as I slowly bike by.

Say's phoebe. Rousing dawn performance, one of three different songs offered nearly every second! (192, 2:41)

So many others here clamor for attention. David bikes on ahead, but I linger beneath an odd robin, soon realizing that he sings far too continuously, without the robin-like pauses and without the higher-pitched *hisselly* notes. These songs are too rich, a "robin with voice lessons," a black-headed grosbeak to be sure. But I've never heard a grosbeak sing so continuously like this. I sing with him, counting off about three phrases every two seconds, 15 every ten seconds. I pick a distinctive sound, three rising whistles I'll recognize as *wheet wheet wheet,* and hear him deliver it again in another minute, after roughly 90 other songs, then roughly a minute later again, in a dawn singing program so different from that of the robin.

Black-headed grosbeak. Robin-like, but richer and purer, now in continuous song at dawn. (193, 10:25)

Spotted towhee. In dawn song, a "mountain" towhee stutters repeatedly on *drink* before offering *tea.* (194, 4:04)

I try to catch up with David, but am slowed by a spotted towhee who alternates two songs, just like his eastern cousin does at dawn. *Drink drink drink drink drink drink teeeeeeeeeeeeee,* six identical introductory notes followed by a rapid trill, is alternated with *drink drink drink te-ee-ee-ee-ee-ee-ee,* usually three introductory notes followed by several distinctively enunciated syllables. True to form, males in this "mountain race" of the spotted towhee begin their songs with multiple *drink* notes, and I'll be listening for these introductory *drink* notes to all but disappear as we approach the Pacific.

I follow the bike trail upstream, through scenic canyons and past vertical cliff faces, enveloped in a chorus of western kingbirds, western tanagers, western wood-pewees, black-billed magpies, Bullock's orioles, house finches, mourning doves, house wrens, even a pair of red-tailed hawks . . . and I'm happy to find David waiting in a large picnic area at the end of the bike path, the dam to the Pueblo Reservoir looming large to the west.

There, perched high in the dead branches of a large tree overhead is one of my oldest of friends, a Bewick's wren. It was 32 years ago, in 1971, when I first heard these Colorado wrens, and they are special. The song overhead now is a full four seconds, with four full-bodied trills among a complex jumble of buzzes and whistles. I listen, marveling at each song, trying to explain to David how I feel when I hear a Bewick's wren. "For my four years in graduate school, these wrens were my life, and the memories are all so good. Let's wait until he switches to another song . . . there. Wow. Let's give him one more switch . . . nice. If we took the time, he'd eventually reveal about ten different songs, but I suppose we should move on."

Leaving the lush river bottom, we head west through arid country with well-spaced junipers. Rocky bluffs rise up beside the road and bridges span impressively large, dry canyons once carved by rivers below.

We climb, soon tackling a 9% grade for half a mile as we rise over a rocky ridge. I love a good uphill, and we've not had a grade like this since the Ozarks. Over my heavy breathing, I listen in the calm of mid-morning to voices from the junipers as we creep along . . . lazuli bunting . . . green-tailed towhee . . . spotted towhee . . . western meadowlark . . . lark sparrow . . . western kingbird . . . Bullock's oriole . . . and the croak of a raven—all good western sounds! Just a few miles to the north looms snow-covered Pike's Peak. Under a cloudless sky, about 75 degrees, David just ahead, western birds, *The West.* How could it get any better than this?

Alexander Barclay would probably agree. According to the roadside sign where we pause, he came here in 1836 in search of silver and gold. Until his death 19 years later, he chased his dream, but survived by trading furs, ranching bison, and smuggling whiskey; he married a most beautiful woman, but lost her to a man with better "prospects." Though his fortune remained unclaimed, the sign suggests "he had reaped other rewards—independence, experience, perhaps even wisdom—that the West conferred in abundance." And much of that happened in the "Hardscrabble" mining district just southwest of here.

We arrive in Cañon City by noon, with about as perfect a morning as I could imagine. Five hours later, having gone nowhere, we're drying out in the Parkview Motel just across from the city park.

David had spent a couple of hours on the internet honing his second mass e-mail message to his friends; I read and waited, mustering patience, knowing that keeping in touch with his friends was as important to him as my dawn rides were to me. Finally, a grocery shopping behind us, we had set out at 3:30 for points west, our destination for the night unknown.

Thunder in the distance didn't deter us, but the wall of water did. Just out of town, along with two rock doves and a flock of house sparrows, we took shelter in a covered shed where the highway department kept its sand and gravel. Yelling to each other over the din of thunder and of rain striking the metal roof, we agreed to wait it out, but eventually donned all of our rain gear and headed back to town, taking the motel room for the night.

DAY 39, JUNE 12: CAÑON CITY TO GUFFEY, COLORADO

It's a long night, thankfully, allowing us to catch up on some sleep, but come dawn, I'm restless. By 4:30 I hear the house finches through the open window; I know what lies just outside of town, and I ache to be listening there, in the quiet solitude of Temple Canyon.

I imagine being there among the junipers, alive with dawn song. In the dark, I sneak closer to a singing juniper, the unperturbed gray flycatcher there soon only an arm's length away. I float through the landscape, gray flycatchers singing everywhere, my first of several truly western *Empidonax*. Spotted towhees awake with *zhreeee* calls and soon launch into a singing tirade. Adding to the growing chorus are Bewick's wrens and black-headed grosbeaks and western tanagers and more. I drift easily among the junipers, the birdsong so pure without the reverberation and echo caused by the broad leaves of eastern forests. In my motel bed I lie in a state somewhere

SONGS OF THE JUNIPER FOOTHILLS

Gray flycatchers. Three neighbors in dawn song. Bird 1 (195, 1:02) Bird 2 (196, 1:03) Bird 3 (197, 0:58)

Black-headed grosbeak. Dawn gives way to day song, still superlative. (198, 5:25)

between full consciousness and sleep, eventually waking well after sunrise to more mundane house sparrows and house finches outside the motel.

Just outside of town we're treated to raw evidence of how the Rockies were born. Sloping steeply up to the west beside the road is the Dakota Hogback ridge, a formation of sedimentary layers that begins in New Mexico and extends through Colorado and up into Wyoming. About 70 to 80 million years ago, a vast mass of billion-year-old granite that had formed deep in the earth was thrust upward, bulldozing aside the sedimentary layers above it. These sedimentary strata were draped and stretched over the protruding mountains like layers of fractured blankets. Erosion took its toll, until today we see only the Rockies and the stubs of a few erosion-resistant sandstone layers aiming up and west here, the angle pointing skyward into thin air and suggesting the heights to which the Rockies once rose.

Only ten miles out of town, David eyes the huge signs advertising the Royal Gorge, and he simply must check it out. So we stop at the nearby campground, where I happily remain with the gear while he cycles the eight miles to the gorge. "I'll be back in 45 minutes," he predicts. It is a sight to behold, I know, as I've seen it: As the Rockies rose, the Arkansas River held its ancient course, cutting its way through solid granite as fast as the Rockies rose above it, so that today the canyon is about a quarter mile deep but only about 50 feet wide at the bottom.

I survey the scene from campsite 36, where I rest. Junipers dot the hillsides nearby, and snowcapped peaks rise above 14,000 feet just to the south, 35 miles away according to my map. To the north are more Rockies, and in the next 30 miles, we'll climb on route 9 to Currant Creek Pass at 9404 feet, rising about 3000 feet from here on an overall grade of about 2%. Then we'll climb for another 48 miles to Hoosier Pass at 11,542 feet, over two miles above the Atlantic. We can do that easily, I tell myself, and remain well above our lowest gear. The Pacific lies "just over there," just to the west, I smile.

I listen to the light breeze caressing the juniper overhead . . . to insects of all kinds chirping and humming in the mid-afternoon warmth . . . to songs of a Bullock's oriole who visits the juniper nearby . . . to a titmouse.

A titmouse? Has to be the juniper titmouse. He repeats himself, not the slowly whistled *peter peter* of the tufted titmouse, but harsher and faster, with eight or nine identical phrases, each phrase itself consisting of three rapidly repeated notes, as if he's stuttering. I work my way across the campground toward him, trying to catch a glimpse, and there he is, a grayish rather nondescript titmouse with the requisite head tuft. His looks are "plain," and until just a few years ago he was the "plain titmouse," with a range from here to the Pacific in California. But the ornithological splitters recently separated this plain bird into two different species, the "juniper titmouse" in the interior juniper and pinyon juniper habitats, the "oak titmouse" in the oak and oak-pine woodlands of California. Very nice—here's yet another western bird.

A vireo sings, almost certainly a solitary vireo, but I remind myself that the splitters have also won here fairly recently. "Plumbeous vireo," I enunciate slowly, the "dull, lead-colored" vireo, one of the three that were once called the "solitary vireo." I count off a song every two seconds and listen for a handle, waiting for an especially unique phrase that I will recognize when he sings it again. He's tough, as none of the songs stand out, but finally at song number 71, I hear a unique, especially burry, down-sweeping phrase, and then it recurs at 73, 81, 83, 84, 108, 110, and off he flies. The first 70 songs contained none of this unique phrase, but the last 40 contained seven of them. His song program is much like that of his close eastern relative the blue-headed vireo, who also favors one set of songs for a while and then another set. I look forward to hearing the third of these vireo species, the Cassin's in Idaho and Oregon, to hear if it is also a package singer.

Juniper titmouse. Pulsed, incisive *tititi-tititi-tititi-tititi*—the tufted titmouse is long gone. (199, 0:49).

Plumbeous vireo

Plumbeous vireo. Like the blue-headed, he sings in packages, but with burry songs. (200, 2:18)

An hour and a half after he departed, David returns. "Ridiculously commercial," he reports, rather disgusted at the theme-park atmosphere of this natural treasure, "but I climbed a fence and got a look at the canyon without paying the $18 entrance fee."

David reloads his bike with his four panniers and I reload my body with 800 mg of ibuprofen to soothe the Achilles tendon and the shin that's been complaining since Virginia. Now already 5 p.m., we are immediately climbing. Junipers own the hillsides, and magnificent rock outcroppings tower above the road on the right; to the left, the valley falls away steeply to poplars lining the stream below. Billowing white clouds grace an intensely blue sky; though rain falls from a few of the clouds, it evaporates in midair and falls short of reaching the ground.

We climb, creeping upward. Grand views of snowcapped peaks emerge to the northeast. The occasional birdsong buoys my spirits, and with a full moon rising, we arrive in Guffey just before the restaurant closes at eight o'clock, just in time for a hearty round of buffalo burgers and fries before calling it a day.

17.
RIDING THE ROCKIES

DAY 40, JUNE 12: GUFFEY TO ALMA, COLORADO

Western wood-pewee

Western wood-pewee. With speed, alternating *bzeeyeer* and *tswee-tee-teet.* (201, 6:12)

An hour before sunrise, sometime around 4:30 a.m., I first hear him just down the road from the hostel in Guffey where we've slept the night away. It's just one down-slurred *bzeeyeer,* then another ten seconds later, the same song I've been hearing throughout the day from these western wood-pewees since Pueblo. I walk in his direction, but by the time I stand beneath him, he is showering me from high in the pine with a blistering pace of his dawn songs: *bzeeyeer, tswee-tee-teet, bzeeyeer, tswee-tee-teet,* on and on he sings. Each song is a mere half second, with a little more than a half-second pause before the next, two songs every three seconds, at a clip of about 40 per minute. Back and forth, he asks *tswee-tee-teet,* rising on the question, and quickly answers with his down-slurred *bzeeyeer*. I close my eyes, listening, taking it all in, now hearing other wood-pewees in distant trees, all on a tear.

Yes, this *is* the West. We last heard the eastern wood-pewee in Kansas, and how fascinating to hear the two of them now singing side by side in my mind. At a leisurely pace during the day, the eastern pewee uses two songs (*pee-a-wee, wee-ooo*), the western only one (*bzeeyeer*), and both birds use this daytime singing behavior briefly just after leaving the roost in the morning. Then, within minutes, the males energize and introduce one extra song, a rising, stuttered affair (*ah-di-dee* for the eastern, *tswee-tee-teet* for the western), and singing rates skyrocket. Their songs and how they use them reveal the song genes within, the similarities offering a glimpse of how their ancestral pewee sang.

A Mosquito Prince

Creeping into consciousness now is another flycatcher who also sings from the top of a pine nearby. It's another Mosquito Prince, another *Empidonax,* but who is he? The songs of other flycatchers in this group since the Atlantic have been easy, the Acadian's *PEET-sah,* willow's *FITZ-bew,* the least's *cheBEK,* and the alder's *fee-BEE-o*. It's not the gray flycatcher, because at dawn he sings low in the junipers, and just two songs, a somewhat harsh *chu-WIP* and a fainter *teeah*.

Here is one of those western species with three different songs. I have saved my cheat sheet for just this occasion, with information gleaned from various sources and from my own listening. Fortunately, although songs of the cordilleran and Pacific-coast flycatchers are essentially indistinguishable, their ranges don't overlap, so I won't worry about the Pacific-slope flycatcher until we cross the Cascade Mountains in Oregon.

Empidonax Cheat Sheet

Species	Song 1	Song 2	Song 3
Dusky	vaguely two-syllabled *pa-EET* rising in frequency	rough, low-pitched *prrdurnt*	clear, high-pitched *PIT-it*
Hammond's	dry, sharp, two-syllabled *SE-pit* delivered briskly*	low-pitched, burry *tsurrt*	rough, drawn out, two-syllabled *SE-luurt*
Cordilleran (like Pacific-slope)	high, sharp *seeet*	thin, high, slurred *ka-SLWEEP*	explosive *pa-TIK,* with first syllable higher than second (reverse of Pacific-slope)

*suggests the *cheBEK* song of least flycatcher

Cordilleran flycatcher. Dawn song: *seeet, ka-SLWEEP, pa-TIK,* unmistakable once one is tuned in to *Empidonax* nuances. (202, 6:37)

That leaves three possibilities. Both the dusky and Hammond's have a low-pitched rough or burry song, but there's nothing like that here. As he sings I stare at the mnemonics for the cordilleran, hearing the sequence of *seeet, ka-SLWEEP, pa-TIK* over and over, the tempo in his songs suggesting that here is his preferred package. Over the next minute I count 30 of the *ka-SLWEEP* songs, which means he's singing about 90 songs per minute. I like him.

As sunrise approaches, first the pewee and then the cordilleran signal a change in mood. Half an hour to sunup, the pewee sings only his "daytime" *bzeeyeer* song, and at half the pace of his dawn singing. Twenty minutes to sunup, the cordilleran drops the *pa-TIK* song and slows to one-third the pace; then he phases out all three dawn songs entirely, now giving a simple, rising *wee-eeet!* "Time to move on with the day," each seems to say.

Meandering through Guffey, I pause to acknowledge others who still sing: robins and chipping sparrows nearby, western tanagers and booming nighthawks in the distance, the rhythmic wing beats of a broad-tailed hummingbird, the calls of red crossbills flying over. In the daylight I now see the sign I've walked past in the dark: "ABSOLUTELY NO TRESPASSERS. VIOLATORS MAY BE SHOT ON SIGHT. SURVIVORS WILL BE PROSECUTED." No, Dorothy, we're not in Kansas anymore.

As the sun creeps over the horizon, I head back to the hostel. David is still curled up in his sleeping bag out in the yard, and just above him in the top of a dead tree is a smallish bird singing a disjointed, rambling series of high-pitched notes, maybe four different notes per second, some of them sounding very familiar. After ten seconds he pauses, then quickly resumes; almost certainly that was a cliff swallow note, a house sparrow there, definitely a robin's call, an unmistakable flicker's *klee-yer!* call. Nothing would do that but the mimicking lesser goldfinch I've read about, and off he bounds in undulating flight, with high, wiry goldfinch-like flight calls.

Broad-tailed hummingbird. Known by the whir of their wings, beating 40–50 times per second. (203, 0:07; 204, 0:12)

By eight o'clock we depart, dropping down from Guffey to route 9, and there, as if to greet us at the beginning of our climb, are four Clark's nutcrackers calling from atop a dead tree. *Shraaaaaaaaaaa,* over and over, a grating, dissonant sound, a bit hollow and buzzy if one chooses to hear them that way, but I hear the sound of a remarkable intelligence. "David, four really smart birds up there. They're nutcrackers, members of the jay and crow family. Each bird caches a mouthful of seeds in the ground in thousands of places each year and has an incredible memory for relocating them. And look up on the hillside there. See those clusters of pines? Planted by the nutcrackers, no doubt, as part of their grand reforesting and gardening program, which results from their not reclaiming every seed cache they make. The nutcrackers depend on the pines for food, the pines depend on the

Lesser goldfinch. Above slumbering David, so fine a song, with oh-so-subtle mimicry. (205, 0:58)

Clark's nutcrackers. Two nutcrackers, *Shraaaaaaaaaa,* the sounds of high intelligence. (206, 0:41)

nutcrackers to disperse their seeds, and everybody's happy." *Shraaaaaaaaaaa, shraaaaaaaaaaa* . . . and they're off, the black and white in wings and tail so showy, now a shorter *kraak*, or *kraaaa*, all the sounds of a superior intellect. Clark's nutcracker, named for William Clark of Lewis and Clark fame, as it was on their expedition in 1806 when these birds were first collected.

We begin climbing, and I listen, to the *bzeeyeer* of wood-pewees, now just the daytime song . . . to song sparrows . . . to the rhythmic drumming of a distant sapsucker . . . to the wailing of a mourning dove . . . to what sounds like a warbling vireo but not quite . . . to cowbirds among the cattle to the left . . . to rock wrens greeting us from every rock outcropping beside the road . . . to a house wren . . . to snorts of an antelope, two adults and three young below us to the left . . . to the distant *quick, THREE BEERS!* of an olive-sided flycatcher . . . to a vesper sparrow just beside us. I stop, focusing on the introductory whistles to his song: two low and then two high over about a second. He then accelerates into musical trills that gradually slow and drop in pitch, the entire song a little over two seconds. I know that each male has just one type of whistled introductory sequence, and I'm curious to learn if vesper sparrows along the way will have dialects in these whistled introductory notes, as I know they do in the Willamette Valley of distant Oregon.

Western wood-pewee. Leisurely daytime songs from on high, a simple *bzeeyeer*. (207, 3:02)

We glide upward for nine miles to Currant Pass, the climb effortless, as if we are being pulled uphill by the extraordinary landscape. Rising over the gentle crest at the Pass, we pull to the side of the road, breathless at what lies before us: The road drops away into a valley, with a panoramic view of snowcapped peaks sweeping the horizon: ". . . unbelievable . . . gorgeous . . . most impressive day yet . . . we flew up here . . . do we have to go? . . ."

David starts downhill first, lifting both hands from the handlebars, his arms stretched high to the heavens. I hear Aaron Copland's stirring *Fanfare for Common Man,* deeply breathe in the brisk air, take one last look at clumps of pines probably planted by nutcrackers and aspen groves and the whole magnificent scene, mark the moment on my voice recorder with "Currant Pass, 10.8 miles on the morning, 9:41 a.m. on June 12, a long way from the Atlantic, closing in on the Pacific," and follow David.

We cruise, slowly, relaxed and in no hurry, alive and happy and enjoying the day, the clouds, the valley we ride through, the forested ridges above, when up ahead we hear a great commotion, with loud, raucous calls by dozens of birds. We're soon beside them, but the flock streams up ahead, the birds flying low through the open pine woodland. Walking the bikes along the road, we keep pace with these jays of some kind. *Krawk krawk*—ah, I know that sound, the deep, hollow call of a pinyon jay. And there, from a bird perched in the top of a tree, a nasal, laughing *yah-ah-ah,* initially rising and then falling away—all classic pinyon jay. How many birds? 100? 200? It wouldn't surprise me if there were more, as the dispersed flock swarms about us on both sides of the road.

While I watch through the binoculars, David runs the recorder, annotating in a genuinely semi-excited voice: "They're going crazy . . . nondescript, bluish gray . . . first time we're heard this . . . very nice sounds from these birds . . . this has been a continuous recording of pinyon jays making a great racket in the trees . . . I'm going to check to see if these are pinyon pines . . ." and he pulls out the pocket guide to trees he bought some time back.

Pinyon jays. Hungry young birds and harried adults, the din of the flock. (208, 6:28)

But what's the racket? *Crrik crrik crrik,* about three notes a second, each one harsh and rising, dozens of birds all calling at once, a din unlike anything I've heard before. We lay the bikes beside the road and walk out into the pines, now among the birds, waiting for some to come near. Landing calmly next to us is a stunning adult pinyon jay, immaculately blue with whitish streaks below. Ah, there's a clue, a gulping sound among the *crrik* calls, and now more. Young birds! Countless numbers of hungry young birds, their harsh, strident, insistent, endless *crrik crrik crrik* their way of encouraging their parents to bring more food. As the flock streams by, we pick out the frantic youngsters, more grayish than blue. I also begin to hear some of their individual voices, the nuances that enable parents to find their own young in so large a flock.

Pinyon jay

"What a surprise this is! I hadn't expected them this far north in Colorado. They're fascinating birds. They live in large year-round flocks, all members of different family clans, and they all breed at the same time with nests distributed among nearby trees. What's so special is how these jays have evolved together with the pines they feed on, much like the nutcrackers earlier this morning. Pinyon pines, the tree after which the jays are named, have especially large, nutritious seeds, but the seeds are wingless, so they're dispersed not by the wind but by the birds. The tree needs the birds to distribute its seeds, and the bird needs the tree for food—a mutualism, as it's called. And like the nutcrackers, these birds have a remarkable memory for caching these seeds in the ground and recovering them later during seasons when the trees don't bear seeds. If seeds are abundant in the fall, for example, the birds can store enough food to begin breeding in the middle of winter, well before other birds who must wait until spring because they rely on finding food as they go." With the open tree guide in his hand, David nods, taking it all in, and fairly confident he's identified these trees as pinyon pines.

Reluctantly, after lingering here the better part of an hour, we move on to the town of Hartsel, where we stop for lunch and to read the Sunday paper. Or pretend to read, as it's hard not to eavesdrop on the couple who sit nearby and comment on passersby: ". . . no doubt who's the father of that one . . . four kids in three years, just poppin' 'em out . . . fast as her mother when I knew her years ago . . . still lightning quick . . . most likely burned the barn down for the insurance. . . ."

Then we're on to Fairplay, marveling at Sheep Mountain and the ridge 3000 feet above to our left. Far less scenic are the mounds of rubble that have been dredged up beside the Middle Fork of the South Platte River, tailings looking as if they've been excavated by some giant mole methodically plowing through the valley in search of gold.

We camp just outside of Alma, leaving for tomorrow morning the six-mile climb to Hoosier Pass. "Never camp on top of a pass," advises David. "Biking down in the cold morning air is not so much fun." We set up camp, eat a leisurely dinner, hang the food well away from the tent and high out of bears' reach, and crawl into the tent as the rains begin. As I recall the dawn singers at Guffey and all the birds heard along the way today, reliving especially the pinyon jays, cresting again at Currant Pass and taking in the spectacular landscape, I understand all the better the rich emotions behind John Denver's "Rocky Mountain High."

DAY 41, JUNE 13: ALMA TO BLUE RIDGE CAMPGROUND NEAR SILVERTHORNE, COLORADO

Come morning, I see from our campsite along Buckskin Creek that a fresh coating of snow has been added atop the mountains to the west. It's cold!

Tsick-a-dzee-dzee. A mountain chickadee offers his husky version of the black-cap's *chick-a-dee-dee,* and how striking his whistled songs: three, sometimes four whistles all on the same high pitch, usually with a couple of tiny whistles at the beginning, *te-te teee teee teee.*

And a ruby-crowned kinglet, his song a masterpiece! He begins in his high, thin voice, squeezing out an accelerating series of *t t tse tse tsee tsee tsee* notes, then catches the unsuspecting listener off guard as he plummets an octave, now accelerating once again and dropping slightly in pitch through a series of low *tew tew tew tw tw tw tw* notes, then galloping into a two-syllable *tee-LETT tee-LETT tee-LETT tee-LETT* chant for the finale, squeaking out a single high *seet* note on the end. I listen and smile, such a loud song for such a mite of a bird, *t t tse tse tsee tsee tsee tew tew tew tw tw tw tw tee-LETT tee-LETT tee-LETT tee-LETT seet.* I puzzle at those *tee-LETT* phrases, as they're a shortened version of the three-syllable *tee-da-LETT* chant I know from the East.

Mountain chickadee. Another whistling chickadee, cousin of black-capped and Carolina. (209, 1:49)

Ruby-crowned kinglet. Exhilarating, from the high thin notes to the galloping conclusion. (210, 4:26)

But other than the chickadee and the kinglet, it is eerily quiet here. One other chickadee whistles in the distance. I walk up the path through the woods, encountering a sign: "Town of Alma Cemetery, this Cemetery was established for the use of Alma residents by President T. Roosevelt on March 21, 1902 . . . Unauthorized burials are subject to exhumation. . . ." It is by far the quietest cemetery I've ever heard.

Wilson's warbler. Simple chatter, crescendo-style, the last note typically faltering. (211, 1:06)

White-crowned sparrow. High in the Rockies, a mountain dialect. (212, 1:32)

I return to camp for breakfast with David, and we're soon ready to depart. I note yesterday's leg of our journey: "53.79 miles for our 40th day on the road, 33.1 mph maximum, 9.8 mph average, time on the bike an easy five hours and 29 minutes; total mileage for the trip, 2460." To capture the details of today's ride, I reset the computer.

We first bike the mile back down the gravel road to Alma, then climb to the north on route 9. Birds are now everywhere, and I acknowledge each in turn: robin . . . flicker . . . cowbird flight whistles . . . Wilson's warbler, first for the trip, the loud, rapid chatter of a dozen or so noisy notes unmistakable, each male with only one song . . . the *whirr* of the broad-tailed hummingbird's wings . . . a junco . . . the local dialect of the mountain race of the white-crowned sparrow. . . .

A violet-green swallow alights on the fence beside us, a stunning combination of immaculate snow white, emerald green, and purplish-violet iridescence, the finery one would expect only on a hummingbird—they're "children of heaven" wrote one early naturalist who was sorry he'd used up all of his superlatives for the lesser tree swallow.

We quickly rise above Alma, the river valley falling below us to the left; beyond the valley, aspen groves abruptly give way to spruce at higher elevations, the barren snowcapped peak of Mount Lincoln across the valley so impressive. With little traffic, we ride side by side, smiling, pointing, laughing, then I fall behind, shifting to my lowest gear as the grade steepens. Twenty minutes later, an hour and a half since leaving Alma, I crest, finding David sitting on the stone foundation for the sign that marks the moment: "Hoosier Pass. Elevation 11,542 feet. Continental Divide." Behind us, water flows to the Atlantic Ocean, ahead, to the Pacific. As if to congratulate us, two Canada jays appear and swoop silently about us, though I know they're merely searching for handouts.

It is a place to linger, to relax and celebrate, to eat an early lunch, to do whatever it takes to delay departing, but heavy clouds materialize from nowhere and the wind picks up. In understated tones, David offers, "Best we be moving along, I think." We quickly put on our rain gear and cover the panniers, and are soon chased off Hoosier Pass, hastily departing the high mountain scenery as we descend through switchbacks in a cold rain, within minutes finding ourselves in Breckenridge, one of the ski capitals of Colorado. Uninterested in the tourist town, we take the fifteen miles on the bike path around Frisco and on to Silverthorne.

There, we take care of business quickly. First, the post office, where we pick up our mail. I'm excited to see my contract from the publisher, Houghton Mifflin, for my upcoming book *The Singing Life of Birds*; in some vague, unspoken way, I see that book as a possible entry into a whole new life beyond the university. Then we head to Wendy's for lunch, where we once again realize that we need to eat and drink earlier and far more often. I leave for the 7–11 to buy dinner, while David takes to the library for the internet. Meeting later at the library, we're ready to depart, but wait for a rain shower to pass.

As the rain lets up, we make a run for it, hoping to cover the eight miles to the Blue River campground before the next storm hits. A mile down the road, the sky darkens and sparks fly from utility wires overhead. Faster, faster . . . and the rain hits just as we pull into camp half an hour later. Taking the only empty site, we quickly set the tent up and crawl inside, counting the seconds between the sharp flashes of lightning and claps of thunder. Five seconds, a mile . . . three seconds . . . two . . . until lightning and thunder occur simultaneously, far too close for comfort as we sit on the inflated pads in the tent and hope for a little insulation between us and the ground.

The Chief Cook is not impressed with the beef stew, spam, cheese, and bread I bought for dinner. After the storm passes and while he reads in the tent, I warm the food, which soon tastes far better than it had looked.

What a spectacular day, we agree, yet there's tension between us. He's tolerated a lot from me in the early morning rides and in the steady pace needed to reach the Pacific before the birds have stopped singing for the year. He's obliged, but the mornings have been a real challenge for his natural cycle, and he'd love to take some days off. But because we arrived in camp late the last two days, I have struggled, because arriving late means it's then difficult to get everything done, get enough sleep, *and* be up early with the birds the next morning. Both our bodies and minds are strained, as we've been on the road a long time since Virginia, and pushing fairly hard. And, I'm his father, not a lighthearted college friend. Life on the road would be richer if another college friend joined us for a bit, especially a girlfriend.

It all boils over with the shoe episode after dinner. Lest the occasional bear dine on our supplies, we need to hang our food from a tree. David wants to tie our rope to his biking shoe, throw the shoe up and over a tree limb, and then pull the food up on the rope. I look up at the high limb and give a firm, fatherly "*No*" to his plan, trying to articulate that the shoe isn't heavy enough for gravity to overcome both the weight of the trailing rope and the friction of the rope on the branch; and when gravity fails him and he tries to pull the shoe back, a crucial piece of his biking gear could be lodged among the branches high up in the tree. I see nothing good in this scenario. The food issue is resolved when the neighbor agrees to store our food panniers in his truck overnight, but the tension remains.

18.
SAGE AND SONG

DAY 42, JUNE 14: BLUE RIVER CAMPGROUND TO KREMMLING, COLORADO

By 4:30 a.m. the robins are singing, and by 5:00 I'm following a nameless singer around the campground. Who is he? I hear about two seconds of rich bubbling phrases delivered with the enthusiasm of a house wren, and the pitch leaps from low to high and back several times within the song. Just when I have his song pattern memorized, he changes his tune, now singing three bubbling phrases in a high-low-high sequence and tacking a single low note on the end. He answers another of his kind just down below along the river, but there's nothing polite about the exchange, as he sings before his neighbor has finished. His haste may sound like an innocent eagerness, but this kind of overlapping is more likely a strong signal, aggressive and dominant. The stakes in this conversation are high, because females are no doubt listening to hear who is top dog in the neighborhood.

Lincoln's sparrow. In Montana I will learn the source of these songs. (213, 2:10)

For half an hour I walk the roads of the campground as I follow these two mystery birds. I listen to when and what and where they sing, mapping their territories and learning to recognize each by his song. I could retrieve my binoculars and look at them to give them a name, but there's a certain satisfaction in not knowing and just listening. Can't be wrens, as I know the songs of wrens in North America. Maybe it's a western warbler that I don't know. A sparrow?

A mountain chickadee calls, a raspy *tsick-a-dee-dee-dee,* then sings like the birds in Alma yesterday morning, three or four whistles on a fairly high frequency, but then he abruptly switches to a perfect *hey-sweetie* of a black-capped chickadee. Remarkable. I knew mountain chickadees had a variety of songs from place to place, but I didn't know they could sound so much like black-caps.

Mountain chickadee. In mountain winds, calling ever so hoarsely, *tsick-a-dee-dee-dee*. (214, 1:14)

I listen on . . . to the songs of a ruby-crowned kinglet . . . to the silence of a junco hopping on the ground. . . . Pretty quiet. Returning after six o'clock to the campsite, I find black-billed magpies gleaning scraps from the picnic table. With the ice I now see on the table and elsewhere, I realize why I am so cold.

David is up and about, but visibly unhappy. "Cold and not enough sleep," he explains. "Are you angry at me?" I ask. "I don't have any good reason to be," he answers, but I feel the hostility. I heat the oatmeal, adding dried milk, raisins, brown sugar, and banana; we eat in silence. I'm eager to meet our agreed-upon goal of 50 miles by noon, but David is struggling to move. I wait, anxiously, but within I'm also smiling, as I know he won't be down for long. He's just not naturally a morning person, I remind myself, though maybe there's more here that needs to be addressed—like, why am I so hell-bent on biking in the cold at this ridiculous time of the morning?

By 7:30 we're following the stunningly beautiful Blue River downhill toward Kremmling, the sunken valley here having been created as part of the "Rio Grande rift" when the North American continent was stretched to the west. I stop on a bridge over the river and search for dippers—none to be seen. And I listen, to green-tailed towhees, Swainson's thrushes, house wrens. David has been keeping his distance well behind me, but now pulls up alongside, somewhat happier with life.

The river suddenly becomes a lake, the Green Mountain Reservoir, and the valley widens. Sagebrush now spans a good half mile to a mile on both sides of the road, with forested ridges beyond. A daytime chorus of Brewer's sparrows and sage thrashers welcomes us, and I promise myself to find a way to be out among them at dawn.

Approaching Kremmling, it sounds as if the world is coming to an end. Sirens of all kinds wail in front of us and far to the left and right, and there's a roadblock up ahead. But pulling up, we see happy faces, and learn that we, just by our good fortune, have arrived during "Kremmling Days"! We're sent on through the roadblock to Main Street, and dead ahead is Engine 409 of the Kremmling Fire Department, an American flag flying high, the driver waving heartily to us. A large float is next, with eight women seated around two tables, a large sign in blue and red letters proclaiming "IF IT'S TUESDAY, it Must BE BINGO." Parading down the street are police cars and ambulances and more fire engines than I've ever seen in one place. Booths of all kinds hawk food and crafts; we're invited to the beer garden, with its live music, and encouraged to participate in the Third Annual Cliff Golf Tournament.

David strikes up a conversation with two young boys, Michael and Ramon. "This is the day where we get an excuse to have a parade. Half of the town is in the parade, the other half watches . . . the parade is only for locals and their families, not for tourists. . . . You see those old ladies on the float? They all grew up here. . . . The owners of the stores work in them—we're a small town. . . ." Michael and Ramon badmouth the town, but in a prideful sort of way.

David is in his element, while I soon despair, as I sense our goal of 50 miles by noon is slipping away. But we come up with a plan: He'll do the grocery shopping and meet me at the laundromat. Far later than I had expected, he arrives with our lunch. "Got to talking with some more middle school kids," he confesses.

As we eat in the laundry, a beautiful young blue-eyed, blonde woman arrives, but she turns down David's offer to share some food. "I'm a stripper," she explains matter-of-factly. "I need a flat stomach tonight. I got two kids but my ex takes them on the weekends when I work. . . . I figured I've never been able to get respect from men in my life, so fuck it, I said. I'm cashing in. . . . It's really easy—you do five minutes of dancing, and then 25 minutes or so of walking around and talking to men. Most of them are kind of sad and down on their luck—that's why they're there at the bar. So you just sit and listen to them talk about their problems, and then they give a big tip. I make 500 to 1000 dollars a night, more money than I ever dreamed of." She's pleasant company but is soon fuming over the "fat woman's panties" in the washing she's doing for her boyfriend. And is that my son trying to console her?

The lunch and laundry done, David and I part, agreeing to meet in five minutes at the sporting goods store. Twenty minutes later he shows up with Keith, the British cyclist we met a week ago in eastern Colorado, and with the Super Soaker 4000 Max-D water cannon he bought, David hoses me down. "It's your Father's Day present," he laughs, "but I'll carry it, as you're not mature enough." How can I not laugh with him?

Finally, three and a half hours after arriving in Kremmling, we depart, but just outside of town David pulls over, remarking that his rear wheel feels out of round. Sure enough, close inspection reveals that the sidewall is damaged, so we head back to town, finding a bike store with one tire in stock. It's simply meant to be, I concede, that we stay here for the night. Resigned to making only 30 miles' progress for the day, I find a motel room in town while David replaces his tire.

I rest, organize my thoughts and gear, and hatch a plan to be out on the road well before sunrise tomorrow. David returns from the bike shop but quickly heads out again, excited about the golf tournament. He's biking to the top of the cliff that rises several hundred feet just to the north of town. The person who hits a golf ball closest to a marker far below wins a prize. Fifteen minutes later, he's back, breathless. "Forgot my Frisbee," he explains. "I want to see how far I can throw it." And he's off again. I smile: David, captain of the national championship Ultimate Frisbee team at Stanford, aims for a personal best by throwing the Frisbee farther than ever. What opportunities, all here at Kremmling Days! I'm reminded again how my stick-in-the-mud, curmudgeon style can smother his, and what a delight this kid is. Kremmling Days is just what we needed.

Fifteen minutes later, I start watching the forms at the top of the cliff through my binoculars. I say "the cliff" and take it for granted, but when I say it again consciously I see a cliff face that consists of the same shale that was deposited in the interior seaway that once covered Kansas and much of the mid-continent. The shale was thrown up here with the rest of the Rockies, but the same shale still lies buried in neat layers thousands of feet deep in eastern Colorado and Kansas.

Finally, yes, there's David—I can tell by his mannerisms a good half mile away. He bends over to tee up a golf ball, then gives a mighty swing. Next, it's the Frisbee. He winds up and lets 'er fly as he did when "pulling" for the ultimate team, and I watch it glide forever, aided by a tailwind, floating down 500 feet, sailing at least a quarter mile, eventually disappearing behind the rooftops in town. Take note, Guinness.

DAY 43, JUNE 15: KREMMLING TO WALDEN, COLORADO

Father's Day! Sunday, 4 a.m., and what a joy on a full night's rest to hear the alarm, to bound out of bed and know that I'll bike through sagebrush heaven this morning. It's 27 miles to Muddy Pass, and I'll take my sweet time, listening to birds all the way. David, with a full day and evening of festivities yesterday, will depart the motel some hours later. We'll meet at the top of the pass, if not before.

Forty-five minutes later I'm outside of town, pausing beside the road. The orangish glow on the eastern horizon hints of the day to come, and the almost-full moon to the southwest is poised above the 2000-foot canyon carved there by the Colorado River. The intense aroma of the sage is intoxicating, as are the sounds of all that I hear. I check the singers off in my mind, robins near the streetlights, swallows singing on the wing, a mountain bluebird; in what must be a wetland just down the road are yellow warblers, song sparrows, red-winged and yellow-headed blackbirds. Snipe winnow overhead.

I set out, slowly, savoring the now. The sage glows in the moonlight, and rising from the glow is a symphony of Brewer's sparrows and sage thrashers. The sparrow territories are small, with four or five birds always within earshot, and they're all in wild dawn song, each male singing continuously, alternating a series of high notes with a long, canary-like ramble at a much lower frequency. Their voices mix and mingle, and at any given instant some are near and loud, some far and soft, some high and some low, some clear and musical, some raspy and buzzy, their collective voices mesmerizing. I pick out an individual beside the road as I approach, hearing him work through high-pitched buzzes and trills and insect-like *zeets*, then cascade down to a long series of varied trills, then rise again for another cycle; his voice grows as I approach, then fades as I pass.

Brewer's sparrow

Brewer's sparrow. A dawn extravaganza
Bird 1 (215, 1:21)
Bird 2 (216, 6:31)

And the thrashers! What remarkable singers, but their territories are far larger, and I pass only one for every ten or so sparrows. I have studied their songs and know how to listen, and I now hear how the next one speeds through his repertoire of hundreds of different sounds (about 700 in one bird I studied carefully), every half second (0.57 seconds to be exact) hustling on to another sound at breakneck speed. And I hear his mimicry, too—how he loves meadowlark and Brewer's sparrow songs, mimicking such a variety of them. Like

Sage thrasher

Sage thrasher. In continuous song, with abundant mimicry, in bounding flight, together with Brewer's sparrows, *wow!* (217, 7:21)

Wilson's snipe. Mechanical songs, the tail singing as directed by the wings. (218, 1:43)

Brewer's sparrow. Remarkable transformation from dawn to simple day songs. Bird 1 (219, 1:09) Bird 2 (220, 1:03)

a robin or rock wren or yellow-breasted chat or plumbeous vireo, he favors one set of songs for a time before moving on to others, though it takes hours (at least eight) rather than minutes for him to present most of what he knows.

Unseen sparrows and thrashers whisk me along, and with no traffic this early Sunday morning, the world is ours. There's a vesper sparrow, *tew tew tee*, beginning with two low whistles and then a high one, his neighbor singing the same. . . a rock wren on the ridge to the right, bouncing among a variety of songs, as he should in the frenzy that is dawn . . . a snipe winnowing high overhead in the distance to the right, *wuwuwuwuwuwuwuwu*, each "*wu*" the sound of one wing beat, the sound of wind directed from the wings through the special outer tail feathers that hum as he dives earthward . . . a western meadowlark bugling in the distance . . . a mourning dove from the utility wires, *cooowaah, cooo, coo, coo* . . . magpies calling in the gulch to the right. Nighthawks call overhead, some near, some far. I stop beside the road again, counting 21 *peeent*s from one who circles above before I hear him dive, and in my mind I can see his wings stretched forward at the moment of truth and meeting the onrushing wind, the feathers vibrating loudly, *VROOM*, sounding, yes, a bit like a bellowing bull or a very short freight train.

It's now 5:10 a.m., still 25 minutes before sunrise, and already the Brewer's sparrows have abandoned their intense dawn singing. Who would know it's the same bird now, their intermittent, simple, two-parted daytime songs such a contrast to their passion at dawn. Just across the fence to my left, I hear *zree zree zrrr-zrrr-zrrr*, to the right *tir-tir-deee-deee-deee*, each male repeating his one unique daytime song as if he knew nothing else.

I bike on, but soon stop beside a sage thrasher who sings on a fence post to the left. He salutes the dawn, facing the orangish-yellow glow on the eastern horizon, and then he's off, singing in flight, dipping into the sage and bounding skyward off some springboard hidden there, circling about his territory in a showy, exaggerated flight as if stitching sage to sky, finally landing on a sage bush in the distance.

I take the advice from the roadside savannah sparrow, *take take take take-it eeeeeeeaaasssssssyyyyyyy*. Ten seconds later, now

behind me, he repeats his wisdom with his lazy, high-pitched, almost insect-like song, with staccato notes at the beginning that briefly accelerate before ending in two drawn-out buzzes, the second lower than the first. As the sun peeks over the eastern hills, an absurdly disproportionate riding companion joins me, my shadow stretching far to the west across the orangish-green glow of the sage. I check my odometer, not surprised that I've traveled only seven miles in nearly an hour and a half since leaving the motel.

A line of pelicans float magically on extended wings high above the lake on the right, and on the water below floats a western grebe—or is it a Clark's grebe? Even at this distance I see grace and elegance, a "swan-grebe" with slender body and long neck, black above and white below, but I cannot see the subtle pattern on the head that distinguishes the two species. I pause and listen for the double *cree-creet* of the western or the single *creeet* of the Clark's, as it is their calls that best give them away. He dives, under water for 41 seconds, pops up for 16 silent seconds, and is then down again. He is who he is, I decide, and doesn't need a name.

Savannah sparrow. Easy and wheezy, so unhurried, *take-it-eeeaaassssyyyy.* (221, 1:38)

My gaze extends beyond the reservoir to Wolford Mountain, where I'm reminded to look for rocks on the move. Yes, I see what I've read about. Trees grow in abundance atop the mountain on ancient granites that have been shoved over far younger, treeless shales below. It's a haphazard, jumbled landscape, and a few miles down the road on Whiteley Peak I will watch for hexagonal columns of basalt from past volcanism.

Another half mile down the road I slow, then stop near a horned lark who still sings with some of the inspiration that would have fired him well before sunrise. He spits out four to five tiny notes each second, every ten to 15 seconds or so culminating in a flourish of a few stuttered, rising notes

Horned lark. Sputtering lower notes followed by high flourish, repeated. (222, 0:34)

that rush to a high, slurred chitter; he then pauses briefly before doing it all over again. He sings erratically, impulsively it seems, his mood swinging between the frenzy of predawn singing and the lazy, occasional flourish of later in the day.

After picking up a lifeless horned lark on the road just yesterday, I'll never look at one the same way again. Head on and just a few inches away, the facial pattern was stunning. I admired the grayish brown crown yielding to a black forehead visor, from which two black "horns" protruded up to the left and right. Just below the black visor and above the eyes, a whitish band circled from the bill around to the sides of the head; below that was a black moustache, extending from the base of the bill and back beneath the eyes, then arcing down to the left and right, the black contrasting sharply with the nearly encircled pale yellow throat and with the outer whitish cheek feathers. And below the yellow-tinged throat was the broad black necklace. All together, the striking pattern was the kind of intimate, bill-to-bill view that larks regularly dazzle each other with.

In the warming rays of the rising sun, I shed layers of clothing. First to go are the rain pants and raincoat, together with the rain booties, all of which were the first line of defense against the cold. Other layers follow. While repacking the panniers, I survey the endless sage brush, marveling at the shades of green that glow in the dawn's changing light. I breathe deeply, absorbing the aroma of the sage. I listen to all the familiar voices, a veritable symphony in the calm of this best of all possible early Sunday mornings.

Horned lark, head on

How out of place—from the *utility wires* up ahead sings a rock wren. Though it's well after sunrise, he still leaps enthusiastically among different songs; I coast to a stop and try to follow along. My task is made easy, because he alternates two different songs, A and B, and then switches methodically to other pairs, C and D, then E and F, never mixing songs of different pairs. I try some mnemonics: *weedle-weedle-weedle-weedle-weedle-weedle*, *jeeer-jeeer-jeeer-jeeer*, *tewtewtewtewtewtewtewtewtewtew*, *techurr-techurr-techurr*, *chuwy-chuwy-chuwy-chuwy-chuwy-chuwy-chuwy*, *jjjrrrrrrrrrr-rrrrrrrrrrr*. What a fine clinic he offers on all he can sing, beautiful examples of the hundred or more songs he knows.

Rock wren. An old friend, alternating songs as he should, A B A B A B. . . . (223, 2:32)

The road beckons and I move on, but now a green-tailed towhee sings from a fence post on the right; I stop and pull the binoculars from my handlebar bag. The sun behind him casts a reddish halo about his head, a greenish glow over his back. I see the stickpin in his clear breast. And his songs—they begin with a few introductory notes, then alternate trills and buzzes and slurred notes that energetically bounce from one pitch and speed to the next. I try to track his program, hearing that he alternates two different songs, A B A B A, then moves on to two others, C D C D C, and then offers still another song before I lose track. A stunning looker, an exhilarating singer. "Thank you, Mr. Towhee," I find myself whispering to him, and after ten minutes we part.

Green-tailed towhee. A bike-stopper, plumage and song haloed by the sun rising behind him. (224, 1:31)

FITZ-bew! The western "species" of the willow flycatcher. He finds a good living in the bushes and small trees that line Muddy Creek, the small river that drains the valley here.

Willow flycatcher. Of the western variety, still with the expected three songs. (225, 7:55)

7:31 a.m., 15 miles on the day's odometer, three hours on the road—a record slow pace. Yes, Mr. Savannah Sparrow, *take take take take-it eeeeeeeaaasssssssyyyyyyy*. Appropriately, another green-tailed towhee and his neighbor capture me for another ten minutes.

Two crows, and what raucous cawing! Just 20 yards off the road to the left, they swoop into the sage and out again, dive bombing something . . . a red fox! When I spot the fox, he sees me, too, and off he runs through the sage. I monitor his progress by watching the crows and two magpies who have joined the chase. Somehow an antelope gets mixed up in this scene, too; he snorts at me from just up ahead before bounding across the road to the right.

MOBBING! American crows, black-billed magpies. Raucous, hair-raising crowd cawing. (226, 3:10) **Black-billed magpies**. An equally fervent clamor. (227, 4:43)

It's a gentle climb to the top of Muddy Pass, and by 9:30 a.m. I pull over beside a grove of aspen trees on the left. Their leaves quake in the wind, the rustling pleasant evidence of the slight tailwind I've had this morning. I admire the aspens; the hundreds, perhaps thousands of trees I see are probably only "one individual," all of the separate trunks rising from the lateral roots of one founding seedling. And this "being" could be tens of thousands of years old. I settle in next to my large, old friend, sitting beneath the Muddy Pass sign that confirms I'm still sky high, at 8772 feet, once again crossing a continental divide.

Half an hour later David and our friend Keith arrive. They're happy for me that I've had such a fine morning, even happier that they slept in and took to the road at their own time and pace.

We're soon gliding off the pass toward Walden, dropping into North Park, still sagebrush country, still cattle country, reminding me of the roadside sign of a few days ago: "Preserve your western legacy. Support ranching in Park County.—Central Colorado Cattlemen's Association." A literal translation seems to be "We got cattle in here once. Just try to get us out." David's sarcastic translation is "What great mountains! Let's fill them with cows!" With barbed wire fences lining the roads and cattle owning the land, David vows, "I'll never eat beef again," but we also remind ourselves not to think those thoughts aloud in this country.

On the downhill, David spots three antelope beside the road and hurries to catch them. They sprint ahead, but their escape is blocked by the fence 15 feet from the road; nearly alongside of them, David pulls out his Super Soaker water cannon and unleashes a spray that, well, falls far short of its mark, but the hunt was good while it lasted.

We'll follow the valley as it opens toward Wyoming to the north, with snowcapped mountains nearly encircling us, the Park Range to the left, the Medicine Bow Mountains to the right, the Never Summer Mountains and the Rabbit Ears Range behind us. And rivers run through it, the Michigan, Illinois, and Canadian Rivers, the wildlife abundant. As we drop into the wetland complex beside Arapaho National Wildlife Refuge, the roadkill carnage is unlike anything we've seen before: a red-tailed hawk, cinnamon teal, mallard, barn owl, prairie dogs, and songbirds galore . . . and a bloated cow rotting beside the road.

Just ahead is Walden, "The Moose Viewing Capital of Colorado" (so declared the Colorado legislature). Though we could camp free at Walden's Hanson Park, I want a sage-country encore tomorrow morning, so I suggest that we once again take a motel. That way I'll have another quick and simple predawn departure without having to break camp.

19.
HELLO, WYOMING

DAY 44, JUNE 16: WALDEN, COLORADO, TO SARATOGA, WYOMING

A little after 4 a.m., I devour the cold oatmeal prepared last night, hoping that the gusting winds and thunderstorms will soon dissipate. By 4:15 I stand beside the road with my loaded bicycle, watching lightning on all horizons, and soon laugh at myself, as I realize that my early morning ride through the sage is not to be. I wheel the bike back to the motel room and, though bed looks inviting, choose to walk around outside to take in whatever happens.

Across the road and beyond the cemetery, a snipe winnows. And there's a bluebird, almost certainly a mountain bluebird, though he's too distant to hear well. "Come closer," I whisper.

Much nearer, a swallow sings nonstop, perhaps from the utility wire along the road, *chi-lip chi-lip chi-lip*, over and over, 10 . . . 20 . . . 30 . . . 40 . . . 50 . . . 55 seemingly identical, two-syllable calls I count in just 30 seconds, at a clip of well over 100 per minute. They're not the harsh calls of bank or rough-winged and not the creaking and rattling of a cliff swallow (wrong habitat, too), not the rich song of a barn swallow, without the variety of a tree swallow . . . perhaps a violet-green swallow. I'm not sure. So little is known of them.

Mountain bluebird. From just above on the motel's roof. (228, 1:36)

The bluebird lands just above me on the roof of the motel. I start my recorder and aim the shotgun microphone at him, but in less than a minute he's flying west toward town, and then he's back across the road at the cemetery again, energetically flying about his territory and singing all the while.

I captured those few seconds, though, and now I replay this track several times, convincing myself that he *was* a mountain bluebird. And just what was he singing? Several series of ten to 20 brief warbles, with one to two seconds between each series, much as a robin would sing his series of caroled phrases. I listen again, picking out one unique phrase, and find it scattered about the performance nine times. Another phrase occurs six times. Based on how often he repeats these two unique phrases, I'd estimate that he has about ten different warble phrases overall, again not all that different from a robin, a distant relative in the thrush family.

Intriguing. I know so little about these mountain bluebirds, and from what I've read, no one knows them well. The most professional accounts, including the famed *Birds of North America* series, admits as much. It is just possible that I am already the world expert on the dawn song of the mountain bluebird.

By 6:15, after a second breakfast and after the storms have dissipated, David and I depart, but it's not what I had hoped for this morning. Though sagebrush still dominates and the cast of characters is much the same as yesterday, the dawn burst is well past, and I hear only snippets of song here and there. But David's happy.

"Is that a thunderhead over there to the right?" I ask. "Nooooo, doooom, dooooom!" he replies, laughing, as he knows that I worry too much. I laugh with him, but I *know* it's a thunderhead; how beautifully the sun peeks around it, and how nice to watch it dissipate as we ride. A savannah sparrow sings . . . Brewer's sparrows . . . a sage thrasher . . . meadowlark, western of course. Ravens perch on utility poles, croaking encouragement as we pass. Blue and yellow flowers, lupines and what looks to me like goldenrod, blanket the valley as we cross the North Platte River about 15 miles out of Walden. Then we climb, and hidden just beyond rock outcroppings on the right, coyotes yip and howl in a chorus, sending us on a mighty descent, hitting speeds over 40 miles per hour as we drop into Wyoming.

More sagebrush. Chattering barks of prairie dogs welcome us, many standing alert on their dirt mounds beside the road. A friendly tailwind whisks us along, but it's a difficult headwind for the first eastbounder we encounter. We pull to a stop on his side of the road, waiting for him. "Started on the Oregon coast three weeks ago," says the rider on his recumbent. "The mountain passes were pretty tough—cold and still lots of snow." Before continuing, David pumps him for information about the road ahead, and he obliges.

Western meadowlark. Ubiquitous; sharp, crystal-clear perched songs; agitated calls; an occasional brilliant flight song. (229, 1:33)

Another mile down the road we stop to check out a large bird on a utility pole. "We got ourselves a hawk. It is about yea big," David announces into my microphone as he looks through the binoculars. With his customary ornithological precision, he estimates the height of the bird by holding his hands about a foot and a half apart. "Oh, he is flying. He is very white underneath, with some dark edges on his wings. He is splotchy on the top . . ."

"Black tip to the wing, leading edge of the wing black," I add.

"Oh, now you're getting all technical . . . kind of lighter brownish tail . . ."

"A red-tail?" I query.

"Might be a red-tail," says David, ". . . Now he's in the field . . . yes, must be a red-tail." Case is settled. This bird is much smaller than the huge, all-dark ferruginous hawk we saw earlier today, about the size of the Swainson's hawks we saw playing in the winds of the shortgrass prairie in eastern Colorado.

Mid-afternoon we are blown into Saratoga and search out the home of my graduate school friend, Jim States. Though Jim is on a consulting job and away overnight, his wife Carol and her mother Gwen treat us like royalty, preparing a meal of barbecued steak (so much for David's "no beef" pledge), baked potatoes, grilled vegetables, and strawberry shortcake.

Talk turns to politics, and coal. David's conclusions: "Perverted politics give Wyoming ranchers all the power to destroy the land; and I hope those vast reserves of coal are never used."

We sleep in their guest house on the bank of the North Platte River.

Red-tailed hawk

DAY 45, JUNE 17: SARATOGA TO MUDDY GAP, WYOMING

After a leisurely start in the morning, and a little over a mile out of town, I come to a screeching halt. Marsh wrens. *Western marsh wrens!!* Dozens of them buzz and rattle and whistle and chatter in the reeds of small ponds beside the road on the right. I remind David that we never heard the eastern marsh wren back in Virginia, but had we been a little farther north of our TransAm route, in Nebraska, we would have crossed a clear boundary between the eastern and western wrens. "Just listen to the variety in these western songs—each male has 100 to 200 different songs, most of them learned from neighbors, and what's best is how they respond to each other with their songs, sometimes matching each other song for song up to ten times in a row. Eastern males simply don't duel like that, and each eastern male has only 50 or so songs." The moment seems to call for some celebration of all that I've learned from studying these wrens over the last 30 years, but the road beckons. We'll listen more later, I promise myself, sometime before we hit the Pacific.

Marsh wren

Marsh wren. The variety of songs endless, the qualities mesmerizing, the singer's energy boundless. (230, 1:36)

Now it is sagebrush again, endless sagebrush, the vistas extraordinary, the birds familiar: sage thrasher, Brewer's and vesper and savannah sparrows, horned larks, and a few lark buntings. I pass the time listening to the variety in the simple daytime songs of the Brewer's sparrow—most are two-parted, but a few have one phrase or three; most songs fall in pitch from the first to the second phrase, but a few rise; and what wonderful variety in the quality of their phrases. Listening to vesper sparrows, I hang on the introductory whistles of their songs, trying to hear how they might change from one place to the next.

In a moment of some trepidation, we climb the on-ramp to Interstate 80, the nation's busiest highway, the link between San Francisco and New York City. We now follow the original TransAmerica route, that laid out by the Union Pacific Railroad just off to the north. It's the only portion of our journey where we cannot get from "here to there" on smallish back roads. Huge eighteen-wheelers rumble past us, pulling us along as if we ride in a giant, open-air wind tunnel with them, the 13 miles to Sinclair and its maze of refineries over in a flash. In moments of silence between the trucks, though, we hear and see the world from our saddles in a way that no one traveling at warp speed on an interstate highway ever can.

We have come to ignore most weather forecasts, unless they are severe, but now, with the hard evidence in our faces, we recall the forecast for a strong north wind today. Sometimes it swings to the northeast for a cross wind, all making for some tough biking.

"What was that you wrote back in Virginia about biking into that headwind out of Wytheville?" I ask David. "It was right on."

"Oh, it went something like this: Everything is ugly with a headwind. It blows right into your head and rattles all of you. The sound deafens thought, and the bike wants to stop. Nature is saying: Don't bike this way, idiot."

About ten miles out of Rawlins we cross the continental divide once again. We hope for a nice downhill into the Great Divide Basin, but the wind would blow us backwards if we didn't pedal. We press on, through this basin of sand and sage that is completely encircled by the continental divide, so that water flows into the basin from the divide but departs only by evaporation. The sign at Grandma's Café in Lamont reads "Now Open." A much-needed food break, we think, but there's not a soul in sight; scattered behind the store is a vast graveyard of machinery and odd gear.

We push into the wind that sweeps unchecked across this open country, soon climbing to the continental divide once again as we exit the basin. We admire the distant mountains, then drop to gravel roads, huge trucks, and road construction as we near Muddy Gap, the narrow pass that Muddy Creek has cut through the ridge line here. Beyond the gap, just past the intersection with route 220, we follow the signs to the Tetons and Yellowstone. With eighty-some pretty tough miles on the day, we stop to inquire about camping at the large sign that declares "Muddy Gap Campers and RVs Welcome."

"You can set your tent here," says our host, "no charge." Next to Muddy Creek itself, with a pink and blue dilapidated mobile home as our windbreak, we happily come to rest for the evening. David prepares a feast of pasta and cheese with carrots, one of his specialties, and we eat heartily. After dinner, David returns to his *Zen and the Art of Motorcycle Maintenance*, which he's been carrying since Virginia, and I'm soon lying in the tent, drifting off, thinking of morning birdsong.

Sometime in the calm of night, he begins singing, roughly around midnight when the moon first glows in the east. I drift in and out of consciousness, taking mental notes and then losing them again, arousing to hear him honking and tweeting and chattering some more, then I drift off again—all night long it seems. A chat. How appropriate the name. Yes, he's named "yellow-breasted" for his looks, and most birds seem to have been named by someone describing a dead bird in the hand. But this bird lives, and it is his chatty nature that so entertains, hence his name. He gurgles and squawks and hoots; pops and rattles and cackles and clucks; he does it all, he "chats."

Yellow-breasted chat. My Muddy Gap bird, with four song packages. (231, 9:19)

In the wee hours of the morning, partly needing to verify that he is in fact real, I dress and walk down to the creek, laying my recorder and microphone down on the ground about ten yards from the willows where he does in fact sing. Punching "record," I walk away, happy that I will take him with me when we leave later this morning.

Back in my sleeping bag and wide awake, I listen intently. Growing familiar with the four sounds he's playing with, I wait, and after three minutes he introduces entirely new sounds, a new package, and again after about eight minutes, now specializing on ultra-brief pips and pops and squeaks, with fewer of the more pleasant longer phrases of before.

I hear imitations of orioles, jays, magpies, robins, and more. "How many different songs do you have, Mr. Chat, and just where do you get them all? You clearly mimic other birds, but do you also learn some songs from other chats, or do you mostly improvise, making up your songs to suit some inner fancy? Are your song packages of four to six songs consistent from one performance to the next, or do you create entirely new scripts as you sing from day to day? And how do you choose which songs to sing as a package, and why? Or why sing in packages at all? Why not just play out all your songs at once in one spellbinding performance and then start over? You are a mystery, Mr. Chat."

20.
THE OREGON TRAIL

DAY 46, JUNE 18: MUDDY GAP TO LANDER, WYOMING

Just after sunrise, with David already caffeinated and happy beyond reason, we're wheeling our loaded bikes out of our campsite, pausing to admire a loggerhead shrike who lands just above us on the sign for the campground. With the mechanical precision of his kind, as if programmed by a Swiss watchmaker, he sings, *gachee gachee . . . gachee gachee*, each song a sharply enunciated pair of two-syllable phrases, successive songs just a second or so apart. After a dozen songs in half a minute's time, he departs.

Loggerhead shrike. So precise, so repetitive, so . . . monotonous? Never! (232, 5:21)

We bike slowly uphill to the north, with oversized gravel and construction trucks lumbering by. The morning sun rises to our right, casting our long shadows on the sage that rises up the hillside to the left, the light green crowns of the sage radiant against the dark western sky. The four of us, our shadows making a crowd, glide up the hill, the Tetons and Yellowstone in our sights.

McCown's longspur. Perching high in the air on hovering wings, in song flight. (233, 5:04)

"The Brewer's sparrows are almost unanimous. Hear their short songs? Almost always a high fast trill followed by a low slow trill, with the occasional dissenter providing some variety." Vesper sparrows—their lazy songs beside the road begin with *weeoo weeoo*, two whistles slurred downward, followed by a single buzzy note, all of them in agreement on the form of the local dialect. Western meadowlarks chuckle and flute. *Two* McCown's longspurs, one on each side of the road, skylark high in the air and sing on the wing, their tinkling songs tumbling back to earth as the birds themselves float down to the sage. Racing down the road is a Bald Eagle, not just an ordinary eagle, but a humongous statue of one perched in the back of an appropriately humongous truck.

After four miles of some of Wyoming's finest, we gain the ridge where the terrain falls away to the north down to the Sweetwater River. There, looming just beyond the river a little over two miles away, is Split Rock in all its splendor. What a sweep of history lies here. The rock itself is a pinkish-red granite, crystallized from molten rock deep in the earth well over a billion years ago and much later thrust high as part a spectacular mountain range. Erosion whittled away the mountains and sediment filled the valleys, so that stubs of former mountaintops now lie on the surface like so many beached whales, an ocean of sedimentary fill all around them. These Granite Mountains, more modestly called the Sweetwater Hills, have aged into knobby mounds, their surfaces wrinkled and deeply fractured. The deep notch in the top of Split Rock itself was a well-known landmark used by pioneers on the Oregon Trail; spotted from far to the east across the wide-open expanses of sagebrush, Split Rock served as a "gun sight" aiming them west.

Our paved route 287 approaches from the southeast, the Oregon Trail from the east, and our paths converge beside the river at the base of Split Rock itself. There, below the current-day roadside overlook, in the meadows beside the river with water so pure and "sweet" compared to the bitter alkaline water found throughout much of the West, there it is easy to imagine the covered wagons in camp for the night, with livestock grazing nearby. From 1812 to 1869, according to the interpretive signs, half a million pioneers pursued their dreams through here as they followed the Sweetwater River on the gentlest of grades upstream to South Pass, about 70 miles to the west, beyond which raindrops and melted snowflakes flowed to the Pacific and to lands full of promise.

"Back in the 1860s," I interpret from the sign for David, "there was a log building and pole corral just down there by the river, a relay station for the Pony Express. Imagine being Buffalo Bill Cody himself, galloping in from the east; a man rushes out to meet you, transfers the mailbag to a fresh horse, and you depart within minutes, off to the next station ten or so miles down the trail.

You and your riding team can deliver mail across the West in 19 days. Perhaps this is even part of your record run of 322 miles in under 22 hours, using 21 horses. And now, just 140 years later, we have FedEx's overnight delivery, and the instantaneous internet."

BESIDE THE SWEETWATER RIVER

American goldfinch. *Po-ta-to-chip* calls, and songs. (234, 16:52)

Birdsong rises up from the bushes beside the Sweetwater River, the same birds the early pioneers would have heard . . . American goldfinches . . . spotted towhees . . . chats . . . a Virginia rail . . . and more.

Spotted towhee. Repetitive, in day-time mode. (235, 3:22)

We bike on, imagining a horse-drawn coach of the Overland Stage along the river just below the road on the right, passengers and payroll and mail headed west, a driver and guard riding shotgun up above. We soon pass Jeffrey City, once a boom town built by the atomic bomb and its need for the uranium that was mined nearby; now it too is largely history, a ghost of a town compared to its heyday during the Cold War. As if a ghost itself in this sea of desert sage, an osprey screams from a utility pole beside the road.

Yellow-breasted chat. A package of three different songs. (236, 2:12)

And *there*, in the distance, just after a small bend and slight rise in the road, lie the famed Wind River Mountains. As spectacular as they are, with their rugged, glacier-carved peaks dominating the skyline to the west, it's hard to believe that they, too, are has-beens. They're the only remains of a massive block of ancient granite, pre-Cambrian like the Sweetwater Hills, that formed deep in the earth over a billion years ago and then slowly uplifted, rising to the surface and up to 11 miles high—so reports a roadside sign. Erosion claimed the earth that once lay above them and most of the mountains themselves, until the highest peaks now lie a mere two and a half miles above sea level.

Virginia rail. *Kick kick ki-dick*, repeatedly, likely an unpaired bird. (237, 15:08)

We pass Ice Slough, just off the river, celebrated by Trail pioneers as the place where they could find ice well insulated by marshy soils into the hottest of summer months. The summer ice is history, too, as grazing cattle have destroyed the insulating properties of the soil.

Just beyond Ice Slough we meet two more eastbound cyclists and we pull over to greet them; they're Michael and Hintsje from the Netherlands, we learn, eastbounders who waited until the Oregon mountain passes cleared so that they could start in the West and have the prevailing winds blow them across the country. I picture them in Kentucky and Virginia during July and August, during the hottest and most humid months, all but devoid of birdsong, but choose not to share the unpleasant image with them.

At Sweetwater Station, we part ways with the river and the Oregon Trail. They turn southwest to South Pass, but we head northwest on route 287 toward Lander, our destination for the night. First, though, we check out "The Station." One building is a Mormon visitor center to commemorate the Mormon pioneers who pulled handcarts across the Sweetwater River here; we're greeted by big smiles from a young woman of unquestioning faith who tells stories of heroic pioneers along the "Mormon Trail." The other building is a bar, and Fox news plays loudly; "most fair and unbiased news source," we're told by the waitress, as "CNN and the other networks are too anti-American." Privately, we celebrate the diversity that is America.

Outside, mosquitoes suddenly swarm about us, getting in their last licks here in the Sweetwater basin. We bike on and I check my speedometer, speeding up a bit, discovering that I need to bike at least 5.6 miles per hour to outpace these bloodthirsty critters.

With 48 miles on our odometers for the day, we pull off the road, poised at the top of Beaver Divide. The snowcapped Winds rise to the west, the Oregon Trail and Sweetwater Hills lie behind, and the road before us drops away to the Wind River Basin a good thousand feet below to the north. What a magnificent basin it is, bounded by mountain ranges on all sides, the basin itself a vast expanse of sagebrush and wind with sediment from the mountains that has accumulated to four or more miles thick on its floor. The Wind River runs through it, of course, the mountains and basin and river all named for those cold winds that rush down the canyons from the high mountain ridges. A horned lark sings continuously just off the road to the right, sputtering away, rising to a flourish every ten seconds or so—a wonderful send-off for a downhill ride into the basin that will be as fine a ride as anywhere on the planet.

Minutes later, we've covered six miles. What a visual and geological feast in the cliff faces and canyons as we plummeted past layers of white and green and gray and red and black sedimentary rock, a textbook written in sandstones and claystones and shales and conglomerates, a text documenting the last 65 million years of the earth's history—all of geologic time since the extinction of the dinosaurs that cleared the way for mammals like us to thrive.

It is time travel like no other I know, hurtling by bicycle through unimaginable expanses of space and time, through global climate changes and continents drifting about the planet and the entire "age of mammals" and the origin of so many *birds*! Time travel continues when, just a few miles before Lander, we cross a black shale that was laid down more than 65 million years ago, during the Cretaceous, when dense, swampy vegetation covered much of current-day Wyoming; the vegetation from the Cretaceous and over the following 30 million years (Cenozoic) turned to peat and eventually was transformed into coal and oil and gas. Just off to the right lies the Dallas Dome oil field, where the first commercial oil well was drilled in Wyoming back in 1884; coal lies conveniently just beneath the surface in most of the basins throughout the state.

In Lander, we pitch our tent in the city park. June 18, I reflect. "A perfect day" David scribbles into his journal. My odometer reads 2840 miles for the trip, with perhaps another 1400 to go. We're two-thirds done, one-third to go, time and distances that I find difficult to comprehend.

DAY 47, JUNE 19: LANDER TO DUBOIS, WYOMING

I lie in the tent, listening and dozing, cozy and smiling, but when I fully awake, I am not sure whether I dreamed that least flycatcher's two-noted *cheBEK*, a voice not heard since Mount Rogers in Virginia. Walking through the park, I'm among western wood-pewees, yellow warblers, yellowthroats, mourning doves, a chat, house finches, and, as sunrise approaches, cowbirds and starlings high in the trees over the tent. A mystery bird sings nonstop high in the canopy, reminding me faintly of a warbling vireo; I listen, feeling the rhythm, the quality, accepting the unknown identity for now.

Returning to the tent, I find that David's up! "Count your blessings," he says, "we're outta here in half an hour." I'm impressed by his energy this morning, and indeed, following a quick breakfast and packing, we soon enter the Wind River Indian Reservation just out of town.

"The reservation was negotiated by Chief Washakie for the Shoshone," David learned in town last night. "Twenty-two million acres of it; Washakie and his Shoshone were 'good' Indians who helped the army defeat the Crow and also helped Lewis and Clark. Then the government moved the Arapaho here, too, who outnumbered the Shoshone three or four to one. Quite a history here, mostly sad."

I soon lead at a moderate pace, and David falls behind. Knowing how he likes to bike alone in the morning, I just try to keep him in view in my mirror.

There! Off to the right, in the cluster of trees beside a house is a *least flycatcher*. Its *cheBEK* is unmistakable, and this morning was likely no dream at all. I check the range map in the Kaufman Field Guide, noting how this

Least flycatcher. *CheBEK cheBEK,* faster than one a second, an "easterner" out West. (238, 2:56)

"eastern" bird, like so many others, stretches across the northern half of the continent up to Alaska. As we bike north into Montana, we'll encounter more of these birds that we've not heard since the far side of the Great Plains. Oh happy day!

As I linger, waiting for David to come into view behind me, I listen to red-winged blackbirds, yellow warblers, the ever-stunning songs of western meadowlarks . . . when a minivan pulls up beside me, a young Native American woman smiling behind the wheel: "Your friend back there has a flat tire and needs the tools you're carrying."

After the appropriate thank you and departing words, I'm alone with my thoughts. I'm shocked that my 45 years of living beyond those boyhood games of "cops and robbers" and "cowboys and Indians" and television programs like *The Long Ranger* (and Tonto) hadn't prepared me better for this moment. Plain and simple, here was an Indian driving a minivan, the image from my youth just all wrong. Rational thought informs me, of course, that this pleasant and helpful young woman was leading the normal life of a human being in North America (or at least in Wyoming). Those rational thoughts take me further, to an awe for the human founders of North America, to Sacagawea, who saved Lewis and Clark, to a proud culture of "First Nations" throughout North America. I race through thoughts of my own heritage, from Dutch farmers in rural western Michigan, back to immigrants from the Netherlands during the mid-1800s, celebrating again the diversity that is now America. I'm grateful for the brief exchange with her—perhaps I've grown just a little from it.

About a mile back I find David beside the road. "I could have gotten a ride to you, but then I wouldn't have biked the entire distance," he explains; "better to have you bike this stretch three times than me not at all." Can't argue with that logic!

Beyond Fort Washakie, so named for the great Shoshone Indian chief and the place where he lies buried, we climb through remote country. Only an occasional car breaks the solitude, and we eventually top out at a breathtaking view. Below us, the Wind River stretches left and right, a lush oasis in what looks otherwise to be a barren, badlands landscape. Far to the northwest and a day's bike ride to where we'll drop off Togwotee Pass into the Tetons, snowcapped peaks line the horizon. White clouds billow high against the bluest of skies, blanketing a quiet, rugged landscape. A sage thrasher comments in the distance, Brewer's sparrows, too, and a raven croaks on the wing over the river.

We stop for lunch in Crowheart and sit on a bench outside the store. As we run up a healthy tab for our insatiable appetites, we admire flat-topped Crowheart Butte across the river to the east. The name tells all, we learn. Back in 1866, not all that long ago, the Shoshone and Bannock Indian tribes fought the Crow for hunting rights here in the fertile river valley. Over several days, the bloodshed was considerable, so Chief Washakie of the Shoshone proposed to Chief Big Robber of the Crow that the two of them fight alone on the butte, the winner eating the other's heart. Chief Washakie won.

Following the Wind River upstream toward its source in the northern Winds themselves, we are mesmerized by the rock formations. Fifteen miles beyond Crowheart, the reds in the massive formations and cliffs beside the road are intense. Naming the red rock as iron oxide, or "rust," seems dismissive, as this is no ordinary rust. These are the iron-rich sands and shales that were deposited in sedimentary layers during warm climates, dating back to the Age of Reptiles; that's during the Triassic and Jurassic periods, as in "Jurassic Park," from roughly 150 to 250 million years ago.

"David, see there above the red layers, the purple and black layers? I've read that they date back to the Cretaceous, about 150 million years ago when *birds were born!* Thought you'd be excited to know that." They were born of dinosaurs, it is now widely believed, with the first birds indistinguishable from the lines of small dinosaurs that preceded them. This lineage eventually distinguished itself, of course, with feathered flight and the remarkable radiation of 10,000 species that we hear today over the planet.

And there's more. Overlying all of these sediments are more recent layers of bright whites and reds, from roughly 50 million years ago. They've been carved and eroded into spectacular badland formations. Cavorting and wheeling and darting about the cliff faces are white-throated swifts, their high twitters cascading down the scale in perfect celebration of all that is.

This surreal wonderland of geologic history continues upstream for another 15 miles to Dubois. Adding to the drama are thunderheads in all directions, and even David watches for possible shelters along the way. I take note of the vesper sparrows—they now begin their songs with three simple whistles, *tew tew tew*, not the exaggerated, downslurred *wee-ooo wee-oo* I've heard most recently, the local dialect having changed.

White-throated swifts. Cavorting and darting through a cascade of twittering *jejejeje-jejejeje* calls. (239, 1:47)

Beyond Dubois about eight miles, I wait for David at Union Pass Road, the left turn up into the forest where we'll stay at the bike hostel overnight. Yes, *forest,* I marvel, as we have gained just enough elevation up the east slope of the Wind River Mountains so that the badlands have yielded to a forested landscape. We're greeted enthusiastically by Dave and Jo-An Martin, generous hosts for any bicyclists who stop here. Before we know it, we are sitting before an extraordinary dinner of meatloaf, mashed potatoes, and vegetables, followed by unlimited apple pie and ice cream, a feast prepared by hosts who know how to win the hearts of any cyclists who wander by.

DAY 48, JUNE 20: DUBOIS TO JENNY LAKE, WYOMING

Eager to hear the forest sing, I'm walking about early, listening, waiting for first song with sunrise over an hour away. A cloud bank hangs over Ramshorn Peak about a dozen miles across the valley to the north, and a half moon shines overhead. Snipe winnow down below over the Wind River wetlands, and I hear *killdeer, killdeer.*

Nearby, a yellow warbler launches into his morning routine, soon working his dozen or so different dawn songs. With eyes closed, I concentrate, hearing swallows (most likely violet-green?), robins, chipping sparrows . . . and one of my mystery birds sings in the underbrush, the same bubbling house wren–like song I heard back at Blue River Campground in the Colorado Rockies. I strain to see him, but he retreats, leaving resolution for another day. There's the slow *yenk yenk yenk* of a red-breasted nuthatch; the raucous calls of Clark's nutcrackers overhead; a mountain chickadee's pure whistles; pine siskins with their raspy, rising *zhhrrreeeeeeeee,* the "zipper call"; a red squirrel chattering at me. And a flycatcher, an *Empidonax* flycatcher; I cringe, knowing what a challenge these birds are, but settling down, I hear the unmistakable *seeet, ka-SLWEEP, pa-TIK,* the three high-pitched songs of the cordilleran flycatcher, the same bird I heard back at Guffey in Colorado.

Pine siskin. Distinctive zipper calls, and high-pitched goldfinch-like calls. (240, 0:23)

Up the road is the distinctive *quick! THREE BEERS!* of an olive-sided flycatcher. I walk closer, and he's soon just above me, in a dead, leafless tree. The *quick* is sharp and brief, and he pauses maybe a quarter second before following with two pure, roughly half-second slurred whistles, the *BEERS* descending and drawn out. He's a close relative of the wood-pewees, in the same genus *Contopus,* but how differently he sings, with only the one song that he uses both at dawn and throughout the day. Walking up the road to the west a little more, I position him so he's silhouetted against the red band just above the eastern horizon, about five minutes before the sun itself will peek over the distant mountains. From there I watch and listen: *quick! THREE BEERS!*, though I increasingly hear it as *hip! THREE CHEERS!*

Olive-sided flycatcher

As I walk the lane back to the hostel, a sapsucker *mews*, and the Lander mystery bird from just this morning confronts me from high in a fir tree. Time to settle this one, I decide. I listen, to the husky voice; I feel, a mildly undulating rhythm in each song, ending low; I measure, each song two to three seconds; I count, 14 songs in the first minute. Positioning the recorder beneath him, I capture song after song dropping down the 15 yards from the canopy, and a good 200 songs later, conclude that this bird must be that "other" warbling vireo who lives west of the Great Plains. His songs have little of the spirited, undulating singsong of the eastern's *If I SEES you, I will SEIZE you, and I'll SQUEEZE you till you SQUIRT!*; these western songs are shorter and choppier, with successive notes rising and falling more rapidly, and without the high, emphasized eastern *SQUIRT* at the end. Their genes differ, too, I've read, revealing that eastern and western warbling vireos have been isolated from each other for some time on opposite sides of the Great Plains. I'm satisfied—no need to fetch my binoculars and try to see him.

Olive-sided flycatcher. *Quick, THREE BEERS!*—whether dawn or day: how curious. (241, 4:00)

Jarring me from my reverie is the sign on a cable across the road: "No Trespassing. Violators will be shot. Survivors will be shot again." Western humor, perhaps—but maybe not. Certainly not Kansas humor.

Warbling vireo. Above bicycle hostel, a new bird?! No, old friend with new voice. (242, 9:21)

Rivaling last night's dinner is this morning's breakfast: pancakes, sausage, scrambled eggs, and juice, with leftover apple pie and ice cream waiting in the wings! David writes a June 20 entry into the hostel's logbook: "11 hours of sleep in your cabin, and I think I could go for 11 more. I'm here with my father, who did not get 11 hours because he woke up at 4:30 a.m. to go bird watching. What a crazy man! David and Jo-An, WOW! Thank you! This has been amazing." And I add: "AND great birds this morning, among them an olive-sided flycatcher in the yard, singing *quick! THREE BEERS!* And chickadees, warblers, vireos, and sapsuckers. Oh Happy Day!" Had I been more forthcoming, I might have noted the challenge I found in listening quietly to Dave extolling the virtues of his Wyoming friend, Vice President Dick Cheney.

Departing late morning, far later than I would have liked given how the wind and weather often act up later in the day, we climb, and climb, the ancient granites of the Wind River Range now back to our left, the high peaks of the volcanic Absaroka Range ahead. With mountain peaks towering over us, we eventually rise to the continental divide at Togwotee Pass by mid-afternoon, with only 20 miles on the day.

Before us in this northwest corner of Wyoming, and in nearby Idaho and Montana, beyond the exciting sign that proclaims "6% grade next 16 miles," are some of the greatest natural areas remaining on the planet. They're set aside as the Teton and Bridger and Gallatin and Caribou and Targhee and Shoshone National Forests. They stretch far to the south and west and north, much of them set aside as wilderness areas. Two of the crown jewels of the National Park system are here, the Grand Tetons and Yellowstone. It's the home of elk and moose, bison and bighorn sheep, mountain lions and grizzlies and wolves, eagles and trumpeter swans. Ten thousand years ago, just after the last ice age, we might have expected the now-extinct American lions, musk oxen, woolly mammoths, American cheetahs, dire wolves, ground sloths, camels, horses, and bears of the long-legged and short-faced types. This is truly big country, with big views on a humbling scale.

If only we could see any of that now! The entire valley is socked in, blackened by menacing clouds hanging low and emptying whatever is in them. With some unease but with little choice, we begin our descent, soon taking cover from the rain and hail among smallish trees beside the road. That storm passes and we resume, only to be nailed again by hail and strong headwinds biting at our faces. We take refuge in a restaurant and refuel while waiting for the storm to pass. David especially enjoys the beautiful young woman who waits on us—it's tough biking with the ole man for a few months, with no life on his own.

Late afternoon we set out again. With a considerable tailwind and a downhill, and in neither rain nor hail, we race along on a collision course with the lightning storm we are all too aware of down below.

At the entrance to Grand Teton National Park we pay our fee but are informed that the campground is full. That means *full,* as in no room for *you,* the attendant emphasized, with little understanding that bicyclists have few options. We press on, hoping for the best, but shortly after taking the turn toward Jenny Lake, our evening's destination, we are hit yet again, this time with numbing, horizontal rain and swirling headwinds so strong that we cannot continue. We retreat, blown back across the highway, where we take shelter in an aspen grove. Kind motorists with trucks offer to transport us north to Colter Bay, but we're headed south; given our goal of biking each inch across the continent, we couldn't have accepted a ride anyway.

For the better part of an hour we are overwhelmed by lightning and thunder and rain and winds the likes of which I have never before experienced, but eventually it all passes and we set out again, now in a calm, gentle rain, the sun shining magically through the parting thunderheads. Rounding the southern end of Jackson Lake, among the fir and spruce, we are buoyed by a symphony of Swainson's thrushes. The songs swirl and spiral upward, each breathy phrase higher and longer and louder than the one before, the entire performance as if lifting up so many spirits to the heavens. I have heard Swainson's thrushes sing before, but never like this. We bike through one territory after another, from one singing male to the next, each of them as if singing "*pass it on*," until we break out into the open sage and grassland on our approach to Jenny Lake itself.

Swainson's thrush

Swainson's thrush. Magical, ascending, swirling and spiraling, heavenly. (243, 2:06)

At the campground, we bike through the section devoted to car campers, confirming that every space is taken with circus tents and big-house motor homes and lots of people milling about. But when we arrive at the "hiker/biker" section, we find all ten sites empty, not a soul in sight.

To the tune of an irrepressible robin singing at the tip of a spruce overhead, and with the day graying away, we pitch the tent and heat the macaroni and cheese dinner that David had bought at the restaurant stop. By 10 p.m., perhaps the latest to bed on our entire trip, with our food secured in the "bear box" nearby, we slip into our sleeping bags, looking forward to our first planned rest day for the entire trip.

21. GRAND TETONS

DAY 49, JUNE 21: JENNY LAKE, THE GRAND TETONS, WYOMING

By 4:30 a.m., from my cozy sleeping bag inside the tent, my slumbering son in his bag beside me, I sense those magnificent yet unseen Tetons looming in the dark just beyond Jenny Lake. On this day off the bike I smile and listen and wait, knowing that first light is racing toward us from the east, over terrain we've biked during the last seven weeks, and the birds here will soon add their voices to the growing chorus sweeping the continent.

I doze lightly, and just a few minutes later our friend the robin offers the first promise of the day, no doubt from the same spire as last night. He repeatedly offers a string of carols followed by a high, screeching *hisselly*, just as a good robin should. I try my usual game of picking a unique carol and listening for him to repeat it, concentrating, but I'm soon distracted by what he's doing with those *hissellys*. As I listen, *hisselly* after *hisselly* is the same, until he eventually switches to a different one. How odd: Other robins rarely repeat a given *hisselly* note like this, but instead race from one to the next. Does such a pattern mean something special in robin-speak? The All-American robin is always full of surprises.

American robin. Teton robin at dawn, with low carols and high *hissellys*. (244, 3:19)

The robin quiets, but replacing him, perhaps even from the same perch, is another one of the Mosquito Princes, another *Empidonax* flycatcher. He's new, not the cordilleran with all high-pitched notes—no, there's that rough, low *prrdurnt*, a dusky flycatcher almost certainly. There's the clear, high *pa-EET* and the rising *PIT-it*, too. For a minute I follow along. . . . Very nice. He has the expected three songs, and they come in bursts of two to five, followed by a longer pause, and I can never really be sure what is coming next. He's a "lifer," one I've never heard before—I should try to see him, but it's dark, and even if it were light enough, all I'd really see is "an *Empid*," who to my eyes is indistinguishable from all the others, *until he sings*.

Dusky flycatcher. Three songs at dawn, *pa-EET, PIT-it, prrdurnt*, the low-pitched *prrdurnt* distinctive. (245, 2:21)

I check in with other birds singing in the distance . . . the rollicking ruby-crowned kinglet and whistling mountain chickadee have been singing for some time now . . . two chipping sparrows sputter through the dawn chorus off toward the car campers, probably facing off with one another on the ground next to someone's lucky tent. Their songs are distinctive, one more buzzy than the other, enabling me to follow their respective roles in the conversation.

Up high toward Jenny Lake is a western tanager in full dawn song. What a fine view he must have with dawn breaking over the lake, the sun's first rays soon to strike the mountains above. He spent the winter somewhere far to the south, perhaps in some tropical paradise in the highlands of Central America, then returned to this temperate wonderland last month. When he looks up at these Tetons, I wonder what registers in his brain. Perhaps he is totally oblivious to them, as my scientific training would advise me to believe, but I admit that I'd like to believe otherwise.

His dawn singing is indistinguishable from that of his eastern cousin the scarlet tanager, except that he replaces the scarlet's *chip-burr* call with his own distinctive *pit-er-ick*. The scarlets from the Atlantic through the Ozarks have long finished their dawn song by now, but the western picks it up and carries it to the Pacific, slowing the delivery of his zippy four-to-six-note daytime song by adding time between the notes and then pausing to call *pit-er-ick* before resuming. I monitor him for a minute, watching the notes dance by in my mind, counting how many he sings between successive *pit-er-ick* calls: 4, 5, 4, 6, 4, 4, 5. I hear two more tanagers fairly near, as if they gather near their shared territorial boundaries to match wits early in the morning, much as the scarlet tanagers seem to do.

Western tanager

Western tanager. Dawn song: slowly delivered song phrases punctuated by occasional *pit-er-ick* call. (246, 6:11)

It's noticeably quieter already, 5:35 by my watch, official sunrise to follow in just a few minutes. I really should leap out of this bag and out of the tent into the world, but it is so comfortable here. The next time I check my watch it is already past seven o'clock, when David is stirring, too.

We think together about the day. A leisurely breakfast of oatmeal, raisins, brown sugar, and powdered milk, still grand after nearly 50 days on the road. A walk around Jenny Lake to the waterfalls. Maybe take the boat back. "Maybe I'll bike around and check the place out," says David. "Not me," I reply, "I plan to rest my bike for the day. Maybe I'll just go sit down by the lake."

Unzipping the tent's fly, we see *The Grand Tetons* for the first time, looming over us to the west. A quick walk down to the lake gives an uninterrupted view of one of the most magnificent and youngest mountain ranges in the Rockies. I breathe deeply, filling my lungs with the mountain air; I take a sweeping look and then close my eyes, trying to burn the entire scene into memory. I'll come back this afternoon, I plan.

Breakfast is served surrounded by our wet gear hanging out to dry, and seasoned with the songs of western tanagers. "Like a robin but hoarser—that's all you need to identify them," I point out to David, "but listen more closely and try to crawl inside their heads." Two birds sing now, both having abandoned their *pit-er-ick* call that was featured in the dawn chorus. Now it's just a burst of song notes, rising and falling over a few seconds, and then a longish pause to the next song.

Western tanagers—so expressive. *Pit-er-ick* calls, with varying enunciation. (247, 2:42) Day song, at double the dawn pace. (248, 1:30) Day song, but the pace varies. (249, 6:50)

But there's more going on here. "Listen how the male by the lake sings in slow motion, with five to six song notes delivered slowly over four to five seconds, almost as slow as during the dawn chorus but without the *pit-er-ick* calls. The male toward the car campers sings on a tear, the song notes ripping from his bill at twice the pace of the other male." A minute later, the lake male picks up his pace, though he's still slow. Had we the mind and ears of a tanager, we'd know what it all means, but we have to settle for just a small window on their singing lives.

After breakfast, we set out on that hike around the lake. At the boat docks we pause and contemplate a touristy boat ride across the lake to our destination of Hidden Falls, but instead choose to walk. A lone barn swallow, perhaps nesting under the docks, flies up into the sunlight on a dead branch overhanging the river. What a fine looker, his back an iridescent bluish black in the sunshine, his forehead and throat a deep chestnut, the tail long and forked; he's buff below, with a dainty bluish necklace across the chest.

In his jumbled bubbling ecstasy of a song, I hear happiness, though I realize it's mine, not his, that I hear. A field guide might coldly describe his song as a bit husky and squeaky, a musical twittering for four or five seconds followed by a half-second rising whine and a half-second dry, creaking rattle. But I hear so much more. I've studied these songs and wallowed in graphs of their songs (sonograms, frequency-time graphs that are a musical score of sorts), slowing the songs down so that I could hear the details that the swallows themselves no doubt hear. In the musical twittering are four to six luscious notes per second, each different from the others; many are only a tenth of a second long, and they're rapidly modulated in roller-coaster fashion over a range of frequencies, all so rapidly delivered and varied that it sounds as if the notes are sparring with each other. That rising whine and the sharp explosive notes in the rattle on the end of the song seem to declare the winner of the bout. In a flash, our swallow is off, *kivik kivik*, calling on the wing, darting out over the lake.

Barn swallow. My happiness expressed in his bubbling song-ecstasy. (250, 0:29)

We hike, or rather walk slowly, enjoying the lake, the trees, the mountains looming above, the total landscape, perhaps mostly just being off the bikes. Swainson's thrushes offer their heaven-ascending arias. Mountain chickadees whistle and call, ruby crowned kinglets chatter, warbling vireos warble in the treetops. A singing junco . . . *yenk*ing red-breasted nuthatches . . . a chipping sparrow chips at us, perhaps because we've wandered near its nest . . . sapsuckers drumming so distinctively—a veritable Teton symphony.

A TETON SYMPHONY

Dark-eyed junco. Same song for nearly six minutes, though he has others. **MacGillivray's warbler.** Typical two-parted, burry, churry song. **Red-breasted nuthatch.** Persistent, weak tin-horn *yenk yenk.* **Red-naped sapsuckers.** Three individuals drumming, in concert. (251, 5:51)

Together with the loads of people arriving by boat, we hike up Cascade Creek to Hidden Falls, then take the return boat trip back to the dock and to camp, where after a quick lunch David goes for a ride. I return to the lake, to sit, and to contemplate life. I have one responsibility for the afternoon, master cook David has reminded me: "Buy tonight's dinner at the camp store."

Perched just above the lake, the mountains beyond, blue sky overhead, I search for some profound thoughts, but settle for just *being*, here, now, and taking in this grand scene. I soon imagine soaring as an eagle, rising above the lake and the mountains themselves, flying the five miles north to Mount Moran and then peak-hopping to the south, past Thor, Woodring, Rockchuck, St. John, and Teewinot just opposite the lake. Then there's Owen, the Grand Teton itself at some 13,000 feet above sea level, Middle and South Teton, Nez Perce, Wister, and Buck, a little over five miles to the south, all listed on my map. A nearby sign explains that a mere nine million years ago the eggshell-thin crust of the earth cracked here, forming two long blocks about 40 miles long. The western block rose to form the Tetons, while the eastern block fell to form the "hole" that the Snake River now drains just to the east.

Everywhere I see evidence of the great ice machines at work. The glaciers in their prime some 25,000 years ago gouged and scraped and clawed at these huge slabs of uplifted rock. What survived are high peaks, some of the finest anywhere, chiseled sharp on all sides by glaciers.

What's missing from this extraordinary scene is even more stunning. I imagine laying a 10-mile-long level atop Grand Teton, almost a mile and a half above the valley floor, and gawk at the emptiness below the level all along the Teton chain. Mountains of rock have just disappeared. Nowhere is the void more dramatic than up the valley now drained by Cascade Creek. Four miles from Jenny Lake the valley forks, continuing for another three miles up each leg of the "Y" to two main headwalls of the glacier that once filled this valley; countless glacial tributaries contributed to the main ice flow, carving out a valley well over a mile deep.

And here I sit, on a forested pile of that very debris, among the gravel and rocks and humongous boulders that the glaciers stole from the mountain and bulldozed here. This "terminal moraine," in fact, encircles the lake, the lake itself about 250 feet deep and impounded by the debris pushed here and abandoned some 12,000 years ago when the glaciers began their big melt at the end of the last ice age. And where's the rest of the debris that's been ripped from the mountains? I glance toward Mount Moran to the north, as at the very top of that mile-high mountain is a sandstone cap that corresponds to the sandstone layer now lying almost five miles beneath the surface of current-day Jackson Hole. Given that the "hole" to the east sank much lower than the mountains rose, it is obvious that the pulverized bits of the mountains are the vast and deep graveyard covered by sagebrush behind me to the east.

A small flock of red crossbills bound across the lake; they bob and weave in flight, much like their relatives in the Cardueline subfamily, the siskins and goldfinches. I try to hear the important details of their calls that I would see in sonograms, but know my ears aren't up to the task. Perhaps there are ten birds, each calling something like *kip kip kip*, or is it *gyp gyp*, or *jilp jilp jilp*, or *jip* or *chip* or whatever, as they disappear up Cascade Canyon across the lake.

Red crossbills. *Kip kip kip* from a Teton flock, but what "species" are they? (252, 0:58)

So which red crossbill did I hear? I smile as I check the range map in Kaufman. The "red crossbill" breeds across

Canada and then south along the Rockies into Mexico; the birds seem to be nomadic, moving about the continent in search of food, not uncommonly breeding in the middle of winter when they find a good cone supply. With all of this flying about, one would think they'd homogenize by breeding with other crossbills from other parts of North America, leading to a single lineage. But detailed analysis of their calls suggests several different lines, each with different calls, sometimes associated with differences in bill shape, behavior, and the species of cone seeds eaten. Kaufman weighs in: "Might represent up to eight species." If these were flycatchers with "innate" calls encoded in the genes, one could understand how these birds might use their fail-safe calls to recognize one another and breed only with birds of their own type. But these simple calls are learned, oral traditions that are passed down from one generation to the next, making it far less reliable for individual birds to identify other birds of their own type. It's all a big puzzle.

I boldly return to the campsite with my purchased spaghetti dinner, only to good-natured complaints that I actually bought spaghetti and not pasta, which is what he meant, and I should have known, so that he could doctor the pasta into some tasty dish. But in the end, we agree that the spaghetti is OK, and hunger satisfied heals all.

"Big surprise for you, Pops; let's get up really early tomorrow so we can be on the road as the sun rises. I've found the perfect place down the road to see the sun rise on the Tetons. Then we can take off for Yellowstone from there."

DAY 50, JUNE 22: JENNY LAKE TO YELLOWSTONE NATIONAL PARK, WYOMING

A good 40 minutes before sunrise and with our bikes fully loaded for the ride to Yellowstone, we have taken a small detour to the south and now pause beside the road about four miles from Jenny Lake. It is about here that the Grand Teton, highest of them all, will be at its finest, we agree, though we're discouraged by a passing rain shower. Worse, the mountains themselves are shrouded in clouds, and a very dark sky lies to the north on our projected route.

In the calm of the early dawn, the Snake River roars to the east, the very Snake that we'll encounter again in Hells Canyon on the border of Idaho and Oregon; to the west Cottonwood Creek murmurs as it drains Jenny Lake. And what a symphony of vesper sparrows! I wander off the road to the east into the sagebrush toward the Snake River, then turn to face the mountains, hoping

for a glimpse of the Tetons should a break in the clouds occur, vesper sparrows now all around me. I have never heard them like this before. And how many can I hear? Hundreds, maybe even thousands, I imagine, if only I had better hearing, as their territories must stretch up and down this sagebrush plain for miles to the north and south. It is their song beginnings that so intrigue me, two downward-slurred whistles followed by two (sometimes one) higher whistles, as in *tew tew tee tee,* the ten or so birds within earshot all agreeing on the local dialect.

I pick the near bird and follow along. He builds in intensity through the four introductory whistles, the first note barely audible, the last loud and penetrating. He then launches into a fine series of four to five trills that gradually trail off, the total effort about three seconds long. I compare the rhythm and feel of successive songs, noting the same *tew tew tee tee* introduction and the same first trill, but then he's off unpredictably to a variety of other trills. I nod, smiling, as this male sings much like the birds that I studied in Oregon over 30 years ago while I was still in graduate school. By his simple introductory whistles he is easily identified as a vesper sparrow, but knowing how he assembles his entire song makes me smile.

Vesper sparrows. Local dialect, from the sage in the Hole below the Tetons. (253, 3:51)

I bike the dirt River Road to the Snake River overlook for a quick look. The road there drops steeply to a second terrace, then down to a third where the river flows in multiple, braided channels through a green, forested landscape perhaps a mile across. The terraces mark successive outwash plains of glaciations in the valley here and far to the north, where an ice sheet covered all of Yellowstone; the river now is just a trickle of the mighty force that one roared through this entire valley when the glaciers melted in earnest between 25,000 and 12,000 years ago.

I suddenly feel very cold, and in spite of wearing layer upon layer of nearly every bit of clothing I have, my toes and fingers are numb. It's close to freezing, and those dark clouds approaching from the north add a chill to the air. I join David on the main road, and with the Tetons obscured in the clouds, we make a run back to the campground. A thunder and lightning hailstorm strikes just as we arrive.

Weather forecast. Strong south winds will blow us to Yellowstone! (254, 0:27)

Our early departure for Yellowstone thwarted, we kill time in the warm visitor center when it finally opens, reading books, writing postcards, and taking heart in the forecast from the National Weather Service that I receive on our small radio. In clipped, staccato voice, we get the news:

> . . . Across the region, rain was falling . . . the local forecast for Jackson Hole and the vicinity: Today, isolated rain showers and thunderstorms this morning, otherwise partly cloudy. Becoming windy this afternoon, highs in the mid-sixties, southwest wind ten miles an hour, increasing early this afternoon to 20 to 30 miles an hour, with gusts to 35 miles an hour.

Southwest wind, 20 to 30 miles an hour, gusts to 35! With some patience, we realize that we could wait out these storms and then be blown north to Yellowstone this afternoon. That sounds like the best option, we agree, though the waiting will be a challenge.

After warming by the fire in the visitor center, I wander the trails nearby, homing in on that whir of hummingbird wings I've been hearing since Guffey in the Colorado Rockies. I confirm that it is in fact a male broad-tailed hummingbird, as I had suspected. He perches now on a dead twig of a small bush just outside the visitor center—his throat patch is a brilliant, iridescent rosy red when the sunlight strikes it just right, such a contrast to his bright green cap and snow-white breast. He fans his tail and flicks his wings, then explodes from his perch. His wings trill loudly as he flies 30 feet into the air, then he dives, seemingly at a dead stick lying on the ground, calling sharply as he passes and then rocketing skyward again where he hovers, as if surveying all that is his. And poof, he's gone, departing in a showy whir of wings. It is the tips of his

Broad-tailed hummingbird

Broad-tailed hummingbird. Calls, yes, but how fascinating the acrobatic flight and wing sounds. (255, 0:16)

wing feathers that "sing," as they're specially tapered to vibrate up near 6000 Hz, and every one of his forty-some wing beats per second produces a pulse of this high tone.

Next to the hummer's dead stick is a scarlet gilia, identified by a small sign next to the flower. For the ornithology classes I've taught, I've read all about this flower, though I've never before seen one. Hummingbirds pollinate the gilia's reddish flowers by day, moths its whitish flowers by night, and the scarlet gilia cleverly provides the appropriately colored flower based on its "expectation" of which pollinator is likely to be about.

"I found the ranger's presentation on bats really interesting," David tells me about the time he passed in the visitor center. "But I lost interest when I spotted the wedding ring on her finger, so I went to read a few of the books for sale."

A little before noon the weather clears and the winds shift, so we set out. In leaving, we're spellbound by the sights that we missed when we arrived two days ago, and we stop at each turnout along the road. First it's a striking view of Cascade Canyon, and then, meandering among and over glacial moraines bulldozed here and there and everywhere, we bike into turnouts at Mountain View, Mount Moran, and the Potholes, where receding glaciers left huge, isolated blocks of ice that eventually melted, leaving behind small lakes. But at every turn and twist in the road it is the view of the Tetons themselves that steals the show.

"We're at 2101 meters, north 43 degrees 52 minutes 21 seconds; west 110 degrees, 34 minutes, 24 seconds." With the finest precision one can find in a GPS unit, our newfound friend with his large motor home at the Willow Flats Overlook tells us exactly where we stand on this planet, and inquires with what I sense to be some disbelief and admiration and even more envy, "Did you really come from Virginia?"

"Yeah, and what a trip it's been!"

"When did you start?"

"May 4."

"You guys were real fat men before, right, now you're just as lean as can be?"

"We weighed 400 pounds apiece!" I assured him, only during his hearty laugh realizing that he himself wasn't far off that mark.

With a hefty tailwind, we continue north, following the shoreline of Jackson Lake on our left, then gliding up the Snake River. The rugged peaks of the Tetons recede into the distance behind us, and we soon stop at the south entrance to Yellowstone National Park.

22.
INTO THE FIRE

DAY 50, JUNE 22 (CONTINUED): YELLOWSTONE NATIONAL PARK, WYOMING

"No bikes allowed. Yellowstone is closed to bicycles. The road is for RVs and motor homes, and not wide enough for you, too. Bears will eat you. Turn around and go home." Those weren't the ranger's *exact* words, of course, but it's what I felt. I'm sure he politely as possible informed us of the hazards in the park, but it came across all wrong. We informed him politely that we would proceed through *our Park,* but we pulled off the road shortly to try to understand the greeting we received here.

We continue about a mile up the road to where the Snake River turns to the east, then follow a tributary, the Lewis River, north along its impressive canyon. Another hour up the road, at Lewis Falls, we pause, not only to admire the 29-foot falls but to reflect on where we stand, on the very edge of the vast cauldron that is Yellowstone.

Were we to climb up the half mile to Mount Sheridan just to the east, to 10,308 feet according to our map, we could look north, and about 37 miles away find Mount Washburn at 10,243 feet, just 65 feet difference in elevation. Connecting these two peaks but circling around to the east is the Absaroka Range, a giant horseshoe of old volcanic mountains extending around the Yellowstone region and opening to the southwest. A mere 650,000 years ago a high mountain range filled what is now a void, the great emptiness having been created during a massive volcanic eruption that lasted a matter of hours or at most a few days. An incomprehensible 240 cubic miles of rock and ash and superheated magma were hurled into the air, leaving a three-inch layer of ash as far away as what is now Iowa. After the cataclysm, whatever solid earth remained sank several thousand feet, filling the emptiness below and creating the Yellowstone caldera.

Cracks in this thousand-square-mile sunken plug soon yielded to pressure from below, and viscous lava poured out over the floor of the caldera. The lava flowed slowly, barely moving, cooling in ridges and creating the topography of many of the scenes we'll cherish. Lewis Falls cascades over one such volcanic flow, and just upstream Lewis Lake fills a basin between two lava ridges. Lava also forms the west boundary of Yellowstone Lake and several ridges we'll bike over as we cross the continental divide three times in the heart of Yellowstone. Tomorrow we'll bike beside the Firehole River, which follows the seam between two different lava flows.

The cause of all this mayhem festers just five to seven miles below us, as a super-hot chamber of magma and partially melted rock about 30 miles across. Feeding this magma chamber is a roughly 400-mile plume of hot rock that extends down to the molten core of the earth. This is one of about 40 such hotspots that can be found over the face of the earth's thin crust.

A hotspot beneath the ocean produces strings of volcanic islands, such as the modern Hawaiian Islands and the chain of increasingly older Midway Islands that trail off to the northwest. As the Pacific tectonic plate slides ever so slowly over the hotspot, the magma in the chamber below erupts to the surface every few million years, the lava building a seafloor mountain that, if large enough, rises to the ocean's surface to become a volcanic island. Here on land, the North American tectonic plate on which we bicycle has traveled to the southwest over this hotspot at the rate of about 1.8 inches per year, or about 30 miles every million years. If we could look with telescopic vision 500 miles beyond the open end of the mountainous horseshoe to the southwest of Yellowstone, through the barren moonscapes of the Snake River plain in southern Idaho and into northern Nevada, we'd see evidence of a string of supervolcanic eruptions during the last 18 million years.

"Unseasonably cool" was the weather forecast for today. "But just imagine being here 25,000 years ago, David, at the peak of the last ice age. We'd be covered with ice, all of Mount Sheridan and Washburn, too, and the entire Yellowstone caldera." We bike on through this wild landscape that was created by fire and sculpted by ice, with fires as recently as 1988 claiming vast swaths of forest. We then climb over the continental divide and drop into the campground at Grant Village, where we're given site 405 in the hiker-biker section. "But beware the grizzly," warns the attendant. "He took down a young elk just a day or so ago and stashed it in the campground; we removed the dead elk, but the bear is still prowling the campground looking for his kill. He might be a little grumpy."

DAY 51, JUNE 23: GRANT VILLAGE TO WEST YELLOWSTONE, YELLOWSTONE NATIONAL PARK, WYOMING

"Heavy snow is predicted later this morning," David learns from nearby campers, "but just here in East Yellowstone and up to our two passes over the continental divide. If we get out of here soon, we can beat the weather and get over to Old Faithful." I first make a quick run to the laundry, where we left our clothes drying late last night, hoping they'll still be there this morning. Up and back, I hear only one bird singing in this 25-degree temperature: a hermit thrush. I follow along, leaping the scales with him, just happy to be.

We shake the ice off our tent, pack up, eat our oatmeal breakfast, and head out, soon climbing through a pristine forest of lodgepole pine, one of the only trees that will grow in these nutrient-poor soils of the Yellowstone caldera. And an actual *forest,* we remark, as this part of Yellowstone was spared in the 1988 fires. Elk lie beside the road, unmoved as we pedal by. We savor two more crossings of the divide. The drainage to the south leads to Shoshone and Lewis Lakes, then to the Lewis River, out the Snake, and on to the Columbia River and the Pacific; to the west and north and east, water flows to either the Yellowstone River or the Madison River, both of which join the Missouri and then the Mississippi on the way to the Gulf of Mexico and the Atlantic. We are on top of the world and feel it, though deflate just a bit as we drop into the tourist basin at Old Faithful.

Hermit thrush. Invigorating energy and beauty at dawn. (256, 3:11)

Weaving among the cars and streams of people headed toward Old Faithful, we join the crowd and fall into line for the next showing. We are impressed by the predictable explosion of water that has percolated deep into the ground and boiled out, throwing thousands of gallons over a hundred feet into the air.

What a mass of people, though, the likes of which we have not seen before on our journey. This is Yellowstone, I remind myself, *Yellowstone National Park.* I whoop within, knowing that one is expected to drive to Yellowstone and experience it quickly by car or RV or truck or bus, yet here we are on bicycles, having come all the way from Virginia. "*We will cross the continent on our bicycles,*" I now say with a confidence never before quite felt, and it feels as if the Pacific awaits just down the hill.

We soon depart to the northwest along the trail through the Upper Geyser Basin, all of which is drained by the Firehole River; once again largely alone, with the masses back near the parking lot and lodge, we regain our enthusiasm for Yellowstone. The two miles to the highway are a riot

of hydrothermal activity, and in these cool temperatures, the entire landscape "smokes" with clouds of steam boiling skyward across an eerie, barren land.

We pause at Castle Geyser, capturing ourselves on film, smiling, happy, giddy even, admiring the castle that has formed around the geyser. "Says here that water gets heated to boiling down below, then the geyser blows off the pressure; when the boiling water evaporates, it leaves behind the dissolved silica to build the castle," summarizes David.

YELLOWSTONE GEYSERS

Ear Spring Geyser. It bubbles, boils, and surges at times. (257, 3:54)

Daisy Geyser. A steam engine as it fades. (258, 5:01)

Grotto Geyser. Bubbling, the real eruption hours later. (259, 2:12)

We leave our bikes and hike some of the trails through the basin, past hot springs and steaming mudpots and fumaroles (steam vents), past geysers with castles and those with none, some erupting, some not. Geysers are the "Giant" one, or Splendid, or Comet, or Daisy, or Spasmodic, Sawmill, Grand, Vent, Grotto, Fan, Riverside, Artemisia, or, one of our favorites, the Baby Daisy. Pools of boiling water bubble and gasp, the entire landscape alive as it sighs and heaves. At the one-eyed Morning Glory Pool, it is as if we peer into the depths of the living earth itself; a burnt-orange iris envelops the pool, the pupil itself shades of bluish green that give way to mysterious black at the center, so that eyeball to eyeball, the earth and we contemplate each other.

"This place would be really special at dawn," I remark, "when the birds are singing." I'd start at Old Faithful an hour and a half before sunrise when good tourists still slumber. During my private showing, a mountain bluebird who owns the geyser will begin circling high overhead, singing on the wing, skylarking, celebrating the moment with me. Robins will be next, and the sora in the wet meadow just below Old Faithful will call *sor-AH, sor-AH, sor-AH,* the killdeer sounding off, too, *kill-deer, kill-deer*. I imagine hearing them all now, the mountain bluebirds and mountain chickadees, yellow-rumped warblers, white-crowned and savannah and chipping sparrows, Cassin's finches and

Mountain bluebird. High above, in dawn song flight. (260, 1:37)

Sora. *Sor-AH, sor-AH*, nonstop; likely unpaired and seeking a mate. (261, 6:00)

Yellow-rumped warbler. As if halfhearted, a two-parted song that builds, then falls, in energy and pitch. (262, 4:41)

Cassin's finch. Stunning mimicry in brief, excited song. (263, 0:18) Less excited, less mimicry. (264, 2:51)

Ruby-crowned kinglet. Beginning wee and high, galloping thunderously by song's end. (265, 4:03)

Savannah sparrow. Beside the Firehole River, *take-it eeeeeee-aaassssssssyyyyyyy*. (266, 6:08)

Mountain chickadee. Classic *tsick-a-dee-dee*, and whistled song. (267, 3:16)

pine siskins, and juncos. From the wet meadows sing red-winged blackbirds, and an occasional *quack*ing duck flushes from the river. There's the whir of broad-tailed hummingbird wings and the *jilp jilp jilp* of red crossbills commuting overhead; a *quork*ing raven flies by as swallows dart about this Upper Geyser Basin. And all this magic is accompanied by the sounds of the living, breathing earth as it bubbles and gasps and sighs and effervesces and blows, all in a most otherworldly scene of steam rising everywhere into the cool morning air. I most certainly will be back—this is just too good.

We bike on the road along the Firehole River, through traffic jams of cars stopped beside roadside bison and elk, but thankfully no grizzly bears (the idea of cyclists being "meals on wheels" is not too appealing), on to the hot springs of Midway Geyser Basin. The centerpiece here is the Grand Prismatic Spring, a good 250 feet across, over 150 feet deep, the deep blue at its center giving way to bluish green and then green itself, with brilliant oranges and golds and reds on the rim. Every minute over 500 gallons of hot water stream down the sides of the raised cone. The colors are a gift of pigmented bacteria in the water, we read; greenish colors are due to chlorophyll, reddish colors to carotenoids.

Bison

In Lower Geyser Basin we find even more geysers and pools and springs and pots and fountains, continuing evidence of the hot bed of magma not all that far below the surface. Taking it all in, we follow the Firehole River another seven miles to Madison Junction, where the Firehole and Gibbon Rivers merge to form the Madison.

Along a roadside strewn with elk and wildflowers and signs of life returning everywhere after the fire of '88, we float downhill beside the Madison, now puny compared to the torrent that once drained the huge ice sheet over Yellowstone just a few thousand years ago. Kingfishers rattle over the river. Just past a large sign asking motorists to stay in their cars lest they disturb the nesting eagles, a woman walks off the road to get as close to the nest with her tiny camera as possible. All this and more, from the south entrance of Yellowstone to Grant Village to Old Faithful to here, all on the supposedly unfriendly roads of Yellowstone that the ranger had warned us about, among kind and courteous drivers of all types of road vehicles, all happy to share the bounty with two adventuresome cyclists.

It is downhill all the way to West Yellowstone, just outside the park and over the line into Montana, where we take a room at the hostel in town. David and I share a room with a motorcyclist who has biked up from South America, and David pumps him for ideas of how he might bicycle through Central and South America. Our British friend Keith is here, too, having stayed at the motel in Grant Village last night; Keith's tent still rides shotgun, pampered and unused since the Atlantic.

"Montana," I say to myself, "our eighth state. In just a few days, you'll be in Idaho, then cross Oregon and find yourself blocked by land's end with nowhere to go but home. . . . *Slow down,"* whispers a growing voice from within.

DAY 52, JUNE 24: WEST YELLOWSTONE TO MADISON RIVER CABINS, MONTANA

So as not to disturb our sleeping roommate, we eat breakfast down below on the first floor, and by 8:30 a.m. we roll our bikes out the door into a cold drizzle, and into ravens! What a chorus of them! It seems that one sits atop every rooftop in sight, each calling the way only ravens can. I imagine they know that in my next life I plan to come back as one of them, and they have gathered here to croak their approval. What a sight I am now, though, the anti-raven beneath the yellow shower cap over my helmet, yellow raincoat and yellow rain pants covering my body; only my rain booties are raven black.

Common ravens. A conversation—if only we knew what they were thinking. (268, 3:00)

Just outside of town, a cow moose and her calf lumber across the road in front of us, and we're soon riding through the song-rich willow flats beside Hebgen Lake, the large reservoir created when the Madison River was dammed downstream. The *sweet-sweet-sweet-I'm-so-SWEET* songs of yellow warblers and the *FITZ-bew*s of willow flycatchers dominate; white-crowned sparrows speak their local dialect. Across the lake to the left, and then across the Madison River itself after we pass the dam, the continental divide meanders from snowcapped peak to snowcapped peak, from Lionhead to Targhee to Sheep, the melt on our side eventually finding its way east to support the Mississippi barges where we crossed back in Chester, Illinois.

Though the rain has stopped, a cold wind howls into our faces, easily 30 miles per hour. We creep along, taking turns in the lead and pulling each other along. "The wind isn't so bad," remarks David sheltered behind me; "The wind is awful," counters David as he takes his turn in the lead. The river flows uphill here, we are convinced, so strong is the wind that it would push us backwards up the gentle road grade if we did not struggle forward.

White-crowned sparrows. Local dialect just outside Yellowstone. Bird 1. (269, 2:29) Bird 2. (270, 0:57) Bird 3. (271, 3:19)

"Quake Lake," the sign reads. Rising directly from the water are lifeless snags, the skeletons of trees that drowned after all hell broke loose here 44 years ago. Just before midnight on that fateful August night back in 1959, with 28 campers sleeping peacefully down below in their RVs and tents, one of the strongest earthquakes ever documented in North America dropped the face of Sheep Mountain onto the campers and into the Madison River Valley below. Back in Yellowstone, geysers were muddied and the damaged Inn, about 40 miles to the southeast, was evacuated. The Madison River quickly backed up, creating a lake six miles long, a quarter mile across, and nearly 200 feet deep. "Man oh man, imagine the tranquil scene, campers out celebrating the earth and all its beauty along the river and lake, sleeping peacefully, maybe a few still around the campfire . . . the earth shrugs and drops the mountain on them. Could be us almost any day along the way, in one way or another."

Overlooking the scene from the parking lot at the visitor center, we gawk at the large scar on Sheep Mountain and the fill in the valley below. In a geological nanosecond, rising and falling blocks of the earth's crust slipped *20 feet* past each other along what is known as the Hebgen fault. Fifty square miles of the surrounding area dropped a good ten feet. As the hotspot beneath the earth blazes its surface effects from the southwest through Yellowstone, it leaves behind a relatively stable area in its direct path but highly unstable fault zones on the two sides; one swarm of faults extends south all the way down to the Wasatch Range in Utah, the other passes right through here to the west.

In just a few more miles, we know we're licked. Fighting the wind and making three or four miles per hour just isn't worth it. With only 32 miles on the day and another 42 to our planned destination of Ennis, we stop at the Madison River cabins for the night. With the wind buffeting the cabin, we lounge and wait, hoping for better weather tomorrow and digging into the stockpile of old movies the owner keeps handy. We see *Some Like It Hot,* the first Marilyn Monroe movie for both of us, made in the very year of the Hebgen earthquake, and a less memorable 2002 *The Four Feathers.*

23. CATERPILLARS MARCHING

DAY 53, JUNE 25: MADISON RIVER CABINS TO ALDER, MONTANA

A tailwind! We have traded the challenge of yesterday's frigid headwind for the joy of an even colder tailwind, with the temperature somewhere down in the twenties. By 6 a.m. we're at the highway and ready for our 42-mile gradual descent along the Madison River all the way to Ennis.

Across the road, propped on the post for the small sign telling of our cabins, is a sapsucker, and he drums on the metal sign itself, *rat-a-tat-tat-tat-tat-tat-tat—tat—tat—tat.* "A tool-using woodpecker," I point out to David. "Why drum on an ordinary ole tree when one gets the special effects of a metal sign?" We linger long enough for a second drum and then, wearing every bit of clothing we have, launch into our promised downhill. When it's too late to turn back, I realize that I so enjoyed the drum that I forgot to see whether he was a Williamson's or red-naped sapsucker.

A wolf lies beside the road, and we pull up, though gradually accept the reality that it is not a wolf but "only" a large western coyote. He's *huge*, the reddish ears, snout, and forelegs adding an elegant beauty to his overall gray body. Nearby is another roadkill, a black-billed magpie, perhaps having come to inspect the coyote; another remarkable creature, in life so vocal and so strong a flier, so smart as corvids are, but now lying here lifeless, head, breast, back, and the bill, of course, ink black; wing patches and belly a snow white; wings and tail with a tinge of bluish-green iridescence.

We're jarred from our roadkill reverie by a wonderfully familiar bugling and trumpeting. Just to the west, with long neck outstretched and long legs trailing behind, a lone sandhill crane flies above the river. Down below a pair answer. The male calls first, his bugle deep and rich, pulsating and resonant, arising deep in the lungs; with precision breathing and muscle control, he exhales past the vibrating membranes in the voice box, the sound waves

Sandhill crane

Sandhill cranes. Part trumpeting, part bugling, mates sound off as one. (272, 0:22)

then further amplified and enriched as they pass through the specially elongated windpipe that coils through the breastbone, the oh-so-unique bugling finally bursting out the bill for all to hear. She quickly follows in kind but on a slightly higher pitch, so well timed that it sounds as if they are one bird, not two. Again they duet, and again, clearly informing the interloper that this space is taken.

More cranes bugle beside the river as we coast down this broad basin between the Gravely Range on our left and the Madison Range on our right. Just to the right we follow along an impressive hundred-foot-high gravel bank; it looks much like the flanks of an impressive glacial moraine, but it's more likely the slope up to the next terrace in this valley that was once the western glacial outwash from the great ice sheet over Yellowstone. Back in the Tetons we saw the same terraces along the Snake River, which was the outwash plain for the Yellowstone glaciers to the south. Halfway to Ennis, we climb up this bank and find ourselves awash in the songs of western meadowlarks; black cattle graze in green pastures, the impressive Madison Range with its snowcapped Sphinx and nearby Helmet as a backdrop.

"Basin and range," I chuckle out loud to David. "I've never really appreciated those textbook words until now. Right now we're riding the valley, or basin, between two mountain ranges—that's obvious. But if you had a satellite view from Oregon south to Texas, you'd see hundreds of valleys like this nestled between mountain ranges, all oriented north-south, much like an 'army of caterpillars marching toward Mexico.'" We are riding the Basin and Range Province of the geology texts. Somehow the earth's crust was stretched east to west, and it broke into long blocks of rock that then slowly tilted. The exposed edges of the blocks became the mountain ranges, the valleys between them the basins that gradually filled and flattened with eroded sediment. Over the next few days in Montana, we will bike up and over a number of these caterpillars and into the basin on the other side.

Two *eastern kingbirds*, unmistakable in flight, harass a raven who carries something in its bill, the two marauding tyrants driving the raven directly to the ground. He sits there in stunned silence. Their deed done, the kingbirds eventually relent and the raven flies to a nearby fence post to recover. It's nice to see eastern kingbirds again, the first since Kansas. Nearby a horned lark bubbles skyward, singing in a tight circle as he flies back to the fence post from which he came.

In Ennis, the "land of 773 people and 11,000,000 trout" according to the welcoming sign, we warm up with a second breakfast, this one of pancakes at a local café. Then, departing the basin of the Madison River, we head west, climbing the better part of a vertical half mile toward the pass between the Gravely Range to the south and the Tobacco Root Range to the north. In the calm, warming air of a fine Montana morning, our bodies rested from plenty of sleep last night and big sky above, we bask in all that is . . . in the sweeping views of the Madison River Valley that grow below us to the east . . . in song after song from vesper sparrows along the way, all of them beginning with the local dialect of two or three short whistles. "Isn't that nice!" I point out to David . . . in the songs of a rock wren who sits on a tall, tattered mullein plant on the bank above the road, silhouetted beautifully against the blue sky; he alternates two different songs in classic rock wren style . . . and we bask in the strength of our well-traveled, complaint-free legs that carry us and our loads up the mountain.

Horned lark. Day song, lacking dawn's fire. (273, 1:18)

Given the headwind, modest though it is, we choose to camp for the night in Alder after another fairly short day. It's taken almost two months, but I think I'm learning. Take smaller bites out of the continent each day, stop a little earlier, shorten the day, relax a little, and the trip will last longer. "You don't want to go back to work," coaches David. We sit and read. We throw the Frisbee over the large irrigated lawn. I think healing thoughts for my university workplace back home and for my Achilles tendon that I've nursed through Colorado and Wyoming.

Rock wren. Over a quarter hour, telling some of what he knows, in classic rock wren style. (274, 17:10)

The campground owner tells of his ancestor Sacagawea, the trusted Shoshone woman who accompanied Lewis and Clark, and also explains the ravaged landscape. "Twenty years after small-time miners had abandoned the stream here, the Conrey Placer Mining Company brought in a huge barge to dredge the gulch. The barge moved slowly forward through the very pond it gouged out in front of itself. It sifted through the rubble it devoured and dumped behind those long mounds of tailings a hundred feet wide. You can see what they did in 24 years. There's a good square mile overturned right here, and the tailings stretch all the way up the gulch to Virginia City." Yeah, we saw that on our ride in. Pretty ugly, pretty sad.

We settle into our "luxury site" beside a tree with two swings and a picnic table, next to the very pond where the gold dredge died about 80 years ago.

DAY 54, JUNE 26: ALDER TO DILLON, MONTANA

Bzeeyeer . . . bzeeyeer . . . a western wood-pewee awakens in a leisurely mood directly over our tent site, then moves to the willow tree nearby. I dress quickly, trying not to disturb David too much, and ease out of the tent. My plan is to depart early this morning to explore the campground and the Ruby Valley; David's plan is to let the sun's early rays invite him gently into the day, and then catch up with me later in Twin Bridges, 21 miles down the road.

I tiptoe over beneath the pewee and attach the microphone and recorder to a fence there so I can capture his dawn thoughts and take them with me when I leave. He sings his two songs at a blistering pace now, at least 40 songs per minute. Most of the songs feel paired, as if he asks a question on the rising *tswee-tee-teet* and then quickly answers with the falling *bzeeyeer*. Every 20 seconds or so he repeats his question, singing two *tswee-tee-teet*s back-to-back. And then perhaps once a minute he does something a little extra, offering an extra note, a soft and plaintive *peee,* mostly on one pitch and much lower than the louder *bzeeyeer*. Is that a rarely used "third song"? Do all western birds have such a song? What's on his mind?

Western wood-pewee. Awaking song, for him and for me, above our tent. (275, 13:26)

Throughout the pewee's performance, a chickadee sings *hey sweetie,* never pitch-shifting. Is he a mountain chickadee with a black-cap-like dialect, or a blueblood black-cap? He finally calls the robust *chick-a-dee-dee-dee* of the black-cap, not the wheezy call of the mountain—case closed. I wander around the campsite, listening to yellow warblers . . . house finches . . . the *wuwuwuwuwuwuwuwu* winnowing of snipe over the wetlands. . . .

Leaving David only the tent and his gear to pack up, I'm soon on the road, admiring the Ruby Caterpillar "marching" on the left, the Tobacco Root Caterpillar to the right as I squeeze between them to the northwest down the heart of the Ruby Basin. The chorus of birdsong in this rich, well-watered valley is exhilarating. I hear catbirds, the first since Missouri, some with fine mimicry, and eastern kingbirds, and so many other familiar voices: mourning doves, yellow warblers, yellowthroats, savannah and vesper sparrows, red-winged blackbirds, willow flycatchers, all singing from the alders and fence posts along the road. A ring-necked pheasant crows out there, cowbirds whistle from the utility poles, and from a dense spruce tree at the entrance to a farm, a billion house sparrows *chirrup*.

Wilson's snipe. Above the nearby wetland, skydiving snipe winnow *wuwuwuwuwuwuwuwu*. (276, 1:34)

A bobolink, the first of our trip. What a remarkable song. He begins with rich, low notes, the pitch rising higher and higher, accelerating and rollicking upward, and then he stampedes for another two or three seconds through an ecstatic jumble of contrasting notes on different pitches, working his two voice boxes for all he's worth. I bubble along with him, effervescing with the rhythm and overall flow of each song, then wait for the next, knowing that he, like all good bobolinks, should have two different songs. . . . Yes, I hear that now, as one of his songs has a series of harsher notes after about a second. I pause, waiting for him to launch into an excited song flight, to sing continuously on the wing . . . and, *yes*, he so obliges.

Gray catbird. An extraordinary mimic, telling of birds he has known. (277, 5:00)

Bobolink

Adding a distinctly western flavor to the chorus are western meadowlarks, black-billed magpies, lark sparrows, Bullock's orioles, and winnowing snipe. Those marsh wrens are western, too. And the irrigation—perhaps a hundred streams of water shoot into the air off to the east where the sun just now peeks over the Tobacco Roots. And the cattle, also "western." I stop beside a pasture, uttering my best "*MOOOOOO*." Surprisingly, they come running toward me from all over the field. A huge herd soon crowds the fence, their mooing intense! I try to take control of the situation, announcing in my best moo-speak, "I am your leader. Listen. People will eat you. Run for your lives." They're oblivious, and I leave them to their fate.

Bobolink. A spirited rollick, a jumbled effervescence, when perched and especially when airborne. (278, 3:12)

Black-billed magpies. In a loud, complaining voice. (279, 2:11)

By eight o'clock I've covered the 21 miles to Twin Bridges. Waiting for David, I gradually shed layers of clothing as the temperature climbs. From my pannier I remove the flicker that I had picked up beside the road earlier. At a distance or through binoculars, a flicker is a pretty bird, but in the hand it's fine artistry. I stretch out the wings and spread the tail, admiring the "red shafts" and feather linings, though they're more of a soft salmon than a red. In turn, I admire the stiffened tail feathers, the down-soft white rump patch, the off-white belly and breast with small black spots and crescents, the black bib, the grayish wash to the head contrasting with the brownish forehead and the intensely red moustache in this male, the grayish-brown back with dark brown bars from side to side, the "zygodactyl" toe arrangement with two in front, two in back. In admiration, I pluck a few tail and wing feathers to travel with me; his body I lay in the bushes beside the river.

Northern flicker

A stream of about 50 cyclists on shiny, skinny-wheel road bikes whizzes by, followed by a large truck; it's a Cycle America group, I learn, a rather cushy tour in which cyclists have a support team to transport luggage, prepare meals, repair bikes, and the like. The word "cheating" first comes to mind, as they're taking an all-too-easy way across the country, but I quickly move to "cheated," as they're missing so much in the group's hasty efficiency to cover the miles.

David eventually shows up with a new-found friend. Al is instantly likable; "I drive trucks," he says, "but only long enough to fund the next bike tour." Al is Johnny Cash in his "I've been everywhere, man" song, and David and Al ride side by side up ahead of me the 26 miles to Dillon. "Tell me about . . . ," and I lose David's voice in the wind. David is as eager to hear of Al's travels as Al is to share them.

I happily follow, catching bits of the conversation, but largely content to just be. Above the Ruby Range to our left I follow the ever-changing cloud patterns. I marvel at the irrigation pipes that slowly sweep over the lush fields, dousing circular plots a half mile across. I watch the meandering Beaverhead River, tracing in my mind how it flows behind us to Twin Bridges, where it joins the Big Hole River and Ruby River to form the Jefferson. All of these western rivers drain the nearby ranges through the intervening basins and, surprisingly, eventually empty to the east into the Missouri, the Mississippi, and the Gulf of Mexico.

We pause at the stump of a mountain on the right, Beaverhead Rock, so named by native Americans because it looks like the head of a swimming beaver. In August of 1805, when the Lewis and Clark Expedition was in some peril, Sacagawea gave them hope at this very spot. She recognized this rock formation and knew she was near her Shoshone people, who would supply the horses needed for the climb over the mountain passes before winter. By the 1860s, with gold discovered nearby, a saloon, hotel, and stage station sprang up here, making this a heavily traveled thoroughfare. We travel through history along current-day State Route 41, celebrated as part of the "Lewis and Clark National Historic Trail."

In Dillon, Al quickly downs two malts after David explains how he's taken to them lately. "OK, so I'm a bit impulsive!" Al admits. Later, at the campsite just outside of town, Al eats his cold chili straight from the can, but my gourmet chef whips up yet another carbo-loading pasta dinner.

DAY 55, JUNE 27: DILLON TO WISDOM, MONTANA

Through the night a great horned owl calls off to the south, and snipe winnow, too. To the tunes of yellow-headed blackbirds and both eastern and western kingbirds, David and I dine on our oatmeal as Al departs. "See you down the road," we call out.

Yellow-headed blackbirds. Harsh and raucous, such laborious, strangled strains, but a delight. (280, 4:32)

"A big day lies ahead. We'll climb two passes and drop into two more valleys, *and* I am going to clip into my pedals today." Since Eads, Colorado, many miles and almost three weeks ago, I have babied my Achilles tendon, but I now swap out my platform pedals for the clipless pedals, raise the saddle a little, and smile at the thought of a more efficient stroke, pushing from the ball of the foot rather than the heel. "Feels great," I yell back during a test ride in the parking lot.

We've left the lush basins of the Ruby and Beaverhead Rivers and are in arid sagebrush country once again, with western meadowlarks and vesper and savannah sparrows abundant. From a willow grove on the right a house wren bubbles; at the next farmhouse a least flycatcher *che-BEKs* and a young crow caws hoarsely as it begs for food; a cowbird sings, then whistles from a utility pole just before flying off to the north toward snowcapped Torrey Peak, only 15 miles distant and towering over 11,000 feet in yet another mountain range, the Pioneers.

Brown-headed cowbird. Villain or evolutionary marvel? Songs, then departing flight whistle. (281, 1:40)

Off to my left, my shadow extends to the far side of the road, and I see myself leaning forward over the handlebars, the bicycle frame and two wheels and four panniers below, tent and bedroll on the rear rack, my legs pumping. It's a beautiful sight, and with a pain-free Achilles, I'm flying! From a passing pickup comes a friendly *beep beep* and a wave.

How fascinating the fence posts here. Several generations of them line the road, metal posts and wooden ones of various sizes and shapes, each style no doubt telling some story of the land. Among the fence posts, two sage thrashers rise up and grapple in song over the sagebrush, one quickly departing in undulating song flight over the small ridge to the right.

In only 14 miles we climb and cross over Badger Pass, past a few small conifer islands that have taken hold among the sage at this higher, cooler elevation, and then we swoop down over a five-mile descent into Grasshopper Valley. For half an hour we bike beside Grasshopper Creek, stopping only to chat with five cyclists who are headed east; they're jugglers, comedians, and magicians, they say, good fodder for a modern-day Chaucer and another *Canterbury Tales*, all on a cross-country journey doing road shows to raise money for Shriners Hospitals. Soon we begin to climb again, more steeply this time, rising in an hour to Big Hole Pass at 7360 feet, slipping between the rump of one caterpillar and the head of the next.

Opening about a thousand feet below is an immense valley, the Big Hole, "The Valley of 10,000 Haystacks" our map proclaims, and beyond it lie the snow-covered Beaverhead Mountains. Eager for what we have earned, we begin our descent, coasting at first and then pedaling gently for mile after mile, 11 in all as we glide down to the town of Jackson on the valley floor. Population, 38, says our map. From the restaurant there, between bites of my pancakes smothered with ice cream and granola and whipped cream, I call ahead to the motel in Wisdom. We hope to avoid camping among the swarms of mosquitoes that have emerged, and I'm eager for a good night's sleep before my big ride the next morning. Success! We get the last room, just after Al, who's already there.

Outside town we are surrounded by cattle that are being driven down the road, channeled by a fence on each side. Together with a few much quieter men, a surprisingly well-dressed woman on horseback whoops and hollers as she encourages the cattle along. David pulls out the rope we use to bear-proof our food at campsites and forms a lasso; he enters the fray by bike, pretending to be part of the drive. "From Virginia, heading to Oregon," I yell to the real cowboy in response to his question. We bide our time, taking in the big scene in the Big Hole beneath the Big Sky, hemmed in by the Beaverhead Mountains to the left and the Pioneers to the right.

The cattle eventually part, and we ride through and among them, thankful for our bicycle fenders that fend off the abundant, fresh cow pies covering the road. And in parting, with a friendly tip of the hat, the last man on horseback smiles, "Have a nice day—eat more beef."

Cattle. *The West,* Home of Big Beef on the Range. (282, 1:16)

This Big Hole looks like just another "basin," but by the best guesses of geologists it's no ordinary basin in this "basin and range" province. Instead, as the North American tectonic plate slid over the Pacific plate, a huge welt formed deep within the earth along the edge of the old continent. Perhaps 15 miles down, over an area about 75 miles wide and 200 miles long, molten granite formed, then rose and solidified about ten miles below the surface. As this giant mass of solid granite (geologists call it the "Idaho batholith") continued to rise, beginning roughly 50 to 70 million years ago, unimaginably huge slabs of rock above it slid off to the side. I look to the right and see one of those slabs, now the Pioneer Mountains, perhaps ten miles deep and covering more than 1000 square miles. In sliding the 50 miles east from its source, this block opened up a huge chasm, the Big Hole, whose bedrock lies almost three miles below us. During and since the Big Move, the valley between the emerging Idaho batholith to the left and the displaced Pioneer Mountains to the right has filled with almost everything one could imagine, beginning with volcanic rocks in the basement, then a mess of mud, sand, gravel, volcanic ash, limestone, even coal, the valley floor from Jackson to Wisdom now flatter than Kansas.

In this broad valley, the abundant snowmelt provides the water to keep things green, and cattle rule. At this time of year, during the season of irrigation, mosquitoes co-rule. We're grateful for the slight headwind, and we keep just ahead of mosquito ground speed. As a test, I turn to bike with the wind and the mosquitoes are on me in a flash. "No flat tire, please," I request of my bike, "and thank you, headwind."

In Wisdom, we find Al at the local bar. Jawing with the locals over a beer is to him what listening to birds is to me. We eat at the bar, too, but I'm eager to get some sleep, because morning for me will come earlier than it has on the entire trip.

24. CHIEF JOSEPH PASS

DAY 56, JUNE 28: WISDOM TO STEVENSVILLE, MONTANA

"3:00 a.m. 6–28"—so declares the lighted watch face when I punch my alarm off. I ease out of bed, leaving David asleep, and a little before 3:30, with sunrise over two hours away, I'm out the door and on my fully loaded bike, heading west out of Wisdom. At whatever speed I choose this morning, stopping where I please to look and listen, I will play my way across the sagebrush from here to the forest 11 miles away, and then ascend through the Beaverhead-Deerlodge National Forest to the top of Chief Joseph Pass, 25 miles to the west. After recovering from his night out on the town, David will awake and leave when he pleases, and if he hasn't caught me by the pass, I'll wait for him there.

With no moon, the constellations burn intensely against the still-black sky, the Milky Way a brilliant path through the heavens. Off to the right lies the North Star, and I trace the Big Dipper, the Little Dipper, and the Dragon, just for starters. My eyes then sweep across the heavens to spot old friends who fly there, especially the Swan. Mars glows red off to the east; Venus and Jupiter are below the horizon.

As I cross the Big Hole River just out of town, a lone marsh wren sings lazily off to the right, a western marsh wren, with all of the harsh buzzes and rattles and whistles one would expect from his more than 100 different songs. What's in his head to make him sing now in the dead of night? Maybe he's a young bachelor who sings for a mate, or perhaps an older male who already has one or two females in his "harem" and is greedy for more. Flashing through my mind are the thirty-some years I have spent trudging through marshes across the continent from the Atlantic to the Pacific. I've recorded the songs of these marsh wrens, caught them in nets so I could put colored bands on their legs, and followed many as

Wilson's snipe

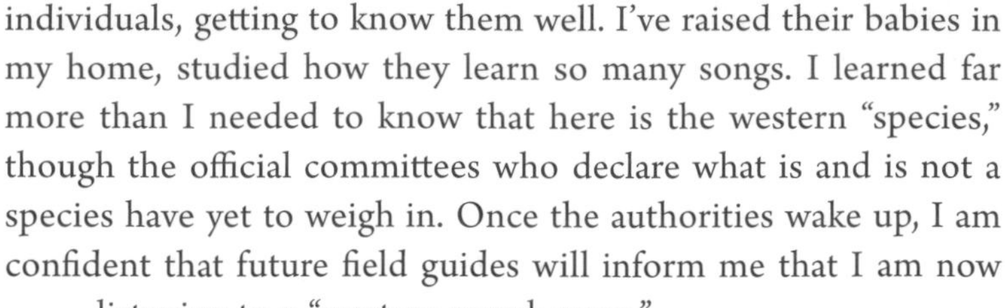

individuals, getting to know them well. I've raised their babies in my home, studied how they learn so many songs. I learned far more than I needed to know that here is the western "species," though the official committees who declare what is and is not a species have yet to weigh in. Once the authorities wake up, I am confident that future field guides will inform me that I am now listening to a "western marsh wren."

Wilson's snipe. Perched, he calls, *tick-tock, tick-tock,* sort of. (283, 5:02)

From beside the road a startled killdeer flushes, *kill-deer kill-deer kill-deer*. High overhead, snipe winnow in earnest, perhaps a dozen of them within earshot, unseen meteors hurtling earthward with the wind whistling past their wings and through their stiffened outer tail feathers, *wuwuwuwuwu-wuwuwu*; some call intensely from the ground, too, *tick-tock, tick-tock, tick-tock tick tick tick . . . .*

To the north, beacons of shimmering, greenish light now dance up from the horizon, engulfing the Big Dipper and all of the Great Bear, rising to the North Star and the Little Dipper, consuming even the Dragon's head. Shifting to the left, then to the right, now a half crown, then a full ghostly halo hovers over the earth's pole, rising, then falling, pulsating as if alive. The northern lights, I smile, the aurora borealis, what a gift now in late June.

The constellation of lights that is Wisdom recedes behind me, and I cruise along at a modest six miles per hour, listening . . . to my breathing, as I'm climbing a gentle grade . . . to the concert of two tires rolling on pavement . . . to the faint squeak of my left shoe on each revolution of the crank arm . . . to the sound of darkness swishing by my ears . . . to my voice whispered into the night, "What a spectacular time of day to ride!"

A wire fence creaks violently on its posts just off the road to the right. A grizzly first comes to mind, but my beacon of light shows an antelope with its horns caught in the fence. I want to help, yet know how dangerous that could be. He'll work it out, I hope.

The sound of sagebrush awaking soon extends in all directions, and a little after four I stop to listen. The sage is awash with the sound of a thousand tinkling bells, horned larks in dawn song; a half dozen sage thrashers add their night songs to the music. Coyotes yip and howl in chorus off to the north. Far away, a lone cow bellows, but something nearby snorts just across the fence. A bit spooked, with my light insufficient to reveal the source, I quickly move on, past a mountain bluebird singing aloft to my left.

With just enough light to see, I pull into the overlook at the Big Hole National Battlefield. In the serenity of this dawn, I strain to imagine the mayhem here on another dawn, on the ninth of August back in 1877. Chief Joseph and roughly 750 Nez Perce Indians are camped down below, where the skeletons of tepees now stand in memoriam along the river. Horses graze nearby in the lush meadows. In a surprise attack, the U.S. Army soldiers first disperse the horses and then advance on the teepees, firing low into them. Fierce fighting erupts, and the soldiers are forced to retreat to the grove of trees that I see just above the river, where sniper fire from the Indians pins them down.

Horned lark. Ode to a Horned Lark, an Autobiography at Dawn. (284, 6:04)

Other soldiers pull a howitzer to the far hillside, intending to rain hell on the Indians below, but the gun is captured almost immediately by the Indians. As snipers contain the soldiers through the day and into the night, the Nez Perce bury their dead and depart, continuing their flight from where they began in Oregon to where they'll eventually surrender just short of Canada almost two months from now.

Those are but a few of the cold facts, but they were anything but cold to Colonel John Gibbon, the army's besieged commander among those trees above the river:

> *Few of us will soon forget the wail of mingled grief, rage, and horror which came from the camp four or five hundred yards from us when the Indians returned to it and recognized their slaughtered warriors, women and children. Above this wail of horror we could hear the passionate appeal of the leaders urging their followers to fight, and the war whoops in answer which bodes us no good.*

Sobered, I bike on over the river, listening to songs that the Nez Perce themselves probably heard on that fateful dawn. Sandhill cranes sound off, he bugling low, she finishing the duet high. Willow flycatchers and yellow warblers add to the chorus, as do song sparrows and white-crowned sparrows. There's the *wich-i-ty wich-i-ty* of the yellowthroat, the caroling of robins, the winnowing of snipe, and the downward-spiraling song of a veery, the first I've heard since the rooftop of Virginia. From all along the stream bottom I savor the first northern waterthrushes I've heard on this trip, each singing his particular version of *sweet sweet sweet swee wee wee chew chew chew chew,* a rapid, emphatic song, always three-parted, always descending from one phrase to the next.

FROM THE BIG HOLE NATIONAL BATTLEFIELD

American Robin. On a tear with carols and *hissellys*, 120 songs a minute! (285, 2:09)

White-crowned sparrow. Unique Big Hole dialect. (286, 2:33)

Veery, Canada geese, northern waterthrush. Symphony beside the Big Hole River. (287, 2:13)

Northern waterthrush, black-capped chickadee, yellow warbler. More symphony. (288, 2:09)

MacGillivray's warbler, with **Clark's nutcracker**. The mere act of Listening takes you There. (289, 0:36)

Just beyond this North Fork of the Big Hole River lies the forest and the beginning of my gradual ascent to the pass. The birds here are in full song, though it suddenly feels so much colder, and I add even more layers of clothing.

How empty a dawn would be without the energy and enthusiasm of flycatchers. I cycle past pewee after pewee, each rapidly alternating his two songs, *tswee-tee-teet . . . bzeeyeer . . . tswee-tee-teet . . . bzeeyeer. . . .* Dusky flycatchers hustle through their three dawn songs, as do their *Empidonax* cousins the willow flycatchers. How odd, then, the *quick! THREE BEERS!* from an olive-sided flycatcher; maybe he sings a little more excitedly at dawn than later in the day, but he's an odd flycatcher with no special song to greet the day. And just how, I wonder, are these songs encoded in the genes so that young flycatchers don't need to learn their songs from singing adults?

Overhead are the repeated *peent*s of a nighthawk with his occasional booming *VROOM*. I walk the bike beneath a dusky flycatcher, recording his effort at close range. Then it's a pewee, a white-crowned sparrow, a northern waterthrush, a song sparrow, a western tanager awaking with his fractured songs regularly punctuated by his *pit-i-tuck* call note. A doe snorts loudly and repeatedly at me, her fawn just off the road up ahead.

And there's that bubbling house wren–like song from the mystery bird that I've not yet identified! I have my suspicions, and walk closer, eventually finding myself face to face with none other than a Lincoln's sparrow. He sings in plain view from an exposed perch beside rushing and tumbling Trail Creek, and is unmistakable with his subtle yet distinctive facial pattern, together with fine black stripes on a buffy breast. I should have known all along, I tell myself, but I cut myself some slack, as I've never heard this sparrow before that encounter back at the Blue River campground in Colorado.

Lincoln's sparrow. Bubbling energetically through his complex song. (290, 1:01)

About 5:15 now, directly down the road behind me I see a sliver of a moon rising, Venus, its companion, just behind. I continue west, past warbling vireos, chipping sparrows, MacGillivray's warblers, waterthrushes, juncos, and a whistled song so sweet but so foreign. I imitate it, trying to burn it into memory, guessing pine grosbeak but unsure. A startled group of elk hustle across the creek and into the woods. Slowly I meander up the road, "the most beautiful place I have ever been," I say to myself. I stop to listen where I will, especially in the first beam of sunshine that reaches me, as I am cold cold cold. I linger and listen, amazed at what a little sun can do for warmth, relishing the red-naped sapsuckers who have chosen to drum and call nearby.

Trail Creek now flows in from the right, draining a valley recognized as part of the Lewis and Clark and Nez Perce National Historic Trail. It's a possible alternate route for us over dirt and gravel Forest Service roads, but I continue on highway 43, soon braking to a smooth stop.

Red-naped sapsuckers. Three calling birds . . . with drums. (291, 5:19)

"A *great gray owl*," I whisper into my voice recorder. "Just ahead on a small post across the road. He's *huge*. Look at that head, so round and enormous; two large facial discs beautifully lined by concentric half circles, with small, beady yellow eyes at their centers; raised, white boomerang-shaped feather tufts separate those eyes. Just below his yellow bill is a black chin, with snow-white moustaches extending out to the side, looking much like a black-and-white bowtie; heavy streaking below, from head to tail; feathered legs, impressive talons emerging to grip the post. He's great and an owl, sort of grayish, but with lots of subtle patterns in whites and browns and blacks. What an extraordinary creature!" As I try to capture the moment, he glares directly at me, no doubt trying to fathom in his own way the odd creature he sees. Perhaps it is only five minutes later, maybe an eternity, when he lifts into the air ever so silently, a phantom disappearing into the forest.

I climb more steeply now, perhaps on a 3% grade. With about two miles to the pass, the higher elevation is announced by mountain chickadees and ruby-crowned kinglets. With less than a mile to go, in a lush fir and pine forest, hermit and Swainson's thrushes offer some of the finest songs on the planet.

I move on, but yet another thrush soon freezes me beside the road—a Townsend's solitaire! From the treetops out of sight to the right, his songs ripple out over the mountain, clear warbled notes delivered in a great rush for seconds on end, and then he rests, but only briefly before he's at it again. Those who tell of this song gush in superlatives: most glorious, most beautiful, the richest and fullest and clearest and sweetest and sparklingest, the finest mountain music; I smile at the understated "long rich warble" that Kaufman dares in his field guide. The solitaire is up there somewhere, perched erect as can be on the very tip of a fir tree, his songs such a contrast to his somber gray appearance. What a fine view he must have from above as he surveys his forest domain of green spires.

Townsend's solitaire. The song gushes in superlatives, accompanied by a chorus of coyotes. (292, 3:48)

Thrushes, I say—what marvelous singers. I tick them off for today: the robins early, a mountain bluebird, the veery back at the battlefield, rising to the hermit and Swainson's thrushes, maybe a varied thrush, and now the solitaire.

Just short of the sign announcing Chief Joseph Pass I am stopped short by a winter wren. "Mr. *Western* Winter Wren," I hail him, "just listen to you go!" I last imagined them atop Mount Rogers in Virginia, but that was the eastern bird; this western bird is an even more remarkable singer, so different from the eastern winter wren that I know they'll eventually be recognized as two species. I said as much back in 1980 when I published my first paper on them and showed how different their songs were.

Pacific wren

Pacific wren. Like a speeding, ecstatic eastern winter wren, from here to the Pacific! (293, 7:33)

And here comes David up the hill! "Buenos días, señor," I greet him, though the words would be an understatement in any language.

"You've been taking your time," he says.

I laugh, and we compare our versions of the 25 miles we've just biked. "Left about seven o'clock, letting the sun ease me out of bed," he smiles, "a perfect morning. The battlefield was a sad place. Great forest." I tell him some of what I've heard, especially in the last mile.

"I heard a hermit thrush," he offers.

"Hear this Pacific wren, in the forest? . . . the high, tinkling sound?"

"Yeah. What's that all about?"

"Oh, the most spectacular song in North America. It goes on and on and on, maybe ten seconds, and he's very creative in taking segments of his song and rearranging them, making all new songs." I describe how the songs are so high and fast that our pathetic human ears can't process the sounds, but slow the song down on a computer and then we can hear how the wren leaps about among his tiny, tinkling notes. We know these wrens can hear the details at full speed, because they imitate each other, so that neighbors have similar songs. "Neat bird. And he's only yea big," as I gesture with thumb and forefinger.

"Cool . . . but how can all birds be the most spectacular?"

"It's just the way it is," I explain fully, "and your best listening is yet to come, in the rain forest of Oregon's Coast Range and all along the Pacific."

Finally, Chief Joseph Pass, at 7241 feet yet another crossing of the continental divide. It's a little after nine; over the last six hours I've averaged about four miles per hour, a brisk walking pace. A most perfect morning. Overhead, a flock of red crossbills call as they bounce along in the sunshine among perfect spires of subalpine fir. Yes, most perfect.

We're joined by Al at the top and soon begin our descent into yet another valley, the Bitterroot. In the 14 downhill miles to Sula, we race through rugged terrain and beautiful forests, braking hard to make some of the sharp turns, dropping half a mile in elevation and stopping only to admire a herd of elk grazing in a meadow high on a hillside to the left. I hear lazuli buntings and spotted towhees again—the first since we entered the Rockies many days ago in Colorado.

Now following the Bitterroot River north, we descend for another 18 miles down the road to Darby, stopping in awe of glacier-carved Trapper Peak off to the left, the high point of the Bitterroot Range. On the other side of this Bitterroot Valley to our right are the sedimentary Sapphire Mountains, another one of those massive blocks of the earth's crust that slid east off the rising granite range that is now called the Bitterroots. And we ride the fissure that opened between the two ranges. In Darby, we brake for a second breakfast, and in the warming weather I tackle a milk shake, together with pancakes and ice cream.

From Darby, it's another 17 miles through the valley to Hamilton, and along the way we stop to swap stories with half a dozen eastbound cyclists. Al will spend the night in Hamilton, as he has a brewpub to visit. We bike the 22 miles further to Stevensville, as I have planned an early-morning outing at the Lee Metcalfe National Wildlife Refuge just outside of town. With nearly 100 miles on our odometers for the day, we check into the Stevensville Hotel, exhausted yet exhilarated.

25.
LEWIS AND CLARK

DAY 57, JUNE 29: STEVENSVILLE TO LEE CREEK CAMPGROUND, MONTANA

By streetlight, I navigate the roads of sleeping Stevensville down Church Street to the Eastside Highway, then to Wildfowl Lane, where my bicycle light shines the way out to the refuge.

Along the lane, I'm spooked by the ghostly image of something large floating across the narrow beam of my headlight. Stopping, I shine my light into the trees, and there, perched on the naked limbs of a dead cottonwood, are *three* great horned owls. They must be young birds, but they're so quiet. "Give me some of your hair-raising screeches," I request quietly, their youthful sounds in the night enough to scare the bejesus out of unknowing listeners—but apparently they're well fed and content.

Fledgling great horned owls

I bike on, and a good hour before sunrise stop beside the cattails near the visitor center. Two marsh wrens duel just off the road, each matching the other song for song—"take that, and that, and how about this," I hear them say, hurling songs back and forth every few seconds, each challenging the other with a bewildering variety of more than a hundred songs that they've learned from their neighbors in this singing community.

Marsh wrens. Matched countersinging, endless song variety, in haste—westerners in all their wonder. (294, 0:86)

What a contrast to the yellowthroat who sings nearby; he's also a songbird, but he learns just one song, his particular version of the *wich-i-ty wich-i-ty* sufficing him for life. Each in his own way, I reflect, each an unqualified success. A lone swallow darts about over the open water . . . bullfrogs everywhere moan and groan . . . two ducks in tight formation

Common yellow-throat. Singing *y-wichy y-wichy*, his personal version of *witchity-wichity*. (295, 2:15)

whirl by overhead . . . deer snort in the open field to the south . . . snipe winnow overhead and call *TIK-a TIK -a* from the marsh . . . a rooster crows in the distance. From the cattails, a sora whinnies, and as he calls I imagine the beautiful flow of sound across a sonogram, tiny whistled notes rising briefly at the outset, then descending sharply and lengthening, the longer whistles trailing off. Surely a Virginia rail is here as well, and I await his full *kick-kick-kick-kidick-kidick-kidick-kidick*, but the sora seems to be alone.

In the open water beyond the cattails, the silhouette of a lone duck circles, calling repeatedly with three whistled notes, the second one emphasized. Perplexed, I pull out the field guide to study the range maps and read voice descriptions, and quickly find the American wigeon, who has a "far-carrying whistle, *whee-WHEEW-wheew*, second note highest." That's a fine sound for a duck.

American wigeon. An airy whistled tune, so fine for a duck! (296, 1:13)

Standing beside the yellowthroat and the untiring marsh wrens, I sweep my ears out over the marsh: red-winged blackbirds sing, *konk-la-reeeeeee* . . . a pied-billed grebe *brays* far to the north . . . an American bittern, also to the north, *pump-er-LUNK, pump-er-LUNK* . . . an osprey, with its loud *keyew keyew* . . . several roosters crow. My eyes linger to the west on the snowcapped peaks of the Bitterroots, now beaming in the first rays of the morning's sun.

Osprey. A loud complaining *keyew keyew*, from near its nesting platform. (297, 2:06)

Around six o'clock I head back to the motel and crawl into bed. Restless and energized, unable to sleep, I assure myself that it'll be a short day, only about 50 miles to our planned stop at the campground just below Lolo Pass, our last stop in Montana.

By nine we're on our way, but just outside of town I stop to admire an intense patch of yellow on the road: a male yellow warbler, a beam of sunshine on black asphalt, his intense rusty streaks streaming down his breast, feathers ruffled, tiny feet clenched, bill open, eyes receded, silent.

Twenty miles down the road we stop at a mini-market in Lolo, only to find our biking companion Al, who has already biked 40 miles this morning from Hamilton. We restock our panniers with calories and then head out, taking the turn to the west, knowing that we'll be rising on a gentle slope toward Idaho for the next 30 miles.

Not far out of town is a roadside sign overlooking Lolo Mountain: "The Lewis and Clark Expedition camped here on September 11, 1805." It's humbling to think how we, with a slight tailwind, almost fly uphill on paved roads along Lolo Creek, up and into the heart of the Bitterroot Mountains, when 200 years ago Lewis and Clark had a far more arduous task.

As we climb, the mountain chickadees confirm our progress, having replaced the black-capped chickadees of the valley. Leaving little doubt that we are now in the "far West," songs of spotted towhees here are a simple trill with perhaps a tiny, barely noticeable introductory note, much as they will be all the way to the Pacific; long gone are the multiple introductory *drink* notes of those mountain spotted towhees in the Rockies outside of Pueblo, Colorado.

"The Lewis and Clark Expedition camped here on September 12, 1805" reads the next wayside sign; "Along Lolo Creek and over the Bitterroot Mountains, Lewis and Clark recorded several animals they'd never seen before: ruffed grouse, spruce grouse, mourning dove, Steller's jay and the broad-tailed hummingbird." Perhaps it is Lewis on the sign who crouches on one knee in the foreground, notebook in hand and looking down at the ground, with an oversized spruce grouse and Lewis's woodpecker looming over him, and with diminutive yet full-grown trees that had been documented along the way in the background. The sign is all out of proportion, the birds comically larger than Lewis and twice the height of the grand fir, but, as I point out to David and Al, "They got it right. Birds are big."

Spotted towhee. With tiny introductory *drink* notes, such simple songs from here to the Pacific. (298, 1:34)

Good scientific information—that's what President Jefferson told Lewis and Clark he wanted out of this expedition, and they carefully described a number of bird species. Several were new to science, at least "white man's science," and the preeminent ornithologist of the time, Alexander Wilson, named some of them after the leaders of the Corps of Discovery. Hence, we have Clark's nutcracker, the very nutcracker who is probably responsible for planting some of the patches of forest high on the slopes above us here, and Lewis's woodpecker, who calls and drums beside the road on occasion.

Past Lolo Hot Springs with just eight miles to Lolo Pass, we encounter special ground: We leave behind the sedimentary landfill that we've been riding on through the Bitterroot Valley and creep out onto the exposed Idaho batholith itself. Here is that very granite that solidified deep in the earth and then rose, pushing aside to the east what we now call the Sapphire and Pioneer Mountains and creating the Bitterroot and Big Hole Valleys.

Lewis's woodpecker. Named after Meriwether of the Corps of Discovery—drums and calls. (299, 1:42)

By late afternoon, at site 22 in Lee Creek Campground, with a secure bear box nearby holding our food, we are nestled among the lodgepoles with Ponderosa pines reaching beyond them to the sky. Al is in the next site over. It was an easy day, as we thought it would be, and we'll save for tomorrow morning the brief climb to Lolo Pass and our first listen to Idaho.

DAY 58, JUNE 30: LEE CREEK CAMPGROUND, MONTANA, TO LOWELL, IDAHO, VIA LOLO PASS

Just a few miles out of the campground we begin a serious climb, perhaps a 7% grade. But a slow climb is good for listening, a fast downhill with wind whistling in the ears bad. And just a little after sunrise now, with Al up front and David behind, I'm all ears.

Pacific wren. A true maestro (300, 5:01), the song especially impressive when slowed 4x for our non-bird ears. (301, 0:32)

"David, hear the Pacific wrens I told you about up on Chief Joseph Pass? They line the stream here, bird after bird."

"Yeah, and hermit thrushes and Swenson's thrushes," he calls back. I let the "Swenson" go—close enough. He's listening, and I smile.

The upper half of the mountain before us has a reddish tinge in the sun's early rays, the lower half dark green in the shade. To the sound of heavy breathing and bikes rolling on pavement, and to the strange sight of brightly multicolored us moving by, a red squirrel barks and chatters from a tree just to the left. Songs of Swainson's thrushes rise to the heavens, as we do, as the trees do. Juncos everywhere trill with such variety, and a Townsend's solitaire rips off a sweet, spirited warble from high overhead.

Nashville warbler. Easterner at western outpost: typical two-parted *seebit seebit seebit titititititi.* (302, 2:34)

An oh-so-familiar song hits me from the right—what is that two-parted song, a sweet, slow trill followed by a faster, lower chatter? Nashville warbler? Yes, it has to be, as there's nothing else like it. At the "WELCOME TO IDAHO" sign, we stop and I quickly check the guide. Yes, Kaufman concurs: "sweet double notes followed by faster trill, *seeba seeba seeba tetetetetetetetetetteh,*" and the range map matches beautifully, showing that we can first expect to encounter Nashville warblers right here on the border of Idaho.

There's a brown creeper, *trees trees see-the trees,* two high notes, two quick low notes, ending high again. And a golden-crowned kinglet, yet another first for the trip! All he offers is his string of really high notes at the beginning of his song, *see-see-see*, but they're a dead giveaway. I feel I should be shouting all this from the treetops, and deep within I do, but outside I simply smile, wave my arms, and understate the situation to David and Al: "Wonderful birds singing here."

"ENTERING PACIFIC TIME ZONE" declares the sign. Next, "LOLO SUMMIT." I imagine Lewis and Clark just off in the woods to the right, crossing this pass back on September 13, 1805. Over the next week, as they struggled with little food over rocky ridges and through dense forest with downed trees littering the ground, they would face some of the toughest and most perilous conditions of their entire journey. We, however, will sail downhill on paved road along one of the most wild and scenic rivers in the world *for the next one hundred miles*.

West of the divide between Montana and Idaho, drops and trickles and rivulets from the watershed here soon merge to form a roaring river that we swoop down to meet. A rainbow flies across the road in front of me, a male western tanager, the red and orange and yellow aglow in the rays of the early morning sun that shines over my shoulder. And a Steller's jay, too, *shook shook shook* he calls, flying so close I see the details, the black crest and head giving way to a sky-blue belly, a grayish back, the deep, rich blue of the wings flashing with each stroke. In this light, he's radiant, iridescent even; the birds here are as colorful as found in any lush rain forest, as if to remind us that we are no longer in the rain shadow of the Bitterroots.

In a flash we cover the ten miles to a stand of old-growth western red cedars that tower over the road. "*THIS IS THE WEST*," they shout, "*the home of trees tall beyond measure, of cedars and redwoods and sequoias and Douglas firs and more.*"

Steller's jay

Steller's jay. Why this particular call now, when he has so many others? (303, 1:04)

And the home of special birds. From somewhere deep in these cedars, a varied thrush sings. I have marveled at these songs on my computer, where I've played endlessly, slowing them down and speeding them up, listening to long sequences and trying to understand the overall effect. I follow along now: the first song is high and buzzy; the next is lower pitched, more pleasing to the ear; next he's high again; always, successive songs are different, the contrast striking. He plays with only four different songs, but each is actually two in one, because he sings two different songs simultaneously with his two voice boxes. "Try whistling and humming at the same time," I suggest to David, "and you get the overall effect of what these thrushes are doing." I listen on, mesmerized, the effect

Varied thrush. Try whistling and humming simultaneously, briefly, then change the pitch, and repeat. (304, 10:22)

all the better knowing that Meriwether Lewis first heard and described these varied thrushes near here two centuries ago.

From up on the hillside above the road, the exact direction difficult to know but not all that distant, deep in my chest I *feel* a ruffed grouse drum. He's atop a fallen log, I bet, one like so many obstacles that made it difficult for Lewis and Clark to scramble through here. He adjusts his position on the log just right, bracing himself with his tail, and flaps those wings at light speed, generating miniature sonic booms and energy at such a low frequency that I feel it resonate in my chest better than I hear it with my ears.

And the rhythm? Yes, I know that rhythm; there almost certainly are two soft initial beats that I'm too distant to hear, then those four louder beats preceding the gradual rise to a climax of over 20 wing beats per second, the drum then tapering off, the rise and fall symmetrical around the climax. He's the same ruffed grouse that I know from back home on New England, but somewhere beyond Idaho, I know, is a different ruffed grouse with an entirely different drum, one that I wager will eventually be described as a second species. I've heard him on Vancouver Island in British Columbia, but how far to the east he drums is unknown.

Ruffed grouse

At the Lochsa Lodge in Powell, the three of us find Keith walking his bike out to the road. While we camped last night, Keith stayed here at the lodge, his camping gear still unused. "You'll catch up and pass me like I'm standing still," comments Keith as he leaves, a parting from him that we've heard before.

Ruffed grouse. The eastern "species": One drum (305, 0:17) or 36 over two hours. (306, 2:02:35)

Tension builds as we wait at the lodge. David wants to stock up on food, given that there are no services until we reach Lowell 66 miles down the road, but the store is closed and we're unsure when it will open. I want to move on, to take advantage of the excellent riding conditions in the morning before heat and wind become a factor. "Too much worrying," David remarks. Al looks on in amusement as David gets out his Frisbee, tossing it to me and trying to lighten the mood. There is a history of food shortage here, I remind myself, as just up the road is Colt Killed Creek, where Lewis and Clark were forced to kill one of their young horses for food. But we have plenty, Al and I come to agree, so we soon take off, relying on whatever calories we carried over the pass ourselves.

Half an hour later we spot Keith biking slowly up ahead, clearly having the time of his retired life. Conspiring, we form a paceline, building speed, and with heads down we race by Keith as if he were standing in the road. We slow, then stop at a pull-off, laughing with the amiable Brit as he pulls up.

We ride on as a foursome, downhill beside the Lochsa River. The federal designation as "wild and scenic" is an understatement, as mile after mile of wild, untamed water flows beside us with continuous rapids for miles on end. Small feeder streams cascade into the Lochsa, which cuts sharply into the canyon here, the river conspiring with gravity to cut its bed through a landscape that is still rising.

At our next stop, a fox sparrow sings beside the river. Given that we are in Idaho, he's of the "slate-colored" group, one of several different species now lurking under the general name of "fox sparrow" in our field guides. What a rich medley of slurred whistles and buzzy elements, sounding much like an especially accomplished song sparrow. He alternates two songs now, first one, then the other, back and forth, one song nicely ending with the flicker's *klee-yer!* call, the other with a robin's *tut* note, those ending notes so typical of many fox sparrow songs. I listen on, expecting other songs, but he's stuck on these two; that's odd, I think, because a three- or four-song performance is more typical, with all the songs delivered in an orderly sequence before he starts over.

Fox sparrow. The slate-colored western variety, like a super song sparrow. (307, 10:05)

Biking on, I hear a new bird, the buzzy nature of the opening elements so much like that of the black-throated blue and

Townsend's warbler. With all the buzziness of his black-throated relatives. (308, 2:14)

black-throated green warblers I know back East. I catch it as *zree zree zree see see*, three buzzy notes followed by two on a higher pitch. A glimpse of one beside the road looks much like the black-throated green: a Townsend's warbler.

In Lowell, after our generous helpings of pie and ice cream at a local eatery, Keith chooses to press on another 23 miles to Kooskia, where he'll find a motel. Rather than fight the headwind that has arisen, David, Al, and I happily pitch our tents in the campground here at Three Rivers Resort, where the Lochsa and Selway Rivers join to form the Clearwater, which in turn eventually joins the Snake.

David swims in the river while I relax on the bank. A spotted sandpiper bobs and teeters across the river, calling *weet weet weet* as it departs. Next, a shower. Then a leisurely dinner with David and Al. Early to bed.

26.
PACIFIC ISLANDS INCOMING

DAY 59, JULY 1: LOWELL TO RIGGINS, IDAHO

To avoid the heat we have been promised later today, the three of us awoke at 3:30 a.m. in the campground here, now departing at 4:25, about half an hour before sunrise. With Al concurring, David's happily on board with this plan.

As we depart, swallows swirl about the campground's lampposts and dine on insects that have been attracted to the lights overnight. Robins carol and *hisselly*, and from seemingly every tree yellow warblers proclaim their "*sweet*ness," though the energy level varies from bird to bird. "The western tanagers and wood-pewees are still in dawn song," I point out to David, "and there's a treat, the first red-eyed vireo since the Ozarks."

Yellow warblers. Dawn singing from Idaho neighbors, so different in intensity. Male 1 (309, 11:39) Male 2 (310, 2:47)

It's miles of bliss down the canyon carved by the Middle Fork of the Clearwater River, but just before Kooskia we enter another realm. We have now coasted off the Idaho batholith, that great granite mass that formed deep in the earth on the edge of the old North American continent and then rose to the surface some hundred million years ago. Ahead lie exotic lands, ancient Pacific island chains and old volcanic islands that were scraped off the Pacific plate as it slid under the westward-moving continent that would become North America. These lands to the west include the towering "squashed volcanoes" of the Seven Devils complex along Hells Canyon, as well as the Blue and Wallowa Mountains of eastern Oregon, all of which were added in slow motion to the continent. Prowl around these mountains and one finds limestone from ancient

Western tanager. *Pit-er-ick*, he begins, another voice recruited to a mighty chorus of dawn singers as the sun sweeps the West. (311, 4:20)

Red-eyed vireo

coral reefs with fossils more like those in tropical, southeast Asia than anywhere in North America.

But there's no immediate evidence here of all those galactic collisions. No, towering over us now near Kooskia in textbook roadcuts are far younger columns of basalt that oozed from the earth about 50 miles to the west only 15 to 17 million years ago; lava filled the valleys as it flowed to its easternmost point about here. As the lava cooled, it formed these black, hexagonal columns that are stacked along and above the road much like giant organ pipes. Somewhere beneath all this basalt lies the old shoreline of what was once western North America.

A little after six o'clock, we turn up the South Fork of the Clearwater River, the gentle tailwind pushing us along, the miles flying by . . . until I yell up to David, "Another flat—the front tire this time," just a few miles after repairing the rear. We have the routine down well by now. I remove my front wheel, use tire irons to slip one bead of the tire over the rim, remove the tube, determine cause of flat by partially inflating tube and finding air leak, mark puncture on tube, give to David to repair, check tire to remove anything sharp that might have caused flat, slip new or previously repaired tube into tire, reinstall tire on wheel, inflate tube, put wheel on bike, and within minutes we're on the road, the tube that David repaired in the meantime ready to reinstall at the next flat.

As we climb over layers of basalt and out of the valley of the South Fork, the country suddenly opens up. Hilltops in the distance are now grassy knobs, with trees growing only in sheltered pockets. The fertile hills look like windblown dunes because they are just that, rich surface silt blown in during the ice ages from glacial outwash plains to the southwest. These rich, rolling farmlands, known as "The Palouse," extend into southeastern Washington, all good wheat and soybean country. With not a cloud in the sky, to the songs of a red-eyed vireo and crickets and cicadas, with the knowledge that the high temperature today will be only in the eighties, and all this with a tailwind, Al sums it up as he often has in a variety of circumstances over the past few days: "It's all good."

California quail. Western *chi-CAH-go!* replaces eastern *BOB WHITE!* (312, 2:20)

Climbing slowly, I give a nod to the black-headed grosbeaks, with us since Colorado, and to a new sound for the trip, the California quail calling in the distance, *chi-CAH-go*. I've missed the friendly calls of quail, the last being the *BOB WHITE!* heard back in Missouri, at Prairie State Park.

John, as he introduced himself, had stopped at an overlook when we pulled up to chat. "Pretty bad saddle sores," he admitted. “Don’t know if I can continue.” We learn that he is from western Massachusetts, back home. We wish him well, then encounter yet other cyclists heading east, first a pack of five, than four more a little later, and we stop again, sharing stories, laughing, enthusing about our journeys. They have about ten days behind them with 60 or so to go; I’m envious, in a way, but can’t imagine their hot and humid, birdsong-free Kentucky and Virginia when they arrive in August. “You were really *blown west* across Kansas?”—that’s their universal disbelief, distraught at the possibility that they’d fight a headwind rather than sail with the prevailing westerlies through the prairies.

In Grangeville we find Keith, and the four of us dine together on hotdogs and lemonade at a corner stand festooned with Fourth of July decorations. Outside of town, I have my *fourth* flat of the day.

An hour out of Grangeville, we’re climbing steeply, then suddenly find ourselves atop the world, delighting in one of the prettiest sights a cyclist ever sees: a sign warning truckers of the long 7% downgrade. The valley to the left lies more than a thousand feet below; we’ll drop over half a mile in the next seven miles, but pause just over the crest at the memorial for the Battle of White Bird Canyon.

Here, on June 17, 1877, was the first battle between the U. S. Army and the Nez Perce, led by White Bird and Chief Joseph, who dared to object that their reservation was shrunk by 90% to accommodate gold prospectors on their lands. With white flags flying during peace negotiations, army volunteers began firing; the dead soon tallied 34 for the army, 0 for the Nez Perce, who continued their flight along the route by which we arrived, up the Lochsa River and over Lolo Pass, up the Bitterroot Valley and over Chief Joseph Pass, to where they would clash again with the army at the Big Hole Battlefield in Montana seven weeks later, where we passed just three days ago. Chief Joseph would eventually surrender, declaring “I am tired; my heart is sick and sad. From where the sun now stands, I will fight no more forever.” (White Bird escaped with perhaps a hundred others to Canada.) As we quietly reflect on our collective roles in history, a lone lazuli bunting sings nearby.

Lazuli bunting. More hurried and shrill than the indigo, but equally persistent through the day. (313, 2:16)

Back to the bikes, we check the brakes to ensure they’re fine, then hold tightly to the handlebars as gravity accelerates us to 30, then 40 miles per hour. With only modest curves and no hairpin turns in the road, there’s no reason to brake, and in a flash we are beside the Salmon River, waving to the river rafters heading downstream to the north as we bike upstream toward Riggins.

Mid-afternoon the birds are quiet, but the canyon scenery is stunning, the geology even more so. The Salmon River has cut through hundreds of feet of 15-million-year-old volcanic basalt to the far more ancient rocks below, revealing there the squashed remains of those oceanic islands that were added to the continent some 100 million years ago. Near the town of Lucile are roadcuts through black island slate, the original fine layering destroyed as these Pacific lands were mashed in the oceanic trench off the North American continent. Just south of Lucile, "There's that limestone," I point out to David, "formed in coral reefs around tropical Pacific islands over 200 million years ago, with fossils like those from southeast Asia. They've traveled a good distance before settling here. So I've read."

Mile after mile we follow the Salmon River upstream, a gentle tailwind pushing us along. We pick up a fifth rider in Stefan, who wears a goatee and a smile, financing his way across the country with credit card debt and earnings left over from waiting tables. In single file, we are a bike-train heading upstream to Riggins, a growing Willie Nelson band of gypsies making music with our bikes, the waving rafters down below making music of their own as they float past.

Late afternoon we pull into Riggins, famished yet exhilarated. After a quick meal and some food shopping, Al, David, and I head to the campground while Keith and Stefan disperse to sites of their choosing for the night. David, champion swimmer that he is, takes on the Salmon River, while I settle for a tame shower.

To the ever-present roar of the river, we call it a day, crawling into the tent and reflecting on what extraordinary country we travel through. Drifting off to sleep, I try to take in the big picture—the Salmon River flows north here and parallels the Snake River in Hells Canyon a mere 13 miles to the west, and rising between the two rivers are old Pacific islands in the form of the Seven Devils Mountains. We sleep somewhere on the old coastline of the Pacific.

DAY 60, JULY 2: RIGGINS TO CAMBRIDGE, IDAHO

At dawn, among the songs of yellow warblers and wood-pewees, above the roar of the river and the whistling of the wind, from the rocky cliffs across the river I swear that I hear the song of a canyon wren. Whistles, successively lengthening, cascading down the scale . . . but just once, and perhaps I only imagined, as I've been listening hard for them.

With no particular plan for the day, we pack up camp and set out. The bikes of Al and Stefan rest outside a local restaurant, so we step inside to say good-bye, with the usual assurances that we'll almost certainly see each other down the road. Keith is just outside of town, where he has stopped at a phone booth to send an electronic file back home to England. "You'll catch us and pass us like we're standing still" I call to Keith.

And he does. The California quail lying on the road has reined me in. He's gorgeous, a study in gray and black and white and rust and more, the source of the *chi-CAH-go* song we've been hearing since yesterday. He lies on his side, his head twisted down, the black comma-shaped topknot reaching out to the pavement. I study the ink-black face boldly bordered by white, the mix of brown and black and buff atop the head, the finely scaled pattern in feathers on the back of the neck, the larger-scaled patterns on the belly, the rich brown and white stripes on the flanks, the fine gray breast and back. As with all too many other roadkills, I lay him to rest in the grass beside the road.

Through the valley carved by the Little Salmon River, we admire the barren hillsides, and the naked roadcuts all containing evidence of those ancient islands that once lay far out to sea. About five miles out of Riggins, shiny, dark green rock faces begin to appear in the roadcuts, serpentinite that formed deep beneath oceanic ridges and somehow in the geologic turmoil squeezed and slithered its way through rock fractures to find its way to the surface here. And gilding these ancient rock faces are those singing canyon wrens I've been waiting for, their whistled notes cascading down to us. "David! Canyon wrens, finally!"

California quail

Beyond Pinehurst, we catch up with Keith, now cycling among stately ponderosa pines and an assortment of familiar birds: western tanagers . . . western wood-pewees . . . warbling vireos . . . lazuli buntings . . . spotted towhees . . . yellow-rumped warblers . . . rock wrens . . . and more canyon wrens.

Canyon wren. An exquisite cascade of whistled notes, leaving one yearning for more. (314, 5:07)

"*Glass!*" shouts Keith up ahead, warning of hazards in the road and jarring me to attention. I automatically check in my mirror for traffic behind, and seeing none, swerve to the left to avoid the brown fragments of a broken beer bottle.

After following the rushing and tumbling Little Salmon River upstream for nearly 30 miles, we break out into a widening valley, home to the town of New Meadows. The wetland beside the road here teems with birds, making a killing field of the road. The imaginary sticker on my rear fender is I BRAKE FOR ROADKILLS, and for bird after bird I stop to take a closer look. Soon a small flock of the finer specimens travels with me in my left-rear pannier, just a select few from the many squashed and mangled ones left behind.

From New Meadows we climb over a ridge to the Weiser River and the town of Tamarack, which seems to consist solely of an impressive sawmill. A little inquiring reveals that they process some of my favorite trees, including Ponderosa pine, Douglas fir, white fir, Engelmann spruce, and western larch (or tamarack, hence the town's name), giving a glimpse of the forests nearby.

"Only 42 miles to Cambridge, David. If we push a little, we can get there by five and you can pick up your mail." He's all for it. He has friends. As for my mail, well, I'm still going cold turkey, just disappearing for the summer to see how it feels. Almost two months out, it feels mighty fine, and increasingly I realize that my academic career at the university is in some happy jeopardy.

The forest gradually gives way to open country well before Council, and we can now see more clearly where we are. Off to the east, ten to 15 miles away, are the West Mountains that are part of the granitic Idaho batholith on the western edge of the old continent. Off to the right and a little behind are the high peaks of the Seven Devils volcanic complex of islands that was reeled in from the Pacific. We ride again on the much younger basalt that oozed out of the Columbia Plateau some 15 million years ago; the shoreline of ancient North America, as it existed up to about 100 million years ago, lies buried in the basalt some distance beneath our tires.

We climb to the town called Mesa, then swing to the west, soon biking along the lush valley of the Little Weiser all the way to Cambridge, a modest tailwind blowing us into town. After picking up David's mail, we check in at Bucky's Motel. Stefan soon follows, and Keith shows up later, reporting, "Al stayed back in Council, but wanted me to remind you of the plan for the five of us to meet in Baker City for a Fourth of July party." We're on board.

David heads to the library to use the internet, and I do a little shopping for tomorrow's supplies before collapsing back at the motel. Yesterday was a tough day of ninety-some miles in the heat, followed by a tough, eighty-some miles today in more heat. I'm grateful my Achilles tendon is holding up, and my shin has healed, but we're pushing a little hard, and have revised our plan for tomorrow to a mere 40 miles, our destination the depths of Hells Canyon.

In the waning light, I summon the energy to check in with my fellow, nonhuman travelers. Pulling them out of my pannier, I lay them side by side from smallest to largest on my rear rack. A black-capped chickadee, black cap and bib, white cheeks, once so curious and, yes, cute. A yellow warbler, a male, the ray of brilliant sunshine with rusty-red streaks on the breast. A

Black-capped chickadee. So simple, so fine, *hey sweetie.* (315, 0:38)

Yellow warbler. *Sweetness* in song. (316, 5:55)

Red-winged blackbird. Energized male calling in the local dialect. (317, 9:06)

Bullock's oriole. Chatter and song. (318, 2:17)

Wilson's snipe. Ground calls. (319, 5:02), aerial winnows. (320, 1:57)

Sora. Whinny (321, 0:22), *SOR-ah.* (322, 6:00)

fledgling red-winged blackbird, plumpish, brown, just barely out of the nest, short-lived. A Bullock's oriole, bright burnt orange below, black throat and cap, large white patch on the wing. A snipe, brownish, well camouflaged in life, so boldly striped above, such an impressively long bill. A sora, ultra-secretive in life, now exposed, short stubby tail, long toes for marsh walking, black and white and rich brown stripes on the flanks, its short yellow bill glowing against black face and throat. I look them all over one more time, listening to the life-voices in each before taking them out behind the motel for a decent burial.

DAY 61, JULY 3: CAMBRIDGE, IDAHO, TO HELLS CANYON

Outside of town, after a leisurely mid-morning start, we will follow Pine Creek to the northwest for 17 miles, climbing about 1500 feet before dropping into the canyon. Along the way, beyond the lush streamside vegetation, the hillsides become as dry and brown as the sky is blue. Already it is warm.

"I wonder why they call it Hells Canyon," I call out to David behind me.

"You'll find out why," comes the simple answer, not from David but from a woman hanging clothes on a line beside her house a good 50 yards off the road. With the wind and road noise in our ears, we hear each other with some difficulty but forget how our voices carry to others listening along the way.

JOY OF HEARING OLD FRIENDS

Veery. *Veeer* calls, and *veerying* songs. (323, 2:46)

Gray catbird. An exceptional mimic. (324, 4:25)

Black-headed grosbeak, warbling vireo. The hurried rush of a vireo, the elegant richness of a grosbeak. (325, 2:33)

Mourning dove. Whistling wings (with western scrub-jay). (326, 1:05)

Having been assured of hell later, I relish the surprising abundance of birdsong along Pine Creek. Both kingbirds are here, offering a nice comparison of the two, and grasshopper sparrows sing for the first time since Kansas. Veeries and catbirds—how nice to hear birds I think of as "easterners" along the stream.

The regulars are here, too, all expected in the dense vegetation and trees along the stream: yellow warblers and song sparrows and spotted towhees down low, warbling vireos and western wood-pewees and black-headed grosbeaks up higher in the trees, red-winged blackbirds everywhere. Territories of these regulars are stacked back-to-back like beads on a string all the way to where Pine Creek peters out, perhaps 30 to 40 yellow warbler territories one after the other in each mile of prime habitat.

There's the occasional red-shafted flicker's *klee-yer!* call, the *chi-CAH-go* of a California quail, the hoots and pops and whistles and squeaks from a yellow-breasted chat, flight whistles of cowbirds, a *per-CHIK'o'ree* flight call of an American goldfinch flying by, the whir of mourning dove wings beside the road, barn swallows flitting by. A lone Brewer's sparrow offers his simple daytime song from up in the sagebrush.

Speed limit 40. By habit, from driving a car, I check my speedometer—7.9 miles per hour. I'm within the law. And not a leaf is stirring in this calm air, so I have a 7.9 mile per hour wind in my face, too.

Soon we're gliding down, through a side canyon carved into layer after layer of Columbia River basalt, dropping 2000 feet in just a few miles until we ride beside the Snake River itself, the same river that drained Jackson Hole back in the Tetons. Or, at least it was once a river here, as now it has been dammed to form the Brownlee Reservoir. Tightly stacked hexagonal columns of basalt rise directly beside the road, and in the heat of late morning it is only the lazuli buntings who sing, one after the other lining the road.

With a mild breeze at our backs, we ride easily beside the reservoir. Clouds have kept the temperatures tolerable, in the mid-80s, and a brief shower is a bonus. At Brownlee Dam we cross to the Oregon side with enough fanfare to scramble the skittish post-9/11 security guards. On the Oregon basalt just above the road, a small herd of mountain goats stand sentinel, and we stop to talk, to each other and to them.

Early afternoon, with a bit of good fortune for this holiday weekend, we settle into the last available site in the campground, our small tent dwarfed by the monstrous motor homes surrounding us. David is soon swimming in the river, while I nap nearby in lush green grass beneath a tree. Down here where the Snake River has cut through enough young basalt to expose those old Pacific islands, I imagine lying beneath a palm tree on a tropical island, waves lapping at the shore, with multicolored fish swimming among coral reefs just beyond.

Later, our neighbor comes over and strikes up a conversation. "Where'd you start?" "Virginia," our standard answer. "Where you heading?" "Oregon, well, no wait, we're here. The Pacific. We're headed to the Pacific." We need to revise our answer to the most oft-asked question.

We ask him about his "camper": "What's it like camping in an 'American Dream'?" There are lots of superlatives in his answer. "How do they drive?" Something in this small, elderly man's answer about "just aiming it down the road" is more than frightening to a cyclist. "How long do your batteries last if you can't plug in?" He'd never camp without electrical hookups, as he was done roughing it.

Dinner is early, and by seven o'clock I'm in the tent, eager to get the rest needed for tomorrow's climb out of the canyon and the eighty-some miles to Baker City. With sunset still an hour and a half away, robins and yellow warblers are singing, and through the tent's mesh I watch swallows dart about. For a more intimate experience with Hells Canyon and Oregon, David will sleep out in the grass tonight.

27.
ASCENDING INTO OREGON

DAY 62, JULY 4: HELLS CANYON TO BAKER CITY, OREGON

In Pacific time now, sunrise suddenly comes an hour "earlier." At 3 a.m. the campground is eerily quiet, and we are the only sign of life as we quietly pack up and eat breakfast.

A little after four, with sunrise still an hour away, nighthawks continue to *peent* and boom overhead, robins are just under way, and we're fully loaded and on the bikes, rolling slowly out of the campground. I am euphoric, as we are *climbing into Oregon.* I run through the reasons in my head. *Oregon is home,* as I lived here while in graduate school from 1968 to 1972, before David was born. We're *on our bicycles,* and I'm *with my son.* The *Pacific Ocean,* intensely blue waters with frothy white surf pounding on rocky headlands, awaits our "discovery." And the *western birds*—it'll be the final hurrah of all the special western birds that we've come to know. It is *the anticipation of all things good,* not the least of which is simply *crossing an extraordinary landscape* and knowing a bit of how it came to be.

Not all that long ago in geological time we'd be setting out to sea here, as what we now call Oregon simply didn't exist on the western edge of the continent. Oregon's story begins roughly 400 million years ago, based on the oldest rocks found in the state, and if I round that off to 365, I can imagine encapsulating Oregon's entire history into the 365 days of one year, from January through December, speeding through the year at a million times the rate at which events actually occurred.

During January (and the December before, in fact) ancient volcanic islands complete with coral reefs and sandy beaches form in an ocean well beyond the coast of what will eventually become the land mass we call North America. Through time, from January to February and on through August, this ocean becomes cluttered with islands the likes of those now in Indonesia, New

Zealand, or Japan, or extensive island arcs like the Aleutians. During June and July, with the breakup of the single, large super-continent Pangaea, the current configuration of continents begins to take shape, and as the North American tectonic plate moves westward over the Pacific plate during August and September (about 150 to 90 million years ago), many of these island systems are scraped off the crust and added to North America. Evidence of these old islands now lies in the volcanic rocks and fossilized sandy beaches and coral reefs found atop and within and beneath mountain ranges throughout Oregon, and in fossil shorelines that crept successively westward.

In late October, thanks to a large meteor that struck the earth, dinosaurs disappear and we mammals begin our rise to prominence. In early November, Oregon's first homegrown volcanoes become active, and we'll bike among their remains in just a few days near Mitchell. By late November, an impressive volcanic range of mountains appears, the Western Cascades, which are currently dwarfed and weathered, lying just west of the mountains we now know as the High Cascades, which arise much later.

At a million times the actual rate, the last half of December is a blur. On the 16th and 17th, lava gushes from 10- to 25-mile-long fissures in far northeastern Oregon, first filling valleys and then smoothing the landscape with basalt up to three miles deep; we've been riding over and through and beneath this basalt since Kooskia, Idaho. By December 20, "basin and range" faulting begins, and from the 27th to the 29th, a huge Lake Idaho larger than current Lake Erie forms in southeastern Oregon and nearby Idaho. Early on the 30th, this lake finds an outlet to the north, the river we now call the Snake then carving out most of Hells Canyon before midnight.

Early on December 31, a new line of volcanoes forms just east of the worn-down Western Cascades, giving rise to the familiar High Cascades of Mounts Hood, Jefferson, Washington, and the like. In the hour before midnight, the Pleistocene ice ages come to a close, but not before huge floods ravage the landscape; the ice dam that blocked glacial Lake Missoula in western Montana (and filled the Bitterroot Valley through which we biked) repeatedly gives way and then rebuilds itself, cataclysmically flushing the Columbia River system some 40 times, and Lake Bonneville (the current puddles of the Great Salt Lake in Utah its remnant) abruptly empties to the north through Hells Canyon. At 11:50 p.m. (just 6,580 years ago), not all that long after humans have arrived in North America, volcanic Mount Mazama blows, leaving behind Crater Lake, and a minute before midnight, as a reminder that Oregon is still a work in progress, lava oozes from craters near McKenzie Pass, our route over the High Cascades.

In this process accelerated a million times, untold numbers of islands race to port at the galactic speed of about 15 feet per hour, three inches per minute, and as large islands and subcontinents collide with North America, with one immovable mass meeting another, the entire state trembles and convulses and heaves; mountains are smashed and then folded or tilted this way and that. Volcanic ranges appear overnight, and entire mountains and canyons and vast lakes are created and disappear in the blink of a geologic eye. Yeah, this is *Oregon*, with the landscape so alive I can almost feel it breathing beneath me.

It's all good and it just keeps getting better, I tell myself, as I float effortlessly upward, out of the canyon and among singing birds beside tumbling Pine Creek down below and among stately ponderosa pines towering above. Dark has given way to gray as we ease into the dawn, and towhee after towhee lines the stream, each rapidly alternating among two or three different songs. Yellow warblers chip sharply between songs, and each song sparrow hustles through the different songs in his arsenal. Western tanagers sputter haltingly through their dawn effort, singing ever so slowly, then pausing to *pit-er-ick*. And lazuli buntings—how odd that last one, as he seems to sing mostly a yellow warbler song, but then ends unmistakably with bunting phrases. It's the peril of being a songbird, as having to learn one's song has its risks, and sometimes the wrong song is learned. I'm flying, wheels barely touching the pavement, and it's only 70 miles to Baker City.

Song sparrow. From Hells Canyon, six different songs, eventually. (327, 9:43)

David's up ahead about 200 yards, and I watch for signs that the spell has broken. "Don't talk to me" are his instructions, "and stay behind." So when he stops, I stop, and when he bikes on, I follow, keeping my distance. His sleeping pad failed him in the night, leaving him on hard ground and sleep-deprived. And it's gawd-awful early. Some caffeine would help, as would a second breakfast. Patience, I tell myself.

I listen, lifted by the odd sounds of yellow-breasted chats along the stream below . . . by pewees in dawn song . . . by nighthawks calling and booming overhead . . . by a crow cawing in the distance . . . by a goldfinch flying by, calling on the wing . . . by tanagers and robins and orioles and kingbirds . . . by a rock wren and a canyon wren singing side by side on a cliff face just to my right. I melt at the sound of any wren and have spent much of my life studying them, but these two are among the finest. The canyon wren repeats one exquisite pattern of the most beautiful descending whistles over and over before eventually switching to another pattern, and maybe he has half a dozen different songs at his command (but it seems that no one knows for sure, as no one has ever listened to all a canyon wren has to say). The rock wren, especially in this dawn frenzy, races from one song pattern to another in a jumbled series of highly contrasting songs; I know he can sing 100 or more different songs, as I've studied them. What delightfully different singing paths they have chosen since the ancestral wren, thanks almost certainly in large part to the females in each lineage who have demanded that the fathers of their young sing in this way and not that. Yes, "it's all good," as Al would say, but this morning is especially good.

MORE FROM HELL

Yellow-breasted chat. Out of darkness, a bewildering variety of voices. (328, 11:04)

Rock wren. Leaping among 11 different songs; with lazuli bunting. (329, 2:03)

Cassin's vireo. Another package singer, just like his "solitary" cousins. (330, 3:16)

And vireos! I've heard plenty of the familiar warbling and red-eyed singers, but a new one sings now, like a hoarse red-eyed vireo. "Has to be a Cassin's vireo," I reason, and I pull up beside him, soon confirming that he sings in packages, playing with a set of several songs in his repertoire before moving on to other sets, just like his two close relatives, the blue-headed and plumbeous vireos, all formerly called the "solitary vireo." And no doubt in the same style as the ancestral vireo that gave rise to these three. How nice to find some order in how they sing.

At 5:27, I glance over my shoulder to admire the red ball of the sun emerging over the Seven Devils Mountains on the far side of Hells Canyon. The Devils: born far out to sea, adopted ages ago by "our" North American continent, covered with basalt just 15 or so million years ago, then carved so deeply by the Snake River on the west side and the Salmon River on the east that the unburdened crust beneath the Devils rose like a boat unloaded at port, sending the Devils skyward.

Half an hour later, just five miles to Halfway, I stop to talk to the horses in the pasture beside the road. All 13 stand and stare, as if questioning, wondering what this roadside beast might be. "Hey horses, how ya' doin'? You're beautiful, all of you. Lucky you, living here in so fine a place—life is good, eh." They politely listen to my monologue, but at the sound of my best whinnying neigh, they sprint to the far side of the pasture. Much smarter than those mindless cows back in Montana, I conclude.

Western kingbirds—an entire family is just to the right, six in all, with four young birds nestled together on a dead, exposed branch and two adults flying about. The young already look just like the adults, and they call and twitter like them, too. These kingbirds tell of a history no less fascinating than the hills here, how millions of years ago (maybe 50?) the lineage leading to songbirds and to flycatchers diverged. These flycatchers retained what was probably the ancestral style of song development: Leave nothing to chance and encode the basic form of the adult vocalizations directly in the genes. Not so, said the songbirds, who rewired their brains and voice boxes to imitate complex songs from each other. It's the big story of two evolutionary paths taken that I hear all about me as I continue up the road.

Western kingbirds, siblings

Western kingbird. Directed by the genes, these sounds are not learned. (331, 4:39)

As I pass David, we exchange just enough words to agree to meet at the first restaurant in Halfway. Once there, I park my bike outside a window where I can watch it from inside, then go in and sit at a table for two. David arrives, parks his bike beside mine, then walks inside past me and into the next room. I linger over pancakes, eventually even drinking some coffee . . . and an hour later the old David emerges from the far room, full of energy and eager to hit the road. I smile, swallowing words that needn't be said.

When we emerge from the fertile valley around Halfway, the landscape dries and the birdsong confirms it. From the sagebrush we hear snippets of meadowlark, sage thrasher, and vesper sparrow, none of the forest birds of earlier. In the six miles outside of Halfway we climb over a thousand-foot ridge; the view back to Hells Canyon is stunning, bounded on the north by the Wallowas, on the south by low desert, and here on the west by the distant Elkhorns. At 9 a.m. on this Fourth of July here at the top of the pass in this high desert, the silence is deafening. I strain to hear something, the stirring of a breeze, anything . . . and the barely audible flute-like whistles of a lone western meadowlark arrive from some distant land.

The descent into Richland is swift, roughly a 6% grade for five miles. We lose *all* of the elevation that we gained since Hells Canyon, as this low oasis of Richland lies on the shore of water backed up into a valley that leads directly to the Brownlee Reservoir in Hells Canyon. After restocking with supplies, we're headed west again, eager to make Baker City by nightfall for the fireworks with our friends.

"Bank swallows," I manage to point out to David in the 95-degree heat of mid-afternoon. "It's the first colony we've seen, even though they occur all across the country. See all their nests in the sandy bank on the right? And look at all the dead ones on the road. Pretty sad. . . . They have the dark necklace across the breast against the all-white undersides—typical bank swallow."

Once again we bike beside the Oregon Trail, the wagon wheel ruts still visible through the sagebrush just off the road. On a hill up to the right is a large covered wagon, and we try to imagine crossing the country with such a vehicle. Another mile down the road is the Oregon Trail Interpretive Center, up on Flagstaff Hill, several hundred feet *up*. Exhausted, dehydrated, sick from the heat, I encourage David to check it out while I rest in the shade of a small entry building at the base of the hill. Half an hour later I join him after a kind soul sees me sprawled on the pavement and offers a ride to the top in his pickup.

Walking among the life-size exhibits, hearing the voice of a mother who lost her child, reading of hardships and the occasional joy, I am struck most by one exhibit: "If you're over fifty years old," I read, you're too old for a trip like this—it's a brutal journey, and one in ten people die. Today, so tough on me because of the heat, I could readily identify with that advice, especially given my fifty-seventh birthday in three days. I'm also humbled—I'm riding a bicycle on paved roads with abundant food and drink for the purchase all along the way, but under infinitely more difficult conditions it took the average pioneer six months to make this journey from Missouri.

After dinner, David and I meet our friends at the pub.

"Would you believe that here we are, in Baker City on the Fourth of July, the only city in the entire country without fireworks tonight?" asks Stefan. We grumble.

"Did you see all those dead birds on the road just outside of town?" asks Al.

"Birds?" responds David, taking a sip of his beer. "Those were *bank swallows*; you could tell by their nests in the bank and the dark band across the chest. As opposed to the cliff swallows that nest mostly under bridges and look like they have a little miner's cap on, and certainly not barn swallows with the long forked tail and rusty throats and pretty songs, and I bet you didn't know that purple martins are swallows, too, the big dark ones—the males, that is, as females are a little lighter—they nest in the bird condos and hanging gourds by people's houses, and you wouldn't believe what goes on in a martin colony, and then there are the tree swallows and violent green swallows. . . ."

At which point he is drowned out by roars of laughter from Al and Keith and Stefan, who see the growing smile on David's face and realize that he's parroting me, recounting all that he's digested on this trip. "Yes, he *was* listening, taking it all in," I think to myself as I join in the laughter. "But, David," I correct him, "they're not *violent* swallows. It's *violet*-green, and if you get them in the right light. . . ." There's no way they'll let me finish this crucial point.

DAY 64, JULY 6: BAKER CITY TO DIXIE PASS, OREGON

At 5:07 a.m. we stand in the motel's parking lot, ready to depart. My headache persists, as does some nausea, but I want to try to push through it. Two days ago, from Hells Canyon to Baker City, we were in the saddle for seven and a half hours, too much of it in the afternoon heat. Throughout the night I felt chills and nausea, and I stayed in bed as much as possible yesterday, sleeping, resting, and thinking healthy thoughts. David spent time on the internet, shopped, delivered food, and exercised patience, reminding me on occasion that "You're just getting old." He accused me yet again of wanting to make the trip last, of not wanting to return home.

We will scale back some today. Already it's warm and the sun's not even up. Dixie Summit is only about 60 miles away, and we aim to camp there at the higher elevation overnight. To a rousing send-off by a pack of noisy house sparrows in the shrubbery next to the motel and to the nasal *caws* of a young crow begging for food across the road, we set out through the streets of a Baker City just coming to life.

House sparrows. *Chirrupping*, as along city streets continent-wide. (332, 1:38)

Out of town we ride through lush Bowen Valley, awash with irrigation, as the sun's reddish rays highlight the top of the Elkhorn Mountains to the west. There are the expected songs along the way, and I acknowledge each in turn, yellow warblers, song sparrows, willow flycatchers, red-winged blackbirds, winnowing snipe, even some savannah sparrows, which we've not heard in some time. But it is the occasional starling along the road that most intrigues me. Being the mimics they are, they can tell us what we're missing, and I'm astounded as they announce *willet* and *curlew* and *killdeer*, together with bugles of sandhill cranes, all picked up from the wetland birds in the valley here.

European starling. Stunning mimics, even imitating two different species simultaneously! (333, 0:45)

The sun has worked its way down to us now, with the sole purpose of showcasing the ponderosa pines along the Phillips Lake reservoir. Compared to other wavelengths, the sun's reddish rays survive best the long travel through the earth's atmosphere at dawn, and the already red bark of the ponderosa pines is aflame; long, deep green, graceful needles droop from the branches, the red and green of the pines brilliant against the deep blue sky. Ravens croak and chortle, urging me that in my next life, when I come back as a raven, I must come to live here among these pines.

We stop briefly beside the lush undergrowth at the entrance to the Powder River Recreation Area. A rufous hummingbird flits nearby among the flowers, and how apt the name, especially in this morning light—I note the bits of green, but it's the reds that astound, in the back and tail, and especially in the throat, where the gorget at just the right angle becomes an intense red headlight. A pewee lands beside us, a rare treat, the first we've *seen* of either eastern or western on this entire trip—shades of olive and gray, lighter underneath, a suggestion of two whitish wing bars, and a *bzeeyeer*. How rich the sights and sounds of dawn.

A dog-sized Bambi lies on the road, lifeless but intact. A tanager sings and a sapsucker drums as I gently grab one leg and pull the spotted carcass to the ditch.

"When nature calls, the breeze becomes a symphony of birdsong." So says the roadside sign that David reads aloud, and in mock excitement he reads off the list of birds that we can expect to hear. Then, "The lake, marshes, shrubs, and forest create a rich and diverse habitat. . . ." Sadly, though, we soon bike beside mile after mile of tailings piled high by the monster dredges that once worked this valley for gold. The entire valley, it seems, was exhumed, sifted, and left upside down in long, caterpillar-like mounds of gravel nearly a hundred yards across. The gutting began in 1913, we learn, and didn't end until 1954, when costs finally outweighed the profits.

Climbing over Sumpter Pass and then Tipton Pass, we rise one more time, to our campground at Dixie Pass. After pitching our tent among the magnificent Douglas firs, David naps while I fuss with maps and recordings.

Soon after dinner I am in the tent, preparing for my big day tomorrow. It's my birthday and I'll be out early walking about the campsite, listening to all that is here and celebrating the disappearance of my headache and nausea on the last climb.

Hermit thrush

Hermit thrush. Outside the dawn chorus, their song can sound mournful. (334, 2:06)

In the fading light, somewhere in the Doug fir above, a hermit thrush begins his evening song. I melt into my sleeping bag, just listening, just being, as I gaze upward through the mesh of the tent's roof into the lower limbs of a tree many times my age. There's none of the intensity of the hermit's dawn singing now; instead, he sings more slowly, as if drawing each song out and deliberating on it at the day's end. Down low . . . then high . . . he still leaps from one pitch to another, but it's different, more gentle, almost tender, bordering on melancholy. Two others answer in the distance, the entire forest now their stage. In the "hush and stillness of twilight . . . the sweetest, ripest hour of the day," I remember John Burroughs writing, with no other birds singing and the last breezes of the day having gone to rest, one experiences in the hermit's song that "serene exaltation of sentiment of which music, literature, and religion are but faint types and symbols." Ever so slowly, consciousness fades.

28.
GEOLOGICAL CHAOS

DAY 65, JULY 7: DIXIE SUMMIT TO DAYVILLE, OREGON

"Happy birthday to me," I hum quietly to myself as I ease out of the sleeping bag and then the tent. It's still dark, maybe 45 minutes to sunrise, and so quiet, except for . . . and I stand still, listening, taking inventory . . . two chipping sparrows, a western tanager, a hermit thrush. I listen to each in turn, to the chipping sparrows sputtering their songs to each other two campsites away, to a single tanager in his halting dawn song above them, and to the hermit just beyond.

Hermit thrush. At dawn, such exuberance, such energy! (335, 3:11)

How different the hermit's tune now from what it was last evening. He's racing, the songs and the pauses between them equally short, and I clock him at 20 songs per minute, twice his evening pace. Walking the camp road until I'm beneath him, I hear how sharp and brilliant he now sounds, so energized, entirely different from last night. Is it just the pace at which he delivers his songs that makes the difference? Or is each song itself now compressed and hurried, more piercing, more powerful, delivered with more enthusiasm? Maybe what I hear is my own mood, how content I am at the end of a good day contrasted with the excitement and energy and expectations at a new dawn. It's also possible that he shares my mood, and says so.

Walking quietly along a logging road, I slowly approach a junco who sings low in a small fir. At five yards I stop, and in the growing light I can see his elegance. He's a "dark-eyed" junco, the field guide would remind me, its range stretched all across North America, but such a name does no justice to this bird. He is an *Oregon junco,* so distinctive in dress that this western race was formerly considered a separate species.

He's immaculate, almost like a miniature spotted towhee, with his dark hood, rufous sides and back, and white belly. And how he sings! Chipping sparrow–like, one might say, with each song consisting of a phrase repeated several times, but there's a lingering ring, a pleasant musical effect to each phrase that a chipping sparrow never achieves. I hang with him, hoping he switches to another of his several songs, but 130 songs and seven minutes later, I lose patience and yield to the invitation of other singers.

Dark-eyed junco. My Dixie Summit junco, 18 songs a minute. (336, 2:30)

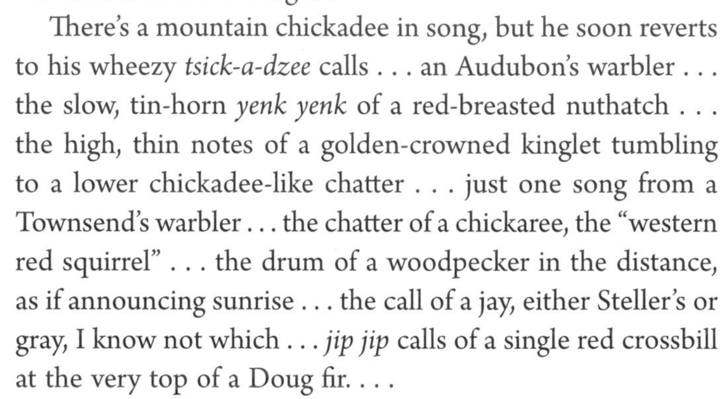

There's a mountain chickadee in song, but he soon reverts to his wheezy *tsick-a-dzee* calls . . . an Audubon's warbler . . . the slow, tin-horn *yenk yenk* of a red-breasted nuthatch . . . the high, thin notes of a golden-crowned kinglet tumbling to a lower chickadee-like chatter . . . just one song from a Townsend's warbler . . . the chatter of a chickaree, the "western red squirrel" . . . the drum of a woodpecker in the distance, as if announcing sunrise . . . the call of a jay, either Steller's or gray, I know not which . . . *jip jip* calls of a single red crossbill at the very top of a Doug fir. . . .

Golden-crowned kinglet. At first so high and thin and wiry, dropping slightly to a lower chatter. (337, 0:35)

Mostly, a little after sunrise now, I listen to the silence of early summer, and I strain to hear anything. July brings the end of the breeding and singing season, and there's no chorus now, only an isolated song here and there, halfhearted efforts by males on the verge of calling it quits for the year. Dominating the soundscape now are not the birds but instead bumblebees, their wings humming as they forage for nectar in the low bushes.

Bumblebees. Pollinating insects, sounds of warmer summer days. (338, 1:17)

Back at the campsite I find David eating breakfast. I join him, and we gradually pack up and ready our departure. "July 7, 7:52 a.m.," I announce into my recorder. "We're about to drop down off Dixie Pass, biking *downhill* for 53 easy miles beside the John Day River, dropping almost 2000 feet to where we'll spend the night as guests at the Presbyterian church in Dayville." To David's playful, muted strains of "happy birthday to you," we mount our bikes, aim them west, and begin coasting.

We drop quickly to the lush John Day Valley, then wind along beside the river, mile after mile falling behind us. There's probably no stretch of country more fascinating to a geologist; a "geological chaos . . . a tossed salad made of stones" was how one put it. All about us are those exhumed bits and pieces of oceanic islands that were scraped off the Pacific plate; these adopted

lands were smashed and stretched, transformed and metamorphosed, then all scrambled to form the landscape here. I eye Canyon Mountain in the famed Strawberry Mountains about five miles south of John Day, and imagine the very top of this 8000-foot peak for what it is, the remains of a magma chamber that once, about 280 million years ago, lay beneath a volcano far out to sea.

Dayville arrives quickly, and with permission kindly granted we're soon camped inside the church. After a quick visit to the nearby general store, David prepares a feast fit for any birthday, topped off with two brownies laden with 57 candles, enough to create a conflagration (wisely, I note, he has a bucket of water nearby, just in case). "Happy Birthday, Pops," smiles David as he unveils my birthday present: A second Super Soaker water cannon to match my Father's Day present, which David still carries, but this new one is entrusted to me.

Before calling it a day, we pull a video from the shelf that tells us how perfectly God created birds and other animals. Happened "just 10,000 years ago" my cousin Bill back in western Michigan once told me. In jest, I referred to that version of creation as an "immaculate deception," a world that was created so quickly and recently but made to appear as if it were billions of years old. But he was serious. The fact that scientists of all disciplines throughout the world were essentially unanimous in their "old earth" theory was irrelevant to Bill: "They're all biased," he said, and went on to tell me how his four years of science training at the local university had been adjusted by his Sunday school teacher. Those were close to my views once, too, I realize, acknowledging my roots in the Dutch Reformed Church of western Michigan, though as a university professor teaching evolution, my view of how the world came to be has diverged markedly from those early teachings.

DAY 66, JULY 8: DAYVILLE TO OCHOCO PASS, OREGON

I ride early again, leaving David to catch up when he chooses. Today we will also cover about 50 miles, but how different the terrain from yesterday, as we'll climb two major passes for a total of 4000 feet.

By 4:35 a.m., I am riding through Dayville by streetlight, past the two antique stores and then the mini-mart on the left. On the right, from high in the trees near the post office, zip code 93825, a single wood-pewee alternates his two dawn songs. A deer spooks beside the road and crashes off through the brush. Just outside of town a western kingbird sputters through his dawn songs, *kip kip kip kip ki-PIP ki-PEEP-PEEP-PEEP-PEEP,* and in the lush riparian habitat along the river I soon hear other birds, too, robins, song sparrows, yellow warblers, house wrens, house finches, lazuli buntings. To the left a California quail offers his *chi-CAH-go,* and high above a nighthawk calls. As night eases into day, I have the world all to myself. It's just as Aldo Leopold once wrote in his *Sand County Almanac*: The land deeds in the local courthouse mean nothing, because at this time of the morning the world belongs to me. With clear skies above and the temperature in the mid-fifties, my bike rolls effortlessly down the valley along the John Day River. I am a rich man.

Lazuli bunting. At dawn, often puzzlingly abrupt and brief. (339, 3:09)

"John Day Fossil Beds National Monument" announces the sign, and Oregon offers yet another geological delight, all explained in the abundant pamphlets provided by the monument. I'm soon riding through Picture Gorge, a thousand-foot-deep canyon carved by the John Day River through the Columbia Plateau basalt flows that originated to the northeast roughly 16 million years ago—the same basalt flows we first encountered back in Kooskia, Idaho, I recall. The layering here is clearly visible, and I count upward from beside the road, trying to identify each of the 17 separate floods of basalt that are visible in this canyon. The layers vary in thickness, and what beautiful symmetry to the canyon, as the layers beside me on the left of the road can be matched perfectly with the layers just across the river to the right. I double back to ride the canyon twice more, slowing to admire the striking hexagonal columns into which the basalt has cooled. I bask in songs of song sparrows and yellow warblers but especially canyon wrens and rock wrens, and eye a silent pair of dippers bobbing for food down below in the fast-moving waters of the river.

The extraordinary terrain here invites me to leave the prescribed bike route and instead follow route 19 and the river north a few miles beyond my turnoff. Sheep Rock towers ahead, 1200 feet above the river, and what a beautiful sight it is, a coffee-table book in itself with innumerable stories told in its face. Layer upon layer of the earth's history lie exposed here, but the layers are no longer horizontal as they were when laid down; instead, they're tilted sharply down to the south toward Picture Gorge, the entire mountain askew. Capping the *very top* of the mountain is a dark layer of the same basalt flows that form the *bottom layers* of Picture Gorge, so much has the land shifted and tilted. Below the 16-million-year-old basalt cap, in shades of reds and browns and grays and whites, are layer after layer of history dating back about 33 million years. It's as if the entire mountain shrugged and the north side simply slid down a hundred feet or so, so that I can follow a particular sedimentary layer up from the south until I come to the slip line, which I follow down to the left until I find that same layer continuing up to the north.

The 1200 feet exposed here in Sheep Rock are just a glimpse of layers *three miles thick* that tell the earth's complete story in this area over the last 100 million years. The oldest and deepest layers are ocean sediments that were laid down in the delta of a large river just off the ancient coastline of North America; they're exposed in Goose Rock just to the north of here. Next most evident are layers from 54 to 37 million years ago, consisting of mudflows, ashfalls, and lava deposits from nearby volcanoes; remarkably abundant and complete fossils in these layers show that the land was warm, nearly a tropical forest, with abundant rainfall. Next up are distinctive layers of ash that fell from the sky during the peak activity of the Western Cascades, 39 to 18 million years ago, when the climate had become temperate, with deciduous hardwood forests; Sheep Rock consists mostly of these layers, as do the stunningly beautiful Painted Hills just up the road, where layers of ancient soils color the landscape. Then come the dark Columbia Plateau basalt flows, dated at 16 million years. Above the basalt are more layers of ashfall from volcanoes to the west, south, and east that were active 15 to 12 million years ago; and above all those layers are sediments that eroded from nearby mountains and filled valley floors in the time since the last major volcanic activity. Three vertical miles of history are written upon the land, three miles of story-telling volcanoes, shifting and tilting lands, and fossils of a life rich and wild beyond our imagination.

Reluctantly, I head back south to the bike route and begin following Rock Creek upstream as I climb to the west out of the John Day Valley. Rising vertically above me on both sides of the road are dark columns of basalt painted with bright yellow lichens. A kingfisher rattles along the stream, and a lazuli bunting sings—no, two buntings sing, back and forth. I follow the cadence of each, smiling, as these two neighbors have identical songs. Given bunting actuarial tables, with only about half of all adult birds surviving from one year to the next, they're most likely a yearling and a two-year-old bird, the youngster having come to match his neighbor just as their eastern cousins the indigo buntings do.

As the canyon opens to a broader valley, cattle graze beside the road, and I stop beside a nighthawk lying on the pavement, noting the white throat and tail band, a male. He's well camouflaged in life, mottled brownish and nicely barred on the sides and belly, but how out of place he is here on black asphalt. I pick him up, stretching out those long pointed wings and exposing the white patch across the flight feathers, admiring the stiffened wing feathers that vibrate in a deep dive to produce the booming *VROOOM*. The bill is tiny, but I open it to reveal the cavernous mouth that he once used to catch insects as he danced about the heavens, and from which his familiar *peent* calls emerged. And how sad those large black eyes that now gaze out into nothingness. "Dad was here"—I scribble my message to David on a piece of paper and tie it conspicuously to a leg, laying the bird on the edge of the asphalt. He can't miss it, I conclude, as how could David not stop to study this remarkable specimen?

Common nighthawk. Ten series of *peent* calls, each followed by a diving *VROOM*. (340, 4:04)

Cliff swallows and kingbirds thrive here, as do towhees. Magpies call upslope, and a canyon wren sings from among the yellow lichens in the basalt, whose black columns, tilted to 45 degrees, are perfectly framed against the deep blue sky. Up ahead, Table Mountain lies to the left, Juniper Butte to the right. Oregon, I smile.

Bushtits! I stop to look and listen to this far western marvel. What odd little grayish creatures, such dainty things with longish tails and smallish wings, half bird and half air, it seems. Airborne, they bounce from one juniper tree to the next, a flock of them—I count 11, then two more follow in acrobatic pirouettes. It's another songbird without an obvious song, but I bet there's not a silent bird in the flock. They twitter and buzz and scrape and chatter a nearly continuous, sparkling conversation among themselves. This baker's dozen, all moving together, collectively weigh a mere 70 grams, two and a half ounces of fluff. They're just as social on a cold night, when they all huddle together to stay warm as they sleep.

Bushtit

Bushtits. Songless, but all atwitter. (341, 1:08)

I continue climbing, listening to vesper sparrows and western meadowlarks in the sagebrush, to an occasional *bzeeyeer* of a pewee from the junipers. At nearly 40 miles on my day's odometer, having climbed about 2000 feet to Keyes Creek Pass, I wait for David. A full hour later he arrives, confessing to my inquiry: "Nope. Missed the nighthawk." After a brief break, we quickly descend the seven miles to Mitchell.

Western meadowlark. Their songs carry in open sage, as do their dry rolling chatters, *vicicicicicicic*. (342, 1:21)

"Oh, I'm considered an outsider," the clerk at the general store admits. "I've only been here 29 years. I'm just helping my wife. She runs the place." And restrooms? "The wooden building just up the road." Not little plastic outhouses, but grand public restrooms in a building devoted solely to the task, right there on Main Street, with carloads of grateful tourists milling about this center of activity.

We drop down out of Mitchell, then start our 2000-foot climb to where we'll camp at Ochoco Divide. Willows line the stream, and the widely spaced junipers on the hillsides give way to sagebrush on drier slopes. It's classic central Oregon, dry, not a cloud in sight, nothing but blue sky overhead, all thanks to the high Cascade Mountains that cast their rain shadow far to the east, sucking all the moisture from the westerly winds.

Also classic is the terrain here, a jumble of this and that. Roadcuts expose fine sedimentary layers of shale and sandstone that were laid down horizontally about 100 million years ago in a river delta just offshore, then brought on land and folded and deformed and tossed about, now resting artistically beside the highway. Dimpling the landscape are giant pyramids, the remains of Oregon's first native volcanoes, those that erupted right here 42 to 52 million years ago and spewed thick layers of ash and lava and mud over land that had once been oceanic floor; the lava stopped plants and animals in their tracks, leaving so many fine fossils that they warranted a U.S. National Monument. We stop at milepost 54 about ten miles west of Mitchell to admire the roadcuts and the sweeping view back to the southeast, with the central plugs of two of those old volcanoes, Black Butte and White Butte, dominating the skyline.

Late afternoon we pull into our campground, set among towering ponderosa pines, taking site 15. It was an easy day in spite of the climbing. We devour David's specialty: pasta, carrots, spaghetti sauce, and tomato paste. For dessert, it's a Snickers bar.

In the dead calm a little before midnight, I stir at the sound of a vehicle that pulls into the vacant site opposite ours. Seconds later, a door opens and the screams of children are soon joined by the loud, angry voices of two adults, a man and a woman, all followed by a clatter the likes of which I have never before heard. Taking up the chorus is a nearby pack of coyotes, their yips and howls then joined by two great horned owls several sites over. She calls, *who-ho-o-o whoo-hoo-o*, and immediately he responds, in his deeper

Coyote

foghorn voice, *who-hoo-ho-o-o whoo*, the she-leads-he-follows pattern repeated several times. How long the pandemonium continues I cannot know, as I alternately doze and am again jarred from my slumber by another surge of verbal carnage or clanging gear. Eventually, calm reigns, and I drift off until morning, amazed that David never stirred through it all.

DAY 67, JULY 9: OCHOCO PASS TO SISTERS, OREGON

"You circle left and I'll go right," I suggest to David during breakfast, "and one of us is sure to see him." From midway up the hundred-foot pine next to our campsite we hear the asthmatic scream of a red-tailed hawk, but it's a little off, beginning well but then becoming a little too asthmatic and breathy. "Maybe a young hawk." Slowly and quietly we circle, to the tunes of a whistling mountain chickadee, a distant hermit thrush, juncos, a chipping sparrow, a pewee, a Cassin's finch, flight calls of crossbills, *yenks* of red-breasted nuthatches, the *tut tut* then *piik piik* calls of a robin nearby . . . and from the far side of the pine our mystery bird flies. "A Steller's jay mimicking a red-tail. Had me stumped. I think they must do that just to play with our heads." David responds to the great adventure with all the solemnity he can muster: "Birds are awesome!"

Steller's jay

Steller's jay. Red-tailed hawk? or . . . it *is* a Steller's jay! (343, 0:31)

Back in camp, we admire the tranquil scene across the way, a happy family quietly greeting the day, preparing breakfast around the campfire beneath some of the most magnificent ponderosa pines we've seen. "And you didn't hear a thing last night? Reminds me of that campsite in Montana just below Lolo Pass. We had the rain fly off the tent, enjoying the good air through the mesh, but in the night I heard the thunder getting louder and louder. Had I not gotten out of the tent and put the fly on we'd have been soaked. You slept through it all, even the thunder and lightning. Then there was the flood in the night back in Prairie State Park—you never stirred there either. Good thing you have me to take care of you." To which our chief cook and mechanic simply smiles and nods his head.

The jays continue to play with my head, uttering calls I never would have thought possible. And it's not just one jay at a time, but two, sometimes three birds simultaneously around the campground, all using the same odd call and matching each other just as the eastern blue jays do. Nice.

Out of the campground, we drop steeply from the pass toward Prineville, soon stopping to put on every bit of clothing we have. "What's that, you say? Never camp at the top of a pass?" David's long-standing advice is driven home in the tough wind chill during our 20-mile-per-hour descent; it's hard to believe that temperatures will soar from the current thirties up to the eighties today, when gravity's gift would be far more welcome.

We bike beside majestic ponderosa pines, along lush and pristine roadside meadows, dropping down nearly 2000 feet into dry juniper and sagebrush again. Past the Ochoco Reservoir and just before Prineville we round a bend, and *there they are,* the High Cascades! I pick out old friends that I came to know when I was a young graduate student at Oregon State some 30 years ago.

Continuing toward Prineville, I survey the barren landscape, far more ancient than the Cascades themselves, and try to imagine what it was like some 40 to 50 million years ago: Ancient volcanoes towered over tropical forests and stretched from here to northeastern Oregon, the remains of a few still evident where we saw them near Mitchell yesterday. Later volcanic activity, roughly 15 million years ago, would add some finishing touches of lava and ash and mud to the landscape. In the calm of this early July morning, it is hard to imagine the inferno and fury that once ruled here.

Beyond Prineville, we bike beside the Crooked River, marveling at glimpses of the canyon this river carved while the Snake River was gouging out Hells Canyon. Across the open sagebrush and junipers the view of the Cascades has widened, and I pick out the peaks, Bachelor Butte (as I knew it in the late 1960s, to become "Bachelor Mountain" when the ski resort took over) and Broken Top to the south; the Three Sisters, South, Middle, and North; Mount Washington dead ahead, Mount Jefferson to the right, and somewhere farther north Mount Hood, all youthful volcanic peaks born of fire within the last million years. Tomorrow, I smile, we will climb up and over that mountain range.

In Redmond, as we await the 11 a.m. opening of a restaurant so we can get our late morning shake, David befriends a young boy on his bicycle. "Flat tire, huh, and a few other things?" In no time David has a friend for life, as he repairs the tire and makes some minor adjustments here and there.

Outside of town, I repair my own flat tire. "Builds character," I'm told. About halfway from Redmond to Sisters, with the lush valleys of the Crooked River and the Deschutes River not far behind, in the dry rain shadow of the snowcapped Cascades that loom on the horizon to the west, with sunburning temperatures already in the eighties but protected by the shade of a fine juniper beside the busy road, I happily build. David looks on as he munches an apple.

Early afternoon we settle into the Sisters City Park on the east side of town, once again among ponderosa pines. While David is off using the internet in the library, four police cars converge on a nearby campsite. I ask a neighbor what's going on: "Yeah, the Indian has been asking campers if he can use their cell phone, and when he's told 'no' he gets upset." But four police cars? When David returns, I warn him of the situation. Next thing I know he's over talking to the "troublemaker," offering him use of his cell phone. Some kid.

Afternoon gives way to a pizza dinner and early to bed. Sometime in the night I'm yet again awakened by a commotion. It sounds like an army of monstrous ants scuffling through the campground, surrounding us and settling in for the night. I listen, soon wondering if it was only a dream, as silence once again reigns, and I'm back asleep beside a son who never stirred.

29.
OVER THE CASCADES

DAY 68, JULY 10: SISTERS TO LIMBERLOST CAMPGROUND, OREGON

It *was* an army that surrounded us in the night! At 5 a.m. we awake to a sea of small, forest-green tents, on the road and paths and everywhere a bare, level surface once was, and sharing our campsite as well. "Firefighters," we learn, as they're already stirring after so short a night. They're traveling from some distant fire and heading to yet another, perhaps the one we've been told we'll see today from McKenzie Pass. I'm struck by how quietly they had moved in the night, and also by my sense of admiration and gratitude for their efforts, and the honor of sharing our small site with them. There will be no complaints about their invasion from the paying campers.

As we pack up, a reluctant wood-pewee and robin offer a few snippets of song, their lone voices declaring that the dawn chorus is rapidly fading for the season. About five miles out of town we stop for breakfast at Cold Springs Campground, yet another Forest Service campground set among ponderosa pines. Dining on cold cereal and raisins mixed with powdered milk and water, all smothered with banana, we're treated to ravens cavorting about the campground—they *croak* and *crock* and *rrock* and *quark* and *quirk* with such a diverse vocabulary, sounding like they have so much to talk about that there just has to be a vast intelligence there.

Common raven

Common raven. A small sample of all that is on his mind. (344, 1:50)

"More intelligent," I think out loud with David, "than the fox sparrow over there who repeats one of his three or four songs with rigid precision? Or more intelligent than the Townsend's solitaire high in the pine who races pell-mell through one of the finest songs imaginable? Or the mountain chickadees, the flock near the outhouse probably mom, dad, and the kids of this year?"

Other quick comparisons run through my mind: the crossbills who fly over, *jilp jilp*; the red-breasted nuthatches who *yenk* high above; a brown thrasher with thousands of different songs; a Clark's nutcracker or a pinyon jay with an extraordinary memory for relocating thousands of food caches; a bobolink who navigates to Argentina and back each year, returning with precision to a familiar little patch of ground in both places. And how about the pewee over there? Unlike the others, he's not a true songbird but instead a flycatcher, so he lacks the special brain and learning abilities of the songbird, his relatively simple songs instead somehow encoded in his DNA.

EACH INTELLIGENT, IN ITS OWN WAY

Fox sparrow. Leaping among four different, exquisite songs. (345, 2:53)

Townsend's solitaire. Rippling, elaborate, melodious mountain music, with calls. (346, 4:22)

How can one be elevated over another? Each of these birds has some native intelligence that has been passed down through countless generations from the beginning of time; each is best at what it does, and each by definition is a success in its own way, each a survivor of a long unbroken lineage of success stories since the primordial ooze first generated life. Measuring something we humans call "intelligence" seems pretty silly, though my inevitable anthropocentric view gives a nod to song-learning songbirds because they learn to sing much the way I learned to speak, and to the loquacious ravens, who seem to have some human-like insight into how things work, how a simple tool might be used, for example, to extract food from a crevice.

Back on the road, we cycle among the ponderosas, with the sun at our backs and the red bark aglow in the early morning light. "Hermit," calls out David from up ahead. Yes, on the right a lone hermit thrush sings; we follow the curve of the road around him, and I follow along with him as he leaps artistically from one song to the next. The ponderosa pines gradually give way to Douglas firs, and everywhere a congested understory of lodgepole pine seems ready tinder for the next forest fire that will sweep through here. The trees grow tenaciously from an ocean of lava, ripples and waves and tsunamis of lava that have been frozen in place, making any venture off the road an impossible scramble. About 12 miles out of town the road skirts a more recent lava flow, rugged and treeless, and the expansive view from this Windy Point, as it is called, offers a close-up of heavily sculpted Mount Washington, which seems to be losing its battle with glaciers. Hermit thrushes, juncos, and chickadees are our companions, and the *hip, THREE CHEERS!* of an olive-sided flycatcher celebrates our climb.

And then it is *only* lava, a great wall of dark, jumbled lava, that we face, our passage a narrow canyon carved by some road crew through the once fiery inferno. A quarter mile later we surface in the parking area next to the Dee Wright Observatory, an observation structure with a well-camouflaged, lava façade in this austere setting. We climb up for an unobstructed view of the High Cascade peaks.

Hermit thrush. In wind-blown ponderosas, bounding among seven different songs. (347, 7:58)

I can't help but reminisce a bit: "This was my backyard for four years, long before you were born. I climbed South Sister with graduate student friends—can't see it from here, though, because it's hidden behind North and Middle, those two cinder cones just to the south. Beyond Washington, smoke from the forest fire hides Jefferson; lots of times we'd hike into the wilderness area at its base and camp around the lakes there, once when your sister was only a few weeks old. It was a wonderful four years that your mother and I lived in Oregon, and we hated to leave."

Olive-sided flycatcher. An unmistakable *hip, THREE CHEERS!* as we scale the mountains. (348, 0:55).

Nearby are the craters that oozed lava over this landscape between 1500 and 3000 years ago. To the north, only two miles away, are the Belknap Craters, and about three miles to the south, Yapoah Crater, the cinder cone that spewed the lava that covers the pass here. On higher ground in this dark, barren sea are a few islands of more ancient lava flows on which trees have taken root. Just a little over 100 miles to the west and a mile lower lies our destination, the Pacific Ocean. "Gotta stop biking then, Pops. We run out of land. Better slow down."

Out of the silence, off toward Mount Washington, rises the simple song of a lone bird; it's just three notes, a pure note followed by two lower, more buzzy notes, floating up from the lava every ten seconds or so. I offer a guess, a gray-crowned rosy finch, as I can't imagine what else might be happy in such a desolate, high-altitude landscape.

We begin our 3000-foot descent, soon among stunted subalpine firs staking their claim to the land, and within four miles we are engulfed in forest once more. Descending gradually at first but then more steeply, we pump our brakes on the straightaways but brake hard through hairpin turns until we stop at a pull-off to let our rims cool, lest their overheating blow out our tires.

Down the road a pack of wildly colorful cyclists approaches, all bunched together and racing uphill on their ultra-light bicycles. David quickly hands me his camera and mounts his bike: "Take my picture." He drops down to the beginning of the pull-off and then, his timing perfect, begins his ascent through the pull-off, arriving opposite me just as the peloton arrives on the road, so that my picture shows David with his fully loaded touring bike leading the pack up to McKenzie Pass. The racers, clearly of a different breed than us, with our heavily loaded touring bicycles, yell friendly words of encouragement to David as they whiz by.

Though I know that the logged clear-cuts probably begin just a quarter mile off the road in this Willamette National Forest, the forest we ride through as we drop off the High Cascades on the way to Limberlost Campground looks untouched, pristine. Spotted owls, I think—this must be great habitat for the endangered owls that environmentalists have used to help save some old-growth forests from logging. Sentiment about these owls runs high in logging country, and I recall the bumper stickers I've seen: "Spotted Owl tastes like Chicken," and "I like Spotted Owls—Fried."

In the campground we pitch our tent beneath western hemlocks and red cedars and Doug firs, beside a rushing river in which David is soon dipping, never mind that the temperature is a bone-chilling 38 degrees. All about us is dense undergrowth the likes of which we have not seen on our entire journey; the westerly winds greedily absorb water as they sweep over the Pacific, but as they rise up and over the Cascades the air cools and relinquishes what it can no longer hold, drenching all that we see. Moss grows between the toes of native Oregonians, I know, because I've lived here through the long, wet winters.

My baseball-throwing arm still a threat, I launch a stick skyward and over the low limb overhead, the trailing rope soon pulling our food supplies well out of reach of any bears who might prowl in the night. To the sound of a rushing river, beneath trees that have been spared so that we might imagine we are in some grand primeval forest, a gentle breeze whispering through the fir needles high overhead, we drift off to the last sharp *chip* notes of a Pacific wren going to roost nearby.

DAY 69, JULY 11: LIMBERLOST CAMPGROUND TO THE WILLAMETTE VALLEY, OREGON

It'll be a long day, perhaps 90 miles to the special place I need to visit, so we arise early and begin packing up in the dark. In about ten minutes the Pacific wren *chip*s from his night roost and then, high up in the Doug fir, begins his morning song.

What fine memories he jogs. "I was your age, David, just twenty-four, when I became intrigued by these wrens. I was a graduate student at Oregon

State then, and always looking for more projects. So I'd come out in the wee hours of the morning like this and wait for the wrens to sing, and then record them for as long as they'd cooperate. They usually sang from high in the Doug firs, and I'd often lie on my back and aim the parabolic mike to the sky. It took months to analyze the songs back at the university. I titled my eventual paper 'A Pinnacle of Song Complexity.' They're amazing singers. Too bad we can't hear the dainty little notes in real time. When I slow these songs down and we follow along in a sonogram, my bird-loving audiences always burst into applause at the end."

There's a flycatcher, one of the Mosquito Princes—I listen, searching for a mnemonic . . . just an occasional single song, all of them the same, two-parted, the first note sharply higher than the second . . . much like the least flycatcher's *cheBEK*, or maybe *SE-pit*, which according to my *Empidonax* cheat sheet would be a Hammond's flycatcher!

Pacific wren. Old friends bring back fine memories. (349, 5:01)

"Hey, it's a new bird, a Hammond's flycatcher, another of that special group we've heard since the Atlantic. Back in Virginia there was the Acadian's *PEET-sah* and the least's *cheBEK*, and most likely the alder's *fee-BEE-o* atop Mount Rogers. We've heard the willow flycatcher's *FITZ-bews*, first the eastern and then the western forms—I'd call them two different species. And in the Rockies we heard the gray and dusky and cordilleran species. Here's the ninth! He's singing just a single note now—probably too late in the season or past the dawn chorus to hear all three of his songs. There's just one more species to go, the Pacific-slope flycatcher, and then we'll have them all. We'll hear him tomorrow for sure." David seems mildly interested in this flycatcher surfeit.

Hammond's flycatcher. The ninth Mosquito Prince—one more to go. (350, 3:59)

Out of the campground, we ride through an evergreen paradise drenched by winter rains, the extra water from snowmelt at higher elevations now tumbling beside us in the McKenzie River. It's all downhill to Eugene in the heart of the Willamette Valley, and the miles fly by as I listen to all that sings and calls from the roadside. White-crowned sparrows sing again for the first time since Montana, and I listen carefully, assuring myself that all are in tune and singing the same dialect. Juncos and robins and Pacific wrens line the road. Songs of Swainson's thrushes spiral upward; those of the "hoarse-robin" western tanagers hold steady. A pileated woodpecker calls off to the left, *kuk-kuk-kukkukkukkukkuk-kuk-kuk*, followed by a higher pitched call across the road to the right, most likely from his mate.

White-crowned sparrow

At 43 miles on our odometers, in the metropolis of Walterville, population 50 according to our map, we happily leave busy route 126 to take the less traveled, back roads around Eugene, population 117,155.

"So what is it about the science and scientists in your field that upsets you?" David asks on the quiet country road. Oh, that's a huge topic, one that I've mentioned occasionally on this trip, but now he asks me to address it head on.

"Maybe it boils down to one simple issue, that I take things too seriously."

"That doesn't surprise me," David says.

"Yeah, right. OK, let me rephrase that. I take *birdsong* seriously. I love listening to birds and all they have to say. I love studying them and trying to understand what they're doing. Even the little discoveries are exciting. Some others who study birdsong aren't quite so hampered by the gravity of it all, and over the last 15 years I've published a few critiques, suggesting how we could improve our science so that we learn what birds actually do. You can imagine I've not made a lot of friends."

"Can you give an example?" he asks. I recoil, not eager to relive the old battles here, reminding myself that it's too nice a day, but David is a scientist himself and persists.

"Well, OK. Here's one example: It's widely believed that the more songs a male has, the better he is at impressing females or fighting other males. You can read it in the textbooks. It's a nice, sexy story, but there's simply no evidence, and it pains me that popular mistruths like this are so pervasive. What's that? . . . you want one reason why it's not true?

Song sparrow

Song sparrow. In 11 songs, he uses two different songs. (351, 2:57)

"Well, take that song sparrow singing over there. He probably has eight different songs, the average for song sparrows around here—I know because I studied them near here many years ago, ten years before you were born. We'd have to believe that a male with nine or ten songs is better than one with eight, but when that song sparrow was a few months old and learning his songs, he practiced perhaps 15 different songs, eventually keeping the eight he wanted and discarding the others. If a female were really impressed with how many different songs he could sing, the male who kept just one of those discarded songs would be a rock star. He'd father more young than the other males, and the race for larger and larger repertoires among males would be on. But that doesn't happen, in song sparrows or any other species."

David presses on, wanting more reasons, and I list them briefly, focusing on the excitement of discovery, how good science is done, on the "utter honesty" and "scientific integrity" of which the physicist Richard Feynman wrote. Someday, I promise David, I'll publish a well-reasoned paper that will disprove once and for all the widely accepted notion that "bigger is better." A very small titer of adrenaline surges through my veins.

"You want another example? I could tell you about my three-wattled bellbird project in Costa Rica, how these bellbirds are close relatives of the nonlearning flycatchers yet learn to sing just like songbirds. It's really exciting, because each singing male changes his songs *continuously* to match what the other males around him are singing, so that our yearly samples during July show remarkable changes from one year to the next. No other bird that we know does that, not even any songbirds, the song-learning specialists! Songbirds typically learn just during the first year of life, and then they're done, and they die after a couple of years; the bellbirds can live 25 years, learning and relearning all the time. But my unequivocal proofs of song learning are dismissed as ridiculous by the pea-brained, know-it-all experts who can't think outside of the little boxes they live in. . . . Did I say that? . . . Yeah. . . ."

I think I am not long for the academic life. By the time we reach the Pacific, I hope to have figured out how to reinvent myself. I'd be happy if the Pacific were another thousand miles away, not just over there beyond the Coast Range.

We bike on, the Willamette Valley that I know so well gradually revealing itself. The Coast Range lines the western horizon, the Cascades the eastern, and we bike through open country, the rich farmland due in large part to fertile topsoil delivered courtesy of Montana, Idaho, and Washington by those Lake Missoula floods during the Pleistocene. Oregon oaks grace the countryside, some singly but often in small groves. In the heat of midday, we dine at a restaurant in Harrisburg, then hang out at the park along the Willamette River. David swims; I lounge on shore. We both know that the route north up the valley will be all the more beautiful in cooler, late afternoon light.

Three-wattled bellbird. Flycatcher relative learns! An evolutionary breakthrough! (352, 0:40)

A young female cyclist wheels up, her bike loaded like ours, the camaraderie instant, especially for David. "Heading to Virginia," she says. "I started two days ago with a guy I met on the internet, but he was so obsessed with speed and miles each day that I told him to go ahead. I'll go it alone," she says with all the confidence she'll need. David is full of advice for her, and as I drift out of earshot to give him some space, he is offering her our Chamois

Butt'r, the soothing skin lubricant that keeps our butts happy for long days in the saddle. Ten, maybe 15 minutes later she leaves, alone, somewhat to my surprise. "Hot" is the one word I catch from David's report.

In time, with lower temperatures (the air, that is) and in softening light, we stow the Adventure Cycling maps in a pannier and head "off-route," now relying on local advice and a local map to find our way toward a special place where I want to be at dawn tomorrow. Crossing the bridge over the Willamette River, we stop to admire the osprey and its nest before heading north on River Road, up the valley known as the "grass seed capital of the world." Through fields of rye grass and fescue, among corn and beans, beside picturesque Oregon oaks silhouetted against a blue sky, past farmhouses and farmers tending their fields, along roads that turn and wind, some paved, some gravel, we find our way north, then eventually west—"the finest hour of the day to bike" says my riding companion. Just before sunset, with about 86 miles on the day and about a half mile to busy route 99W, we agree that it's time to find a place to slip off the road and bandit-camp for the night. Rounding the next curve, we spot a couple of RVs in a yard, then a big sign, "BUNDY BRIDGE Campground," as if magically placed there just for us.

"Yes, you can camp back there in the grass, under the trees next to the picnic table," we're told. "The ten dollars gets you a shower, too." We set up camp, shower, and then dine simply on bread and chili directly from the cans, listening to robins and goldfinches, and to the pleasant rhythm of a brown creeper's song in the oak above: *trees trees mur-mur-ing trees*. I can picture the sonograms I've made of the songs recorded here in the valley, with two high notes, a couple of faster lower notes, then another high one again. *Trees trees mur-mur-ing trees,* to the accompaniment of a gentle breeze rustling the leaves of the spreading branches of the Oregon oak overhead. With light fading, the landscape turns first yellowish, then reddish yellow, and streams of irrigating water rocket skyward from the valley floor as if venting some vast reservoir below.

Brown creeper. *Mur-mur-ing trees,* from the Willamette Valley. (353, 1:08)

With the sun down, I'm in the tent, setting the alarm for 3 a.m. and readying myself for the big reunion tomorrow. For four years during my graduate student days, from 1969 to 1972, my field home was the William L. Finley National Wildlife Refuge, just a few miles up 99W from here, where I studied the songs of Bewick's wrens and spotted towhees and other birds. Tomorrow, in just a few hours, I will return to see and smell and hear this special place once again. I struggle to comprehend the simple fact that after Finley, by nightfall tomorrow, we'll be on the Pacific.

30.
A HOMECOMING

DAY 70, JULY 11: WILLAMETTE VALLEY TO NEWPORT, OREGON

David slumbers on as I find my way in the dark to route 99W and then head north. How eerie to ride alone in the dead of night on an otherwise busy highway. I am King of the Road, claiming both lanes as I weave into the oncoming lane and back to mine, until I meekly yield all pavement to an approaching truck.

How far is it to the refuge's entrance road? I strain to see Winkle Butte on the right and the big refuge sign on the left, but it's so odd to be approaching from the south; as a graduate student, I arrived from Corvallis to the north, and through countless early morning visits knew just where to turn, almost driving the route in my sleep. I worry that I have missed my turn, but the large and familiar Finley sign soon looms out of the darkness on the left, welcoming me home.

Down gravel Finley Road I bike the mile to the west, then the quarter mile to the south, pausing at a prairie observation platform where there was none years ago. Following the curves in the road, I head another quarter mile west and *I am here,* at the very location where I often waited at this time of day for the towhees and wrens to begin their dawn singing. Though I know exactly where I am, it feels so different, as the trees have grown so tall. In the darkness, a little after 4 a.m., the moon having just set, I wait as I so often waited here thirty-some years ago.

The towhees will be first, each male awaking with a call or two, *zhreeeee, zhreeeee,* then launching into his chosen song to greet the day; he'll begin slowly and repeat that song several times, but then more excitedly alternate two or three different songs. I should be able to hear four male towhees here, each on his own territory, each heatedly dueling with his neighbors for a good half hour, though probably for a much briefer period this late in the singing season.

Spotted towhee

These towhees fascinated me, and over several mornings I recorded the four males here. At the university, I graphed the songs on equipment that was so cumbersome compared to software programs that now do instantly what I command. Studying the songs, I learned that each male had seven to nine different songs he could sing, and these four neighbors shared most of their songs with each other. Intrigued, I recorded other towhees in the area, and discovered that over fairly short distances all the songs changed. I caught the four focal males in a mist net (a fine net strung between two poles and used to catch birds), and slipped colored bands onto their legs; and I learned over the following year that these birds were not migratory, but instead resident, each staying on his chosen territory for life.

How satisfying, then, this glimpse into towhee life: A young male towhee leaves home in search of a vacant territory, as young birds do when about four or five weeks old. When he finds a territory, perhaps by two months of age and within a mile of his birthplace, he settles in for life, learning there the particular songs that identify the micro-dialect of his singing neighbors. How satisfying, too, to see this small research project become my first publication; the paper appeared in 1971 in a journal called *The Condor,* a small token that my work and I had been reviewed and accepted by the professional ornithological community.

The Bewick's wrens will sing here, too, and their story is much the same, but what enormous effort I put in to show it. It seems a lifetime ago that I wrote that central chapter of my doctoral thesis, the work of which I am perhaps still most proud. The ruling idea of the day, based on song-learning studies in the laboratory during the 1960s, was that a young male learned the songs of his father (most mothers don't sing), but I doubted. For the birds I knew, such as the towhees and these wrens, the songs changed so rapidly over short distances that a young male had to be flexible, I reasoned, and instead had to learn the songs of the particular little dialect where he settled *after* leaving his singing father. These local dialects are widespread among songbirds but could not exist if a young male rigidly retained his father's songs after dispersing from home.

Bewick's wren

The plan to collect the necessary information was simple, though not so simply executed. I needed to compare the songs of fathers, their sons, and the adult males in the neighborhood where the sons settled. Parting words of wisdom from my faculty advisors were something like "Good luck finding the kids after they leave home and disperse all over creation."

In the spring and summer of 1970, I netted hundreds of birds all over this refuge. The dads were easy. I'd set up a mist net between two poles, play a few wren songs over a loudspeaker from the base of the net, and *vroom*, the master of his territory would attack the intruder and soon be entangled in the net. In a few minutes he was released unharmed, though perhaps a bit ruffled, with a unique combination of colored bands on his legs so that I could then identify him through the binoculars.

The kids were more difficult. I waited until they were out of the nest and moving about in a family flock; when I'd find a family of these hungry and noisy youngsters, I'd imagine myself as one of them, trying to guess where they would move to next. I'd hustle there, perhaps between two bushes down a fence row, set up the net, and then try to herd them in that direction. With the perfect setting, I'd catch the entire family.

After recording the fathers of these families, I scoured the surrounding area for their sons over the following months. One by one I found many of them, beginning at about two months of age when each was on the territory that he had selected for his life; most had dispersed a considerable distance, some over a mile. Each young male was already babbling and practicing his songs, so I recorded him and his neighbors that year and, for all surviving birds, again the following spring. The results were conclusive: A young male clearly rejects his father's songs and instead learns the songs of the small neighborhood where he settles.

Since these two early projects, I've always listened to birds as individuals. "Just who are you?" I find myself asking. "Where are your parents, your grandparents? How did you get here? What songs are you singing? Where did you get them, and from whom?"

A towhee—he awakes by calling *zhree . . . zhree . . .* , then sings, soon playing with three different songs, jumping from one to the next, alternating them, just as he should at dawn. A Bewick's wren is next, offering one song several times, then switching to another, and eventually another, but I don't recognize any of them. Thirty years ago I came to know every song that the wrens sang here; I numbered them, and as I say "1" I still see the sonogram and hear that song in my mind. It's the same with "2" and "3" and all the way through "16," as each of these males had about that many different songs and they shared them with each other in this local dialect. The songs I am hearing now are different, none of them the old familiar tunes. Over 30 generations of wrens, the local dialect has changed. Perhaps a song or two was added or dropped each year, and some songs gradually morphed into something

Spotted towhee. Awaking at Finley Refuge; with calling Swainson's thrushes. (354, 4:03)

unrecognizable. The changes over time were no doubt gradual, just as the songs change gradually over distance.

Among the bullfrogs a yellowthroat sings, *wich-i-ty wich-i-ty wich-i-ty wich-i-ty*. More robins have joined the chorus, and a Swainson's thrush, too. And how familiar the smell, so distinctive, yet I've never bothered to figure out exactly what it is—it's the grasses, I think, perhaps curing in the summer heat.

Though I yearn to linger here, I must move on, to visit other birds in other favorite places. In the bottomland forest along Muddy Creek, I park my bike and walk, adding to my growing list of renewed acquaintances: a western wood-pewee in full dawn song, and a western tanager, too; wood ducks flush from the creek, the females squealing *oo-EEEK oo-EEEK* in flight. And a black-capped chickadee . . . *yes!* It is as I remember, four, sometimes five intense whistles all on one pitch, and he alternates that song with the most beautiful descending series of whistles imaginable, finer and purer than even a canyon wren. It's the local chickadee dialect, strikingly different from the *hey-sweetie* songs that I know from most of North America. It seems that the High Cascades isolate these resident chickadees, and a string of unique local dialects can be found all the way up to northwestern Washington, where the *hey-sweetie* songs again predominate in British Columbia and make a solid run across the continent to the Atlantic.

Bewick's wren. Four different songs of the "new" local dialect.
Song 1 (355, 0:38)
Song 2 (356, 0:52)
Song 3 (357, 0:56)
Song 4 (358, 0:54)

Black-capped chickadee. A special dialect of whistled songs, not the usual *hey-sweetie.* (359, 5:39)

Another Bewick's wren sings—following old habits, I start to raise my binoculars to check for colored bands on his legs, but I stop short, knowing that my old friends are long gone, their leg bands buried beneath three decades of leaf litter. I settle in on a song sparrow, enjoying the tone and rhythm of his song, hearing him deliver it nine times before he switches to a song so different one might think it came from a different species. A pair of great horned owls hoots up by the old barn, triggering a long-forgotten memory; in the barn's loft, I had collected the pellets that these owls' ancestors had regurgitated, and then I dissected out the bony bits of shrews and mice, just to see what the owls had been eating. An orange-crowned warbler offers the only song he knows, a series of high, rapidly repeated notes that then shift abruptly to a lower frequency. "One song does me just fine," he might say, "as I don't need the two-song system that many of my warbler relatives have." "But why not?" I want to ask.

Great horned owls. The entire family—mom, dad, and the kid calling from the very barn I once haunted. (360, 1:32)

Orange-crowned warbler. Loose trill growing in pitch and volume, then fading, as if the singer loses interest or energy. (361, 2:16)

Pacific-slope flycatcher. From the foothills of the Coast Range, the last of the Mosquito Princes! (362, 1:58)

Tree swallows. Airborne twitters heard at dawn. (363, 1:18)

To my surprise, David arrives a little before six, and we settle in for breakfast at the picnic table near the headquarters. I want to tell him how it feels to return to this special place after 31 years, seven years longer than he's been alive, but I settle for pointing out a few of my friends who sing around us.

"Yellow-breasted chat," I call, pointing off to the north beyond the headquarters. We listen to his strange program of hoots and squeaks and the like.

"Swainson's thrush," he points out, "and a robin."

"Yeah, and a willow flycatcher. Hear his three songs? . . . And if we were yonder, a little over a mile away in the lush Doug fir forest in the foothills of the Coast Range, we'd be listening to the tenth and last of the Mosquito Princes. Over there, right now, the Pacific-slope flycatchers are singing just like their close relatives the cordilleran flycatchers did back in Colorado and Wyoming."

I want to tell him about the Pacific wrens who sing beneath the flycatchers in the tangle on the forest floor, the Steller's jays who call *shook-shook-shook* overhead, the band-tailed pigeons who clap their wings as they flush from the treetops, but I resist, letting David enjoy a quiet breakfast. I continue eating in silence as well, just listening, acknowledging to myself what comes to ear: American goldfinches fly by, *per-CHIK'o'ree, per-CHIK'o'ree*; a lazuli bunting sings along the lane to the west; a warbling vireo is high in the trees above the headquarter buildings; cedar waxwings *seeee* on the wing, somewhere above; a house wren bubbles and gurgles up by the barn, from the same place where 32 years ago I found a special house wren who had surprisingly mimicked many Bewick's wren songs. Tree swallows dart about everywhere, their aerial dawn singing long abandoned. *Zhreek, zhreek,* harsh and sharply rising, the signature call of a western scrub-jay, such a familiar sound throughout the valley. Just yesterday I stopped to admire a scrub-jay roadkill, rolling him over in my hands, studying his blue necklace on immaculate white below; the bright blue head, wings, and tail; the grayish triangle on the back; his dusky cheeks and long tail. In the hand, dead, the way most birds were best known 100 years ago, before modern field guides and binoculars and "birders" were born. I prefer modern.

"Hey, a purple finch, just behind you. I'm sure they were atop Mount Rogers in Virginia, but this is the first we've heard them. Listen—there," and I lower my finger to point when he sings. We hear a rich, fast-paced warble, a series of tumbling musical notes, no two the same, rising overall before dropping at the end—very nice. He's a good looker, too. "Peterson in his field guide says he's like a drab sparrow dipped in raspberry juice."

"Red-tailed hawk," David announces, and from the open fields to the north arrives another *kee-eeee-arrr*, the unmistakable, sliding asthmatic scream of the red-tail.

"I'd love to find an Anna's hummingbird around here someplace. They're special; males have a complex song that is learned and sung in dialects just like songbirds. It's a high-pitched, sputtering, six-second wonder, beginning with *bzzbzz-bzz* notes and ending with something flashy, like *chur-ZWEE dzi!dzi!* Our lumbering ears can't parse the details, but we know that the birds can because they learn those details from each other."

Western scrub-jay

Western scrub-jay. From this know-it-all, an unmistakable rising *zhreek*, and a pounding *shreck-shreck-schreck*. (364, 0:17)

In the early morning quiet, before other visitors and refuge personnel are about, we have the world to ourselves. My mind swirls about the refuge, visiting places I came to know so well many years ago. In Cabell Marsh just beyond the headquarters here in the center are the pied-billed grebes and herons and, in winter, untold numbers of Canada geese. On Pigeon Butte far to the south the lazuli buntings have an unfettered view of the valley. Circling clockwise, in the rich bottomland forest of Maple Knoll, red-breasted sapsuckers drum. Past the Pacific-slope flycatchers among the

A Finley symphony (365, 11:30) **Lazuli bunting**. Rapid, brief, repetitive songs. **Purple finch.** So rich a rising and rousing warble. **Orange-crowned warbler**. Colorless (so say some) trill that swells, then fades. **Wilson's warbler**. A staccato chatter, given earnestly. **Warbling vireo**. A young male finding his voice. **Evening grosbeak.** Another songless songbird, with two distinctive calls.

Red-tailed hawk. Asthmatic screams drift lazily over Finley's open fields. (366,1:11)

Red-breasted sapsucker. The Maple Knoll Drummers got fine rhythm. (367, 1:59)

Doug firs is the oak woodland with its black-throated gray warblers. On the grassy knoll of Bald Top I first realized how songs of vesper sparrows occur in dialects throughout the valley. I swing around to Brown Swamp on the north and then to the west, following Muddy Creek bottomland south all the way to McFadden's Marsh, teeming with marsh wrens. All over this map are bright memories of special places where I caught families of Bewick's wrens and later discovered young males on their first territories, where I stood eye-to-eye with a bobcat as we studied each other for a good half hour in a winter rain, where the finest blackberries are come August, where the big bucks hang out.

"Back when I was your age, just twenty-four years old, this refuge was my life. I banded and recorded the wrens all over this place, and started projects on towhees and song sparrows and vesper sparrows and house wrens and more. You would be born years later after we moved to New York." How different our lives: I was so focused and busy at his age, starting a career and a family, but he is single and taking two and a half months to bike with his father across the country. When my father was eighty, he still loved his bicycle, but I don't recall ever biking with him. I'm sad about that now.

As we ride on, past the open grassland on the flanks of Bald Top, a male western bluebird perches atop a refuge sign, nervously watching us bike by; *feew, feew,* he calls. By the maintenance sheds, right where they always are, the clown-faced acorn woodpeckers give us a raucous, laughing *whack-a whack-a whack-a*; in my ornithology lectures, which I just might never give again, we talk about mate-sharing and group sex in these breeding groups, how two or three females lay their eggs in the same nest and how each female does her best to have her own eggs succeed. And there were lectures on foraging, how a group of these woodpeckers drill shallow holes in trees to store acorns, an entire tree becoming a large granary with thousands of such caches. My thoughts are interrupted by a California quail, who touts *chi-CAH-go,* as if he were misplaced in this western geography.

We exit the western end of the refuge where the land beyond rises to the foothills of the Coast Range, a mere 35 miles from the Pacific. I turn on my voice recorder: "As if offering a hearty welcome to his coastal domain, a wrentit sings *pit-pit-pit tr-r-r-r-r-r-r-r-r-r-r*. I can almost smell the salty sea spray, almost hear the waves crashing on black basaltic headlands." Again he sings, beginning slowly, abruptly accelerating into a rapid trill. I listen for his mate to add her extended *pit-pit-pit-pit-pit* song without the faster trill, but not today. We leave them to skulk in the underbrush, sight unseen as is their habit, my parting mental image of this phantom a reddish-brown, wren-sized bird assigned a tail that seems far too long for its body; as if still confused about it, the bird cocks its tail comically this way and that at odd angles, still trying to figure it all out.

Black-throated gray warbler. He sings in Z's, *buzZEE buzZEE buzZEE ZEE*. (368, 1:23)

Heading north along the refuge boundary, we are offered one last Finley gift: A Hutton's vireo sings from an abandoned apple orchard off to the right. He sings such an unlikely program, offering the same simple song over and over, about one per second in a seemingly never-ending series of identical songs, much like a loggerhead shrike. *Chew . . . chew . . . chew . . .* We stop to wait, and eventually he switches to another song, *zu-wee . . . zu-wee . . . zu-wee. . . .* How many different songs he has is anyone's guess, as no one has had the patience to hear him out.

Acorn woodpecker

Acorn woodpeckers. With laugh-like calls so suitable for clown-faced birds. (369, 0:51)

My mind sweeps across the vireo-rich continent, listening to red-eyed and white-eyed and blue-headed and yellow-throated and warbling vireos in Virginia and eastern forests, to Bell's vireos in thickets of the Great Plains, to plumbeous and Cassin's vireos in the mountainous west, and now this Hutton's. In most other species, songs are also brief and simple, but males flaunt their repertoires, the back-to-back variety in songs clearly important to them. Only the white-eyed vireo repeats himself anything like the Hutton's, but each of the white-eye's songs is complex and multi-parted, each one "exciting," and he waits a reasonable five seconds before repeating himself, the overall performance never generating the strain on our ears that the Hutton's does.

Hutton's vireo. So relentlessly repetitive, so unvireo-like. (370, 2:04)

With Finley behind, we head north along the western edge of the Willamette Valley, and soon spot Mary's Peak to our left, the highest mountain in the Coast Range and towering over 4000 feet. High up on that peak, I've seen the ancient basalts that once formed on the floor of the Pacific. A big hunk of the Pacific plate that "should have" subducted under the North American plate "decided" to float up and add itself to North America, thus shifting the subduction zone further west. Continued subduction out there jacked up Oregon's new terrain to modest heights, forming the Coast Range.

At Philomath we head west through the village of Wren, my all-time favorite town name, and follow route 20 toward Newport. Even though it's mid-July, birds are still singing, and in the fields and fir forests along the road I register Pacific wrens, goldfinches, juncos, white-crowned sparrows, and the simple call of the Pacific-slope flycatcher. And there's a good listen to the piercing calls of evening grosbeaks, together with their more grating call, the apparently complete grosbeak repertoire from a small flock atop a Douglas fir. Evening grosbeaks, I smile—on my short list of songbirds who have lost their song.

The ride is as a last day should be, blissful, with singing birds, a wide shoulder on the road, not a care in the world, everything perfect . . . until we reach a stretch of road signed as a "Safety Corridor." My first thought is "fantastic," as it sounds vaguely good, but I soon learn that it means "Watch Out For Your Lives." The road narrows and curves sharply through rolling terrain, with "No Passing" zones stretching forever. Larger and larger trucks and campers and full-sized homes thunder by, Airstreams and Gulf Streams and Winnebagos and *Intruders* (our least favorite by name alone), all without apparent drivers, all seemingly oblivious to something so insignificant as a bicyclist. We spend more time pulled off the road than on our bikes, a perfect day no more.

"Did I tell you about that trucker in the restaurant back in Cambridge?" David asks. "He described how the limp body of a deer bounces off his truck. 'You drive around a corner and then suddenly there is a deer. Pop!—it flies off the road and I don't feel a thing. I suppose a bicyclist would be exactly the same.' He then continued to drink his beer."

At first chance, we detour, finding our way to the ghost town of Elk City and then the road along the upper stretches of the Yaquina River to Toledo. In the ten miles to Newport, with about 4300 miles of the continent behind us, we bike in a gentle rain beside Yaquina Bay, and if there's a finer stretch of country on the entire TransAm route, I'm hard pressed to identify it now. The bay gradually widens to a mile across the broad estuary at Newport. Gulls perch and call atop buoys and docks and abandoned pilings that the sea now reclaims. Song sparrows sing from roadside bushes, and marsh wrens in their unique western style bubble energetically from the bulrushes lining the bay. Sailboats and fishing boats and pleasure boats and others—some motor or sail about, but they're docked by the hundreds as we approach Newport, with the Yaquina Bay Bridge shrouded in fog just beyond.

In Newport, we celebrate at Mo's, everyone's favorite seafood restaurant on the coast, filling up on clam chowder and as much seafood and fries as we can handle. South through town we face a brisk headwind, crossing a Yaquina Bay Bridge that seems magically suspended from the clouds above and floating above the unseen bay far below. In the gray mist, we can *hear* the Pacific and the foghorn, but I had imagined that we would arrive under blue skies with white-capped waves on a blue Pacific. It wasn't supposed to end like this.

Marsh wrens. Dueling in matching songs and countersongs, as only western marsh wrens can. (371, 6:15)

Just south of Newport at South Beach State Campground, on a busy Saturday we pull into the "FULL" campground, the *hundreds* of sites fully occupied, but we know better. At the nearly empty hiker-biker section, we set up our tent on a platform beneath the stunted pines among cyclists who are biking the coastal route south to California. We mill about some, swap stories with other cyclists, find the restrooms and showers, gawk at campsite after campsite with fancy camper rigs, and walk the trails to the Pacific, standing on the dunes and listening to the waves, the wind, the foghorn, the gulls, but seeing little. Anticlimactic, yes. Lewis and Clark managed "Great Joy" when they saw the Pacific. We had thought about this moment, thinking that we should have something profound to say for posterity. In the fog, David musters "It's cold."

Crawling into the tent, we agree that our journey can't end in a foggy, crowded campground. Tomorrow will be different, we declare: In our personal forecast of bright sunshine and a tailwind, we will sail south along sandy beaches and rocky headlands with frothy white waves crashing down below, the rolling blue Pacific extending as far west as we can see. Tomorrow will be a more fitting end.

31.
LAND'S END

DAY 71, JULY 12: THE OREGON COAST

Swainson's thrushes. Awaking, a chorus of calling birds. (372, 2:03) Awaking calls, then song. (373, 4:31)

Spotted towhee. From our campground, with the roar of the Pacific and the droning of a foghorn. (374, 5:19)

Alone, a good hour before sunrise, I walk the deserted drive through the campground. "The End," I think abruptly, "this is it." Morning after morning for 70 days I've celebrated the dawn from sea to sea, and here is my last.

For now, all is quiet, except for a few sighs and groans of the ghostly metal homes on wheels lined up at site after site, and except for the distant roar of the ocean. I turn toward the Pacific onto a sandy trail into the stunted pines and instantly I'm transported to another world, my world.

A Swainson's thrush awakes, calling at first by running through an exquisite assortment of notes, some rising, some falling, some held steady, some long, some short . . . and finally he sings. Within ten yards of him, I listen to song after song, each a masterpiece of his two voices spiraling upward to the heavens. Gradually emerging to my listening ears are his three different songs, their delivered sequence consistent, A B C A B C. . . . I sweep the continent and revisit other Swainson's thrushes, imagining them atop Mount Rogers in Virginia, bicycling among them in the calm after a brutal storm in Wyoming's Tetons, ascending with them to Chief Joseph Pass in Montana, now here beside the Pacific.

Even closer to me a spotted towhee awakes, calling . . . *zhreeeee . . . zhreeee . . . zhreeeee,* then sings. Song after song is the same, a simple, rapid buzz. He switches abruptly to another, this one beginning with an emphatic down-slurred whistle followed by a slower trill; I linger, and he offers still other songs, six different ones in all, singing 20 to 30 of each.

In his voice I hear the singing season winding to a close. It's mid-July, after the females have already decided who will father their offspring, and his lust for the dawn has passed; he sings at a leisurely pace, repeating each song many times instead of frenetically engaging his neighbors with two or three rapidly alternating songs. "The end," he echoes.

Beyond the towhee, I wander the trails among the pines, listening to voices . . . a song sparrow, a Bewick's wren, a yellowthroat, a junco, a yellow-rumped warbler . . . and stumble upon a small flock of chestnut-backed chickadees. They're largely finished for the season, too; mom and dad have raised their family, and here they all are. They flit among the branches beside me, stopping to look me over, as I do them. Face-to-face, we eye each other. They are ten-gram wonders with gorgeous chestnut backs; each of them eyes a being 10,000 times its size, the upper half a fluorescent yellow, supported below by two longish black stumps, perhaps nothing wondrous about me at all should they care to comment.

Dark-eyed junco. Repetitive, dry trills with just a trace of typical junco jingle; with Pacific and foghorn. (375, 2:27)

But they do comment, with high, scratchy *tsick-i-see-see* calls, their distinctive version of the *chick-a-dee* call. "The Chickadee Who's Lost His Whistle," I think, this chestnut-back only distantly related to the whistling Carolina, black-capped, and mountain chickadees that we've heard across the continent. On this last morning, to the tune of the foghorn off Newport's harbor, the final piece of the chickadee puzzle falls into place. Nice.

Chestnut-backed chickadee

The dawn chorus continues in slow motion, muted and with few participants. Back at the campground entrance, between the cluster of refuse bins and the four-lane highway guiding campers in, a lone white-crowned sparrow sings the coastal dialect. From tall evergreens to the east arrives a *quick, THREE BEERS!*, telling of an olive-sided flycatcher there. Goldfinches call overhead, *per-CHIK'o'ree, per-CHIK'o'ree.* An orange-crowned warbler begins high, then abruptly shifts his two-parted song to a lower pitch.

Chestnut-backed chickadee. Songless, but with assorted calls, including characteristic nasal *tsick-i-see-see.* (376, 0:23; 377, 1:28)

Announcing sunrise is the flight whistle of a cowbird perched atop a tree nearby. But how odd—I listen intently to whistles that sound fluid and unformed, nothing like the distinctive, piercing, repeatable whistles that I'd expect from a seasoned adult. It must be a young bird learning the local dialect and just starting out in life. I whisper those words out loud, "just starting out in life," and smile apprehensively. My thoughts are as fluid and unformed as his whistles, an inner battle raging between a small voice that says "The End" and growing voices that proclaim "The Beginning!"

Still another voice speaks up, a whistling chickadee of the black-capped variety. I hear two longish whistles on one pitch, like *hey hey*, over and over. Knowing there's more, I listen on, and a second song eventually arrives in the form of four brief whistles, up to seven sometimes, all on the same pitch, *he-he-he-he-he*. He soon returns to the *hey hey*, cycling through these two songs. That's nice—here's yet another song dialect of the black-capped chickadee, at land's end. How many different dialects are there in western Oregon and Washington, I wonder, all sheltered and isolated by the High Cascades from the *hey-sweetie* masses of the rest of North America?

Nearby atop a pine perches a crow, among the wisest of creatures. He has his say, then launches, cawing on the wing, his wisdom fading into the distance. I yearn to follow, wherever he leads.

American crow. Behind the simple *caw* is a creature of intelligence; with spotted towhee. (378, 1:27)

By seven o'clock on this Sunday morning, the campground is coming to life, but not David. "Best not to get up before the sun"—I recall his words from day 1, way back in Yorktown, Virginia, and for him the trip ends as it began. I crawl back into my sleeping bag, warming to the idea that a little more sleep on this fine day wouldn't be all bad.

We eventually arise to yet another oatmeal breakfast, still so fine after ten weeks on the road. Other cyclists stir in this quiet biker section of the campground, and word spreads that we have just completed the Big One, the Granddaddy of all Bike Tours, the TransAm, descendant of the 1976 transcontinental Bikecentennial. Visitors stop by, some singly, some in pairs, all of them cycling a leg of the Pacific Coastal route; as touring cyclists do everywhere, we swap stories, we compare gear, we celebrate what is, and it's all good.

After breakfast, David and I walk the trails together. The birds are silent, as if nonexistent, yet I hear their unspoken voices. We soon stand beside the Pacific Ocean itself and breathe in its salty air, watching gulls swoop and call in the wind, admiring the sandy beaches that have been pulverized by the elements from some ancient seafloor sandstones, eyeing an ocean that has been our destination for 70 days. "Great Joy?" I ask, questioning why I didn't feel as Lewis and Clark did when sighting the Pacific. "It's still cold," replies David.

The morning disappears, and after a leisurely lunch we find our way to the bikes and aim south. The fog has burned off, and our final ride is even more perfect than imagined: The warming sun shines brilliantly in a cloudless sky, and the wind has shifted, now coming from the north and gently pushing us along. We linger first at Holiday Beach, then Lost Creek, then Ona Beach, and roll past other sandy beaches until we come to Seal Rock, where stubborn headlands of black basalt resist the sea's onslaught, where gulls have gathered in a large colony to rear their young. "This cannot end," I say to myself, but I reflect on how the antecedent for "this" is unclear.

Just south of Seal Rock, a lone cyclist approaches, the overall visage familiar . . . it's John, our British friend, whom we haven't seen since Kansas! After hearty greetings we return to Seal Rock and celebrate our journeys over hot dogs and a large bucket of Oregon black cherries. "Mark and I met in a cycling forum on the internet," John tells us. "We did pretty good together, but in Wyoming Mark just wanted to fly, so he went on ahead. I've not seen him since, though I heard he did 180 miles the day he left me." Mark was also "arrested," we had learned long before John tells us now, for jaywalking in Missoula; news travels among cyclists, whether passed beside the road, left in log books at hostels, or just whispered into the breeze, it seems.

Western gulls. "Long Calls" from Seal Rock headland, above the surf. (379, 6:45)

Beyond Waldport, David stops for a swim at Beachside State Park, undeterred by a water temperature that prevents anyone else from even wading. And in the tradition of cross-country cyclists, I dip my front wheel into the Pacific, trying my best to celebrate the moment that marks the end of the road. We take a closeup picture of each other, our faces weathered but radiant. We linger, admiring the ocean that has been in our sights since early May. And inwardly I continue to reflect on how the finest 70 days of my life could transform all that remain.

"It's after five and we've done only 17 miles," I say in mock disgust. We laugh, realizing that we've set an all-time record for lethargy. David studies the map: "Let's do another 15 miles to Washburne Park. Looks like a quiet campground with a nice beach. And, here's a bonus for you—it's the best time of the day to bike," he reminds me.

Just north of Yachats, rocky headlands rise again, and to the tunes of Swainson's thrushes and Pacific wrens we climb the headland that is Cape Perpetua, a mass of resistant basalt that was formed on the ocean floor and has now risen here for all to see. In the softening, early evening light, we pause beneath weathered Sitka spruce to gaze at the landscape before us. Below, waves froth on tongues of basalt that reach out to sea, the tide pools there no doubt teeming with urchins, anemones, starfish, and the like. To the south, wave after wave caresses a dark, sandy shore, the beach there

Pacific wren. The stuff of superlatives, delivered with remarkable vehemence for so wee a bird. (380, 7:33)

consisting not of weathered sandstone but instead of tiny grains and pebbles of black basalt. To the west is nothing but blue sky and blue ocean that fade into the grayish horizon.

Down below the Cape, we stop at the Devil's Churn and Cook's Chasm, where the ocean has exploited cracks and fractures in the basalt and gouged out hollows where it continues to pound. The basalt here has none of those regular hexagonal columns that form when it cools slowly, like back in central Oregon's Picture Gorge; instead, these pillow-sized hunks formed when the basalt cooled quickly as it was spewed directly into the ocean. The texture of the rock tells its story as surely as if it could talk. Later, having pitched our tent at a rustic Washburne that is happily devoid of all "recreational vehicles," we head down to the beach for the trip's end:

> Well, here we are, it is 8:32 p.m. on the evening of July 13, and we are down by the ocean listening to the roar. . . . Sunset is within half an hour, and we're going to eat a pasta and carrots and cheese dinner. This is the first time we've had this!

"Are you recording this?" asks David, perhaps amused, perhaps a bit annoyed that I'm commenting on his favorite dish of the tour, which by now he should know I love.

I nod yes, and continue:

> The light and the ocean during the last 15 miles have been absolutely extraordinary. 'Best time of the day to bike.' We did lots of pictures.

> Roar roar roar goes the ocean. We heard lots of Pacific wrens along the way, and wrentits, and Swainson's thrushes. But the ocean was really spectacular, especially around Cape Perpetua. There's nothing quite like being high up above the ocean, with blue sky and blue ocean and white waves lining the beaches and crashing on black, rocky headlands.

While we savor his gourmet meal, the sun sinks into the Pacific, and David stands his bike in the sand so that we can watch the sun disappear through the silhouette of his bicycle frame. Half an hour after sunset, we wait for the full moon to clear the coastal hills to the east.

Later, David is still at the beach, calling friends on his cell phone. He's thinking the same thing I am, that "this cannot end," but his "this" is a little more concrete than mine. He's hatching a plan for biking from the Bay Area down to Tierra del Fuego at the tip of South America. Sometime in the near future, his ride will continue. Back in the tent, nestled in a coastal rain forest of pine and spruce and fir and cedar and hemlock, among rhododendron and huckleberry and salal and manzanita, with the muffled roar of the Pacific in the distance, I try to call it a day.

My mind races up and down the Oregon coast, imagining the sounds by day and by night of all that is special here, Leach's storm petrels on offshore islands, marbled murrelets finding their way in the dark to inland old-growth forests where they nest, gulls and cormorants and murres and guillemots nesting on rocky headlands, oystercatchers and purple sandpipers and turnstones at the shoreline, wrentits in coastal scrub, the local song dialects of black-capped chickadees and white-crowned sparrows, all who comment here within earshot of the Pacific's roar.

I retrace our route, starting 71 days ago on the Atlantic, first in a big state-by-state sweep of the country from Virginia to Oregon, then day by day, lingering on the truly special ones and listening to voices of all kinds along the way. The most memorable voices surge to the fore and then fade in a wave that sweeps from the Atlantic to the Pacific in much the same way that dawn's first light and a wave of birdsong sweep the continent every day.

Wrentits. At Land's End, a male-female duet: the Voice of the Pacific Coastal Scrub. (381, 1:47)

But it's over. My first summer off work since a wee kid is history. Our free-wheeling discovery of America has been abruptly halted by the Pacific. Time to go home. Totally rejuvenated, with my threshold for tolerating bullshit in the workplace sky high, it's time to immerse myself once more in the life I left behind. . . .

Hah! I laugh out loud at the very thought. *Who am I kidding? No way* am I going to transition to that work life back home at the university, teaching a few more classes, publishing a few more research papers, all the while being strangled by university politics and other professional nonsense. Not me. Not after experiencing the freedom of the open road like this. Who was that person with my name who started this trip ten weeks ago on the Atlantic? A complete stranger. He disappeared somewhere mid-continent, lost in the sea of prairie grass.

A new life has just begun. Professor Kroodsma no more. We just held the commencement ceremony on the beach, over a pasta feast, to a chorus of full sun setting and full moon rising. *"I guess it comes down to a simple choice, really. Get busy living or get busy dying."* So said Andy in the Shawshank Redemption—my favorite movie line of all time. I'm making my choice, long overdue. It's life, full time, and birdsong full time, celebrating it from the rooftops with all who have the ears to listen.

EPILOGUE—WHERE ARE THEY NOW?

JANUARY 2016

Soon after finishing this TransAm trip, David began planning his ride from the Bay Area in California to Tierra del Fuego, the tip of South America. He journeyed as much for the adventure as to educate the world about climate change, publishing a book to chronicle his efforts: *The Bicycle Diaries: My 21,000-Mile Ride for the Climate* (see *www.RideForClimate.com*). Since then he has been a voice for climate science, attending world summits on climate change (such as in Copenhagen during 2010 and in Cancun during 2011) and attempting to educate the public on these critical issues. As a "data journalist," he writes for the likes of the Huffington Post and Climate Central, was a research analyst for the Skoll Global Threats Fund, and recently completed a ten-month ride (with his wife Lindsey) across Asia. And sometimes he stops to listen to a bird sing.

Don is *free*. He left his professorship at the University of Massachusetts just a few weeks after returning from this 2003 TransAm bicycle journey. Searching for ways to share the magic of birdsong with a wide audience, he published his first book *The Singing Life of Birds* in 2005, which was awarded the John Burroughs Medal for natural history writing, a second award soon following from the American Birding Association. In 2008 his two *Backyard Birdsong Guides* were published by Chronicle Books, followed by *Birdsong by the Seasons* from Houghton Mifflin Harcourt in 2009. These days, he's somewhere listening to birds, wondering how best to capture the experience so it can be shared with others. He always stops to listen to a singing bird.

NOTES

Page

PROLOGUE

xv **from east to west or west to east**. See also Ikenberry (1996).

xvi "**westward I go free**." Henry David Thoreau, p. 266 in *Walking, Excursions*, published after Thoreau's death in *Atlantic Monthly*, June 1862.

CHAPTER 1. BEGINNINGS

2 **A robin begins to sing**. For more on robins and many other species, see my other books, *The Singing Life of Birds* (2005), *Birdsong by the Seasons* (2009), and the eastern and western *Backyard Birdsong Guide* (2008a, b). All birdsong can be heard on the website *ListeningToAContinentSing.com*. See page vii for how to explore this website.

3 **wood thrush . . . his two voice boxes**. See the magic in his two voices as illustrated in sonograms on pp. 237–45 in Kroodsma (2005). A wood thrush, like other songbirds, has *two voice boxes* located at the base of his windpipe. With the left voice box he sings the low voice, the right the high voice.

3 **a grand evolutionary tree**. For more "tree thinking," see Dawkins (2005).

3 ***Mitakuye oyasin***. "Man is just another animal . . . Even the tiniest ant, even a louse, even the smallest flower . . . they are all relatives. We end our prayers with the words *mitakuye oyasin*—'all my relations'—and that includes everything that grows, crawls, runs, creeps, hops, and flies on this continent," says Jenny Leading Cloud of the White River Sioux (Erdoes and Ortiz 1984).

4 **homemade soda-can stove**. See *http://www.wonderhowto.com/how-to-make-soda-can-stove-189957/*.

6 ***VIC-to-ry! VIC-to-ry! VIC-to-ry!*** Some field guides suggest *TEA-ket-tle!*, but any three-syllable mnemonic for a loud, ringing song would suffice.

CHAPTER 2. PEACE, AND WAR

12 **the little mini-disc recorder**. For this trip, we used primarily a Sony MZ-N707 mini-disc recorder with a Sennheiser ME66 short shotgun microphone.

12 "**incredible variety of Carolina wren songs.**" Here is the actual sequence of songs that this male sang during the four and a half minutes that David recorded: A A A B C B C B C B D B D B D D D D D D B D B D D D B D B D D C D B D B D B D B D B D B D B D E D E E E D E E E E E E E E E E E E E E E.

12 **Hoping our field guide will help**. Lightweight, fit for a cross-country bike trip, our chosen guide was Kaufman (2000).

13 **rarely do neighbors have similar songs**. For how chipping sparrows learn from each other, see pp. 313–20 in Kroodsma (2005).

17 **"Over five thousand dead and wounded men."** As reported on a sign entitled "The Last Day" atop Malvern Hill.

18 **Sixty thousand . . . Union losses . . . Confederate losses.** *http://www.civilwarhome.com/casualties.htm.*

CHAPTER 3. LEMONADE

23 **Jonathan Livingston Catbird**. Jonathan Livingston Seagull was the super gull described in Bach (1970).

27 **"brown thrasher . . . could sing over 2000 different songs**. See Kroodsma (2005).

28 ***Flight Calls of Migratory Birds.*** Evans and O'Brien (2002).

CHAPTER 4. BLUE RIDGE DAWN

32 **"June Curry. The Cookie Lady."** Sadly, June passed away at the age of 91, on June 16, 2012.

41 ***fire fire where where here here run run phew***. Mnemonic from Sibley (2000).

CHAPTER 5. A VIRGINIA HIGH

45 **all the way to Tierra del Fuego**. And he did it! See *RideForClimate.com.*

45 **"threshold for bullshit."** Pp. xi–xii in Hays and Hays (1995).

49 ***A Birder's Guide to Virginia***. Johnston (1997).

50 **flycatchers had backseats**. Chapman (1939).

CHAPTER 6. APPALACHIA

56 **power to determine who belongs and who doesn't.** Puckett (2000).

57 **one more important ingredient.** Montgomery (2000).

57 **sparrows with non-local songs.** Beecher et al. (2000).

57 **cowbird . . . is therefore not a vagrant.** Rothstein and Fleischer (1987)

64 **Kentucky warbler . . . raise or lower the frequency of his song.** Morton and Young (1986).

65 **song genes and songs of eastern and western . . . [willow flycatchers] . . . differ.** Sedgwick (2001). In look-alike flycatchers, the DNA can be a better indication of species than plumage.

CHAPTER 7. BOONE COUNTRY

69 **eastern kingbird . . . ". . . a miserable stuttering failure of it."** Walter Faxton quote from Bent (1942).

70 ***Zen and the Art of Motorcycle Maintenance.*** Pirsig (1975).

73 **passenger pigeons so abundant.** Blockstein (2002).

CHAPTER 8. A RIDE IN HEAVEN

77 **Robert Fulghum's cemetery plot.** Fulghum (1991).

80 **starlings . . . not protected by law**. If you kill any other bird (except during certain legal hunting seasons), or even if your cat does, you could be fined $5000 or

imprisoned for six months, as dictated by the Migratory Bird Treaty Act (Sections 16 U.S.C. 703, 18 U.S.C. 3571).

CHAPTER 10. ON THE ROAD AGAIN

88 **dickcissels have the same song**. Schook et al. (2008).

CHAPTER 11. DAWN SWEEPS THE SHAWNEE

97 **Humans can play much the same mating games**. In a semi-nomadic society of Namibia, 17.5% of all births were fathered by a male other than the marriage partner, and 31.8% of women had at least one such birth (Scelza 2011).

98 **being here some 500 million years ago**. Frankie (2004).

99 **has to be the songs from the summer tanager.** See also Elliott (2006).

CHAPTER 12. THE OZARKS

107 **"What's that?" ". . . a male red-winged blackbird."** Beletsky et al. (1986).

CHAPTER 13. A PRAIRIE GEM

119 **recalling the few facts I know**. One actual sequence that I have recorded from a Bell's vireo was as follows; each row is a minute, with songs **A**, **B**, **C**, and **D** boldface, showing how he returns to these four songs after about three minutes of singing songs E–H:

A B C A D C A C A E F E **C** F E G F E F E G
F E G F E G F E G F E F E F E G F E G F
E G E F E G F E G F E G F E G F E G **D**
F E G **D** H **B** H **B** H **B** H **B** H **B** H **B** H **B** H **C B** H
B H **B A C A B** H **C A B C A C A**

120 **leap from his left to his right voice box.** Allan and Suthers (1994).

122 **"robin with voice lessons."** Peterson (2008).

CHAPTER 14. KANSAS OCEANS

125 ***Hesperornis***. Ancient birds are typically known by their Latin name.

CHAPTER 15. SHORTGRASS PRAIRIE

141 **ornithology classes over the years**. See Rising (1983).

142 **His song flights . . . as he crisscrosses his territory**. With careful listening, I come to appreciate that this Cassin's sparrow has four versions of his songs, and over ten minutes, he sings them in the following minute-by-minute sequence: D B C B D, A C B D, C C B A, D C C B B, C D C A C, D C C B B C C, D D A A B B B, D D C C A A, C D D B B, C C C A A.

143 **barn owl . . . Hidden there are the two ears**. See Knudson (1981).

143 **owl knows instantly the exact location of its prey**. Humans have ears symmetrically placed, of course, and we can locate sounds in the horizontal plane (to the left and right) about as well as a barn owl can, but we are poor at determining where sounds are in the vertical plane (up and down).

144 **research papers on orioles**. For a more general treatise on hybrid zones, see Rising (1983).

145 **eastern house finches . . . repeat one song several times.** Western and eastern birds have different singing behaviors; Bitterbaum and Baptista (1979), Mundinger (1975).

CHAPTER 17. RIDING THE ROCKIES

156 **vesper sparrows . . . dialects . . . distant Oregon.** Kroodsma (1972)

160 **violet-green swallow . . . "children of heaven."** Dawson (1923).

CHAPTER 18. SAGE AND SONG

163 **overlapping is more likely a strong signal.** See Naguib and Kipper (2006).

163 **top dog in the neighborhood.** Male song may be directed at males in dominance contests, but females in a number of species seem to be the primary reason for the contests themselves. For eastern kingbirds, see Sexton et al. (2007); for chipping sparrows, Liu (2004); for superb fairy-wrens, Dalziell and Cockburn (2008).

CHAPTER 20. THE OREGON TRAIL

179 **The Oregon Trail.** Sources for information on geology include the Wyoming State Geological Survey (*http://www.wsgs.uwyo.edu/*), Lageson (1988), and roadside signs.

180 **like so many beached whales.** Lageson (1988).

185 **the winner eating the other's heart.** When asked later about eating his enemy's heart, Chief Washakie is said to have replied, "Youth does foolish things."

CHAPTER 21. GRAND TETONS

190 **Grand Tetons.** Geology as told by Good and Pierce (1996).

196 **vesper sparrows . . . I studied in Oregon over 30 years ago.** See Kroodsma (1972). One bird that I studied in graduate school had two favorite trills that immediately followed the whistles, but he had a repertoire of at least 43 different trills. In 400 songs, I documented 218 different songs (i.e., different trill sequences), with 175 of them occurring only once. He was a fine singer, as all vesper sparrows are.

CHAPTER 22. INTO THE FIRE

199 **Into the Fire.** Geology of Yellowstone area as told by Good and Pierce (1996).

199 **try to understand the greeting we received here.** That was 2003, and our particular ranger was perhaps just a little grumpy, but his warnings were appropriate. Bicyclists are welcomed in Yellowstone, to be sure, but are urged to beware the hazards of narrow roads, the large RVs with long-reaching mirrors, the bears and bison and elk (*www.nps.gov/yell/planyourvisit/bicycling.htm*).

203 **I most certainly will be back [to Yellowstone].** In June of 2009 I do return, taking in an early morning of birdsong in the Upper Geyser Basin. It was good!

CHAPTER 23. CATERPILLARS MARCHING

208 "**caterpillars marching toward Mexico.**" So said Clarence Dutton, Union officer in the Civil War and well-known American geologist, according to a roadside sign.

212 **"I've been everywhere, man."** Sung by Johnny Cash, written by Geoff Mack.

212 **Beaverhead Rock.** See pages 261–62 in Ambrose (1996).

215 **This Big Hole . . . no ordinary basin.** Alt and Hyndman (1986).

CHAPTER 24. CHIEF JOSEPH PASS

219 **Colonel John Gibbon, the army's besieged commander.** From the "Guide to Trails at Big Hole National Battlefield."

222 **eventually . . . recognized as two species.** And so they were, in the fall of 2009, now the Pacific wren and the winter wren; see *http://www.aou.org/checklist/north/suppl/51.php.*

222 **[winter wren and Pacific wren] . . . how different their songs were**. See Kroodsma (1980).

CHAPTER 25. LEWIS AND CLARK

229 **varied thrush sings . . . the contrast striking.** As with hermit thrushes, varied thrushes may choose their next song so that it is especially different from the one just sung. See Whitney (1981).

230 **Meriwether Lewis . . . described these varied thrushes**. See Ambrose (1996).

231 **[Lochsa River] . . . landscape that is still rising.** Alt and Hyndman (1989).

CHAPTER 26. PACIFIC ISLANDS INCOMING

233 **Ahead lie exotic lands**. Geology from Alt and Hyndman (1989).

235 **Battle of White Bird Canyon**. From information at the memorial.

CHAPTER 27. ASCENDING INTO OREGON

242 **knowing a bit of how [Oregon] came to be**. An extraordinary book on the geology of Oregon is Bishop (2003), and most information on Oregon geology comes from that source.

244 **islands race to port.** Imagery is that of Bishop (2003), who is a scientist, photographer, and poet.

245 **like a boat unloaded at port**. The poetry is that of Bishop (2003).

250 **I remember John Burroughs writing**. In Bent (1949).

CHAPTER 28. GEOLOGICAL CHAOS

252 "**tossed salad made of stones.**" Bishop (2003), p. 44.

CHAPTER 29. OVER THE CASCADES

267 **Someday . . . I'll publish a well-reasoned paper**. We did it! See Byers and Kroodsma (2009).

267 **pea-brained, know-it-all experts**: The full story is long and convoluted, revealing an underbelly of academic politics that is not so pretty, and involving a healthy dose of "payback" for my general critiques of birdsong studies. I would eventually prevail, publishing the bellbird work to much acclaim (Kroodsma et al. 2013), winning a best-paper-of-the-year award from the Wilson Ornithological Society (*http://www.wilsonsociety.org/awards/edwardsprize.html*), and simultaneously being honored with a lifetime achievement award (*http://www.wilsonsociety.org/awards/nicemedal.html*).

CHAPTER 30. A HOMECOMING

271 **small research project . . . my first publication**. Kroodsma (1971).

277 **wren-sized bird . . . cocks its tail comically**. Dawson (1923).

278 **forming the Coast Range**. See Alt and Hyndman (1978).

CHAPTER 31. LAND'S END

282 **young [cow]bird learning the local dialect**. See Rothstein and Fleischer (1987); also O'Loghlen and Rothstein (2003).

REFERENCES

Birds of North America. *http://bna.birds.cornell.edu/bna*. Best source of information for most species, which I consulted frequently; citations for each species consulted are not listed separately.

Allan, S. E., and R. A. Suthers. 1994. Lateralization and motor stereotypy of song production in the brown-headed cowbird. *Journal of Neurobiology* 25:1154–66.

Alt, D. D., and D. W. Hyndman. 1978. *Roadside Geology of Oregon*. Mountain Press Publishing Company, Missoula, MT.

Alt, D. D., and D. W. Hyndman. 1986. *Roadside Geology of Montana*. Mountain Press Publishing Company, Missoula, MT.

Alt, D. D., and D. W. Hyndman. 1989. *Roadside Geology of Idaho*. Mountain Press Publishing Company, Missoula, MT.

Ambrose, S. E. 1996. *Undaunted Courage: Meriwether Lewis, Thomas Jefferson, and the Opening of the American West*. Simon & Schuster Inc., New York, NY.

Bach, R. 1970. *Jonathan Livingston Seagull*. MacMillan, New York, NY.

Beecher, M. D., S. E. Campbell, J. M. Burt, C. E. Hill, and J. C. Nordby. 2000. Song-type matching between neighbouring song sparrows. *Animal Behaviour* 59:29–37.

Beletsky, D. L., B. J. Higgins, and G. Orians. 1986. Communication by changing signals: Call switching in red-winged blackbirds. *Behavioral Ecology and Sociobiology* 18:221–29.

Bent, A. C. 1942. *Life Histories of North American Flycatchers, Larks, Swallows, and Their Allies*. Smithsonian Institution United States National Museum Bulletin 179, Washington, DC.

Bent, A. C. 1949. *Life Histories of North American Thrushes, Kinglets, and Their Allies*. Smithsonian Institution United States National Museum Bulletin 196, Washington, DC.

Bishop, E. M. 2003. *In Search of Ancient Oregon: A Geological and Natural History*. Timber Press, Portland, OR.

Bitterbaum, E., and L. F. Baptista. 1979. Geographical variation in songs of California house finches (*Carpodacus mexicanus*). *Auk* 96:462–474.

Blockstein, D. E. 2002. Passenger Pigeon (*Ectopistes migratorius*). In *The Birds of North America Online*. A. Poole, ed. Cornell Lab of Ornithology, Ithaca, NY.

Byers, B. E., and D. E. Kroodsma. 2009. Female mate choice and songbird song repertoires. *Animal Behaviour* 77:13–22.

Chapman, F. M. 1939. *Handbook of Birds of Eastern North America*, 2nd ed. Dover Publications, New York, NY.

Dalziell, A., and A. Cockburn. 2008. Dawn song in superb fairy-wrens: A bird that seeks extrapair copulations during the dawn chorus. *Animal Behaviour* 75:489–500.

Dawkins, R. 2005. *The Ancestor's Tale: A Pilgrimage to the Dawn of Evolution.* Houghton Mifflin Harcourt, Boston, MA.
Dawson, W. L. 1923. *Birds of California*, vol. 2. South Moulton Company, Los Angeles, CA.
Elliott, L. 2006. *The Songs of Wild Birds.* Houghton-Mifflin Co., Boston, MA.
Erdoes, R., and A. Ortiz. 1984. *American Indian Myths and Legends.* Pantheon Books, New York, NY.
Evans, W. R., and M. O'Brien. 2002. *Flight Calls of Migratory Birds: Eastern North America Landbirds.* Old Bird, Inc., Ithaca, NY.
Feynman, R. P. 1985. *Surely You're Joking, Mr. Feynman!* W. W. Norton & Co., New York, NY.
Frankie, W. T. 2004. *Guide to the Geology of Ferne Clyffe State Park and Surrounding Area, Johnson and Pope Counties, Illinois.* Illinois State Geological Survey, Champaign, IL.
Fulghum, R. 1991. *It Was on Fire When I Lay Down on It.* Ivy Books, Raleigh, NC.
Good, J. M., and K. L. Pierce. 1996. *Interpreting the Landscapes of Grand Teton and Yellowstone National Parks: Recent and Ongoing Geology.* Grand Teton Natural History Association, Moose, WY.
Halle, L. J. 1988. *Spring in Washington.* Johns Hopkins University Press, Baltimore, MD.
Hays, D., and D. Hays. 1995. *My Old Man and the Sea.* Algonquin Books, Chapel Hill, NC.
Ikenberry, D. L. 1996. *Bicycling Coast to Coast: A Complete Route Guide Virginia to Oregon.* The Mountaineers, Seattle, WA.
Johnson, S. 1984. *The Precious Present.* Doubleday, New York, NY.
Johnston, D. W. 1997. *A Birder's Guide to Virginia.* American Birding Association, Inc., Colorado Springs, CO.
Kaufman, K. 2000. *Birds of North America.* Houghton Mifflin Company, New York, NY.
Knudson, E. I. 1981. The hearing of the barn owl. *Scientific American* 245:112–25.
Kroodsma, D. E. 1971. Song variations and singing behavior in the rufous-sided towhee, *Pipilo erythrophthalmus oregonus. Condor* 73:303–8.
Kroodsma, D. E. 1972. Variations in the songs of vesper sparrows in Oregon. *Wilson Bulletin* 84:173–78.
Kroodsma, D. E. 1980. Winter wren singing behavior: a pinnacle of song complexity. *Condor* 82:356–65.
Kroodsma, D. E. 1990. Using appropriate experimental designs for intended hypotheses in song playbacks, with examples for testing effects of song repertoire sizes. *Animal Behaviour* 40:1138–50.
Kroodsma, D. E. 2005. *The Singing Life of Birds: The Art and Science of Listening to Birdsong.* Houghton-Mifflin Co., Boston, MA.
Kroodsma, D. E. 2008a. *The Backyard Birdsong Guide: Eastern North America.* Chronicle Books, San Francisco, CA.
Kroodsma, D. E. 2008b. *The Backyard Birdsong Guide: Western North America.* Chronicle Books, San Francisco, CA.
Kroodsma, D. E. 2009. *Birdsong by the Seasons: A Year of Listening to Birds.* Houghton-Mifflin Harcourt Co., Boston, MA.
Kroodsma, D., D. Hamilton, J. E. Sánchez, B. E. Byers, H. Fandiño-Mariño, D. W. Stemple, J. M. Trainer, and G.V.N. Powell. 2013. Behavioral evidence for song learning in the suboscine bellbirds (*Procnias* spp.; Cotingidae). *Wilson Journal of Ornithology* 125:1–14.
Lageson, D. R., and D. R. Spearing. 1988. *Roadside Geology of Wyoming.* Mountain Press Publishing Company, Missoula, MT.
Leopold, A. 1966. *A Sand County Almanac.* Oxford University Press, New York, NY.
Liu, W. C. 2004. The effect of neighbours and females on dawn and daytime singing behaviours by male chipping sparrows. *Animal Behaviour* 68:39–44.

Montgomery, M. 2000. Myths: How a hunger for roots shapes our notions about Appalachian English. *Now and Then: The Appalachian Magazine* 17:7–13.
Morton, E. S., and K. Young. 1986. A previously undescribed method of song matching in a species with a single song "type," the Kentucky warbler (*Oporornis formosus*). *Ethology* 73:334–42.
Mundinger, P. 1975. Song dialects and colonization in the house finch, *Carpodacus mexicanus,* on the east coast. *Condor* 77:407–22.
Naguib, M., and S. Kipper. 2006. Effects of different levels of song overlapping on singing behaviour in male territorial nightingales (*Luscinia megarhynchos*). *Behavioral Ecology and Sociobiology* 59:419–26.
O'Loghlen, A. L., and S. I. Rothstein. 2003. Female preference for the songs of older males and the maintenance of dialects in brown-headed cowbirds (*Molothrus ater*). *Behavioral Ecology and Sociobiology* 53:102–9.
Peterson, R. T. 2008. *Peterson Field Guide to Birds of North America*. Houghton Mifflin Co., Boston, MA.
Pirsig, R. 1975. *Zen and the Art of Motorcycle Maintenance: An Inquiry into Values*. Bodley Head, London.
Puckett, A. 2000. On the pronunciation of Appalachia. *Now and Then: The Appalachian Magazine* 17:25–29.
Rising, J. D. 1983. The great plains hybrid zones. *Current Ornithology* 1:131–57.
Rothstein, S. I., and R. C. Fleischer. 1987. Vocal dialects and their possible relation to honest status signaling in the brown-headed cowbird. *Condor* 89:1–23.
Scelza, B. A. 2011. Female choice and extra-pair paternity in a traditional human population. *Biology Letters*. Published online 6 July 2011.
Schook, D. M., M. D. Collins, W. E. Jensen, P. J. Williams, N. E. Bader, and T. H. Parker. 2008. Geographic patterns of song similarity in the dickcissel (*Spiza americana*). *Auk* 125:953–64.
Sedgwick, J. A. 2001. Geographic variation in the song of willow flycatchers: Differentiation between *Empidonax traillii adastus* and *E-t. extimus*. *Auk* 118:366–79.
Sexton, K., M. T. Murphy, L. J. Redmond, and A. C. Dolan. 2007. Dawn song of eastern kingbirds: Intrapopulation variability and sociobiological correlates. *Behaviour* 144:1273–95.
Sibley, D. A. 2000. *The Sibley Guide to Birds*. Alfred A. Knopf, New York, NY.
Whitney, C. L. 1981. Patterns of singing in the varied thrush: I. The similarity of songs within individual repertoires. *Zeitschrift für Tierpsychologie* 57:131–40.

INDEX

NOTE: Page numbers in **bold** indicate the location of QR listening codes.

Absaroka Range, 188, 199
Adventure Cycling Association, xv, 6, 33
Alice Lloyd College, 61–62
antelope, 156, 171, 172, 218
Appalachian Mountains
  bird breeding ranges, 39–40
  geology, 55, 58
  local dialects, 56–57, 66–67
  northern birds, 49–51
  See also Kentucky; Virginia
Appalachian Trail, 46
Arapaho National Wildlife Refuge, 172
Arapaho people, 183
aspen, 171
Atchison, Topeka and Santa Fe Railroad, 135
Audubon, John James, 73
aurora borealis, 218

Backyard Birdsong Guides (Donald Kroodsma), 287
Badger Pass, 214
Bannock people, 185
Barclay, Alexander, 149
Beaver Divide, 182
Beaverhead-Deerlodge National Forest, 217–22
Beaverhead River and Mountains, 212–14
Belknap Craters, 263
bellbird, three-wattled *(Procnias tricarunculatus)*, **267**, 293 (p. 267)
The Bicycle Diaries: My 21,000-Mile Ride for the Climate (David Kroodsma), 287
Bicycle gear, techniques, 18, 23, 37–38, 40, 72–75, 139
Big Hole National Battlefield, 219–20
Big Hole Pass and Basin, 214–16, 227
Big Robber, Crow chief, 185
BikeCentennial, xv–xvi, 5, 32–33, 282
Biking Across Kansas, 135–37
Birds of North America, 174
Birdsong by the Seasons (Donald Kroodsma), 287
bison, 116–18, 121, 188, 203–4
bittern, American *(Botaurus lentiginosus)*, 132, 226
Bitterroot Valley, River, and Range, 223–24, 226–29, 243
blackbird
  red-winged *(Agelaius phoeniceus)*, **78**, 89, 120, 211, 226, 240, 248
    calls, **78**, 78–79, **107**, **239**
    local call dialects, **107**, **239**
  yellow-headed *(Xanthocephalus xanthocephalus)*, 132, **213**
bluebird
  eastern *(Sialia sialis)*, 66, 67, 108, 130
    dawn singing, 62, **62**, 87
  mountain *(Sialia currucoides)*, 167, 222
    dawn singing, 173–74, **173**, 202, **203**, 218
    repertoire, 174
  western *(Sialia mexicana)*, 276
    calls, 276
Blue Mountains, 233
Blue Ridge Parkway, 33–43
bobolink *(Dolichonyx oryzivorus)*, 262
  repertoire, 211, **211**
bobwhite, northern *(Colinus virginianus)*, 10, 89–90, **91**, 120
Bridger National Forest, 188
Brownlee Dam, 240, 246
buffalo birds. *See* cowbird, brown-headed
bumblebees, **252**
bunting, 123
  indigo *(Passerina cyanea)*, 7, 18, **41**, 73, 89, 93, 101, 122–23
    dawn singing, 62, 69, 87, 98, 113
    learning to sing, 93
    mini-dialects, 76, 77
    yearling songs, 76, 77
  lark *(Calamospiza melanocorys)*, **138**, 176
  lazuli *(Passerina amoena)*, 123, 148, 223, **235**, 240, **245**, 274–75, **276**
    dawn singing, 244, **254**, 274
    learning to sing, 244, 255–56

Burroughs, John, 250
bushtit *(Psaltriparus minimus)*, **256**

call learning
  blackbird, red-winged, 107, 239
  crossbill, red, 194–5
  jay, blue, 7
  jay, Steller's, 258
  raven, common, 205
Canyon Mountain, 253
Cape Perpetua, 283–84
cardinal, northern *(Cardinalis cardinalis)*, 18, 60, 64, 66, 69, 89, 101, 108
  dawn singing, 20–22, **21**,59, 62, 99, **87**, 112, 119
  learning to sing, 20
  matched countersinging, **21**, 22, 62
  repertoire, 12–13, 21, 24
Caribou National Forest, 188
Cascade mountains, 243–44, 255, 259, 261–69, 273, 282
catbird, gray *(Dumetella carolinensis)*, 27, 31, 69
  dawn singing, 119
  improvisation, **23**
  mimicry, 23, 211, **211**, **240**
Cenozoic era, 183
Chapman, Frank, 50
chat, yellow-breasted *(Icteria virens)*, 62, **181**, 240, 274
  dawn singing, **69**, 70–71, 87, 112, **119**, 244, **245**
  night singing, 68, **177**, 177–78
  repertoire, 69, **70**, **177**, 178
cherry, 34, 47, 75, 108
chickadee, 123, 262
  black-capped *(Poecile atricapillus)*, **52**, 115–117, **220**, 238, **239**, 281
    calls, 127, 210
    dawn singing, 51, 147, 210
    local dialect, **273**, 282, 285
    pitch-shifting songs, 51, 64, 133–134, **134**, 147
  Carolina *(Poecile carolinensis)*, 3–5, 61, 108, 115–117, 127
    calls, **3**, 61
    dawn singing, **3**, 13–14, 62, 113
    shared repertoire, dialects, 8, 88
  chestnut-backed *(Poecile rufescens)*, **281**
    calls, **281**
  hybrid zone birds, 115–17, **116**, 127, 133–34
  mountain *(Poecile gambeli)*, **164**, 193, **203**, 222, 227, 258, 261, 281
    calls, 163, **164**, 252
    dawn singing, **159**, 186, 191, 252
    repertoire, 163
chickaree, 252
chickens. *See* fighting cocks
Chief Joseph Pass, 217, 222–23
choruses and discussions. *See* matched countersinging
chuck-will's-widow *(Caprimulgus carolinensis)*, **98**, 99, 112, 113
church signs, 61–62
Civil War, 16–19
Clark, William, 156, 227. *See also* Lewis and Clark expedition
Clearwater River, 232–34
Coast Range, 223, 267, 274, 276–79
cockfighting, 58–59
Colorado
  Tribune, KS to Eads, 136–39
  Eads to Bob Creek, 139–40
  Bob Creek to Pueblo, 140–45
  Pueblo to Cañon City, 146–49
  Cañon City to Guffey, 149–52
  Guffey to Alma, 153–59
  Alma to Blue River Campground (Silverthorne), 159–62
  Blue River Campground to Kremmling, 163–66
  Kremmling to Walden, 167–72
  Walden to Saratoga, WY, 173–75
Columbia Plateau basalt, 234, 236, 238, 240–1, 243, 245, 254–55
Continental Divide, 160, 171, 177, 188, 200–1, 205, 223, 241, 243, 245
cowbird, brown-headed *(Molothrus ater)*, 5, 70, **70**, 102, 113, 120, **121**, 156, **213**
  flight whistles, **89**, 160, 211, **213**, 240
  learning songs and calls, 281–82
  local dialects, 57
  male displays, **89**, 121
coyote, western, 174, 207, 218, **222**, 257
crane, sandhill *(Grus canadensis)*, 208, 249
  mate duets, **208**, 219
  vocal physiology, 207–8
creeper, brown *(Certhia americana)*, **52**, 228, 268–69, **269**
Cretaceous era, 125–26, 183, 185, 243
creationism, 131, 253
Crooked River, 259
crossbill, red *(Loxia curvirostra)*
  calls, 155, **194**, 194–95, 203, 223, 252, 258, 262
  multiple lineages, 194–95
cross-country route. *See* TransAmerica Trail
crow
  American *(Corvus brachyrhynchos)*, 69, 100, 113, 213, 244, 248
    intelligence, 282, **282**
    mob-scene, **171**
  fish *(Corvus ossifragus)*, **7**

Crowheart Butte, 185
Crow people, 183, 185
cuckoo, yellow-billed *(Coccyzus americanus)*, 57, 119
night singing, 29, 33, 98
curlew, long-billed *(Numenius americanus)*, 249
Currant Pass, 156, 159
Curry, June "The Cookie Lady," 32–33, **33**
Cycle America, 212

Daisy Geyser, **202**
Dakota Hogback ridge, 150
Daniel Boone National Forest, 71–73
dialects, absence of. *See also* dialects, presence of
flycatchers, 21, 30
robin, American, 20–21
thrush, wood, 20–21
warbler, chestnut-sided, 39
dialects, presence of; 7, 84, 92, 271–73. *See also* dialects, absence of; song learning
Appalachian Mountains, 56–57, 66–67
blackbird, red-winged (calls), **107**, **239**
bunting, indigo, 76–77
cardinal, northern, 13, 54, 88
changes over time, 267, 272–73
chickadee, black-capped, 88, **273**, 282, 285
cowbird, brown-headed, 57, 281–82
dickcissel, **88**, **120**
human, 92
Curry, June, 32–33, **33**
Haupt, Charles, 15–16, **15**
High school students, 79–80, **79**
Love, James, 89–90, **90**
Napier, Mary Lou, 66, **67**
Owens, Terry, 56–57, **56**
Smith, Billy Rays, 47, **47**
Snodgrass, Bert, 54, **54**
hummingbird, Anna's, 275
jay, blue (calls), **7**
oriole, Baltimore, **131**
parula, northern, 42, 67, 71–72, **72**, 84
sparrow, song, 57
sparrow, vesper, 156, 179, 185, **196**, 209, 276
sparrow, white-crowned, **160**, **205**, **220**, 265, 281, 285
towhee, spotted, 271
warbler, chestnut-sided, 39
wren, Bewick's, 271–73, **273**
wren, Carolina, 13–14
dickcissel *(Spiza americana)*, 79, 93, 103, 127, 138
dawn songs, **120**, 130, 132
dialects, **88**, 107, **120**
dinosaurs, 125–26, 182, 185, 243
dipper, American (*Cinclus mexicanus*), 254
Dixie Pass, 249–52
dogs, 33, 62, 63, 65–66, **65**, **71**, 75
dogwood, 20, 34, 42
Doppler effect, 36–37
dove, mourning *(Zenaida macroura)*, 89, 134, 138, 148, 156, 183, 227
dawn singing, **87**, 168, 211
whistling wings, **240**
drumming
flicker, northern, **134**
grouse, ruffed, **230**
sapsucker, red-breasted, 275, **276**
sapsucker, red-naped, **193**, **221**
sapsucker spp., 156, 207, 249
woodpecker, Lewis's, **227**
woodpecker, pileated, **14**, 60, 70
woodpecker, red-bellied, 101, **106**
duck, 203, 225–26
wood *(Aix sponsa)*, 273
duet
crane, sandhill, 207–8, 219
thrush, wood, 21
wrentit, 285

eagle, 86, 179, 188, 193, 204
Ear Spring Geyser, **202**
Elkhorn Mountains, 246, 248
encoded songs. *See* genetically encoded songs and calls
English sparrow. *See* sparrow, house
evening singing
thrush, hermit, **250**
thrush, wood, 27
See also night singing
evolution, 3, 20–21, 37, 65, 122–23, 129–30, 141–42, 144, 153, 155–56, 158, 175, 191, 262
extinct species, 73, 110, 121, 182, 188

females
songless, 21, 113
with multiple mates, 47
blackbird, red-winged, 78
bunting, indigo, 93
dickcissel, 93
human, 97
martin, purple, 97
Feynman, Richard, 267
fighting cocks *(Gallus gallus)*, **58**, 58–59
finch
Cassin's *(Carpodacus cassinii)*, 202, **203**, 258
mimicry, **203**
gray-crowned rosy *(Leucosticte tephrocotis)*, 263
house *(Carpodacus mexicanus)*
eastern variant, **92**, 145, 292nc

finch (cont.)
range, 91, 145
western variant, **145**, 146, 148–50, 183, 210, 254, 292n6
purple *(Carpodacus purpureus)*, 52, 275, **276**
Finley National Wildlife Refuge, 110, 269–77
Firehole River, 200–204
flicker, northern *(Colaptes auratus)*, 144
eastern yellow-shafted variant, 25, 70, 77, 134, **134**, 140–42
hybrid zone birds, 141–42
western red-shafted variant, 134, 141–42, 212, 231, 240
Flight Calls of Migratory Birds (Evans and O'Brien), 28
flycatcher, 51, 290 (p. 65)
Acadian *(Empidonax virescens)*, 90, **96**, 154, 265
calls, 51, **99**, 112
dawn singing, 51, **99**, 112
night singing, 96
alder *(Empidonax alnorum)*, **51**, 154, 265
calls, **51**
cordilleran *(Empidonax occidentalis)*, 155, 190, 265, 274
dawn singing, **154**, 154–55, 186
dusky *(Empidonax oberholseri)*, 154, 265
dawn singing, 190, **191**, 220
evolution, 246
genetically encoded songs, 21, 25, 118, 123, 195, 220, 246, 262, 267
gray *(Empidonax wrightii)*, 149, 154, 265
dawn singing, **150**
great crested *(Myiarchus crinitus)*, 5, 7, 102, 127, 130, 140
calls, 87, 98, 102
dawn singing, 21–22, **22**, 87, **98**, 99, 128, **129**
Hammond's *(Empidonax hammondii)*
dawn singing, 154, **265**
least *(Empidonax minimus)*, 154, 183–84, **184**, 213, 265
dawn singing, **50**, 50–51
range, 50, 183
olive-sided *(Contopus cooperi)*, 52, 156, **187**, 220, 262, **263**
dawn singing, 186–87, 281
Pacific-slope *(Empidonax difficilis)*, 265, 274, 275–76, 278
calls, 278
dawn singing, 154, **274**
scissor-tailed *(Tyrannus forficatus)*, 126
dawn singing, 128–30, **129**
willow *(Empidonax traillii)*, 154
eastern variant, **65**, 122, 265
dawn singing, **119**
western variant, 122–23, **171**, 205, 248, 265
dawn singing, 211, 219–20, 274
fox, red, 171
Fulghum, Robert, 77

Gallatin National Forest, 188
genetically encoded songs and calls
flycatcher, 21, 123, 195, 220, 246, 267
meadowlark, western, 131, **133**
phoebe, eastern, 25, 118
wood-pewee, 262
geological events
Appalachian Mountains, 55, 58
Columbia Plateau basalt, 234, 236, 238, 240–1, 243, 245, 254–55
Grand Tetons, 193–94, 196–98
Hebgen Lake earthquake, 205–6
Idaho batholith, 215, 227–28, 233
inland ocean of the Great Plains, 125–26
midwestern sandstone, 98
Montana's basin and ranges, 208, 215
Oregon's origins, 233–39, 242–45, 252–55, 257, 262–63, 267, 277–78, 282–84
Pleistocene ice ages, 10, 98, 243, 267
Rocky Mountain origins, 150
Wyoming, 180–83, 185
Yellowstone caldera and hotspot, 199–204
geysers, 201–4, **202**
Gibbon, John, 219
gilia, scarlet *(Ipomopsis aggregata)*, 198
gnatcatcher, blue-gray *(Polioptila caerulea)*, **18**, 102
dawn singing, 62, 100
goldfinch
American *(Spinus tristis)*, 61, 89, **181**, 268, 278
flight call, 89, 120, **181**, 240, 244, 274, 281
lesser *(Spinus psaltria)*, **155**
calls, 155
goose, Canada *(Branta canadensis)*, 28, 54, **220**, 275
Goose Rock, 255
grackle, common *(Quiscalus quiscula)*, 7, **89**
calls, 89
Grand Tetons, 188–99, 208, 240, 280
Granite Mountains, 180–81
Grasshopper Valley and Creek, 214
Gravely Range, 208–9
Great Divide Basin, 177
Great Plains. *See also* Kansas
hybrid zones. *See* hybrid zones
inland ocean origins, 125–26
shortgrass prairie, 136–45
tallgrass prairie, 115–35

grebe
Clark's *(Aechmophorus clarkii). See* grebe, western
pied-billed *(Podilymbus podiceps),* **131**, 226, 275
intelligence, 132
western *(Aechmophorus occidentalis),* 169
calls, 169
Green Mountain Reservoir, 164
grosbeak, 123
black-headed *(Pheucticus melanocephalus),* 122, 149, **150**, 234, **240**
dawn singing, **147**
evening *(Coccothraustes vespertinus)*
calls, **276**, 278
pine *(Pinicola enucleator),* 221
rose-breasted *(Pheucticus ludovicianus),* 122, **123**
Grotto Geyser, **202**
grouse
ruffed *(Bonasa umbellus)*
eastern variant, 227, 230, **230**
western variant, 230
spruce *(Falcipennis canadensis),* 227
guillemot
pigeon *(Cepphus columba),* 285
gull, 279, 282, 283, 285
laughing *(Leucophaeus atricilla),* xiii, 7
western *(Larus occidentalis),* xiii, **283**
calls, 283

Halle, Louis, 27
Haupt, Charles, **15**, 15–16
hawk, 66, 108, 126
ferruginous *(Buteo regalis),* 175
red-tailed *(Buteo jamaicensis),* 80, 108, 148, 172, 174–75, 258, 275, **276**
songbird alarm call, 50
Swainson's *(Buteo swainsoni),* 175
Hebgen fault, 205–6
Hells Canyon, xiii, 195, 233, 236, 239–46, 259
heron, great blue *(Ardea herodias),* 275
Hesperornis, 125–26, 291 (p. 125)
hisselly songs (American robin), **2**, 20–21, **50**, 99, 147, **190**, **220**, 233
Hoosier Pass, 150, 159, 160
human dialects. *See* dialects, presence of, human
hummingbird
Anna's *(Calypte anna),* 275
broad-tailed *(Selasphorus platycercus),* **155**, 160, 197–98, **198**, 203, 227
calls, 197, **198**
ruby-throated *(Archilochus colubris),* 102
rufous *(Selasphorus rufus),* 249
hybrid zones, 123
chickadees, 115–17, **116**, 127, 133–34
flickers, 141
flycatchers, 123
meadowlarks, 142
orioles, 144
songbirds vs. flycatchers, 123

Idaho
Lee Creek Campground, MT to Lowell, 228–32
Lowell to Riggins, 233–36
Riggins to Cambridge, 236–39
Cambridge to Hells Canyon, 239–41
Hells Canyon to Baker City, OR, 242–48
Idaho batholith, 215, 227–28, 233, 238
Illinois
Utica, KY to Cave-in-Rock, 90–95
Ferne Clyffe State Park to Chester, 96–105
Chester to Pilot Knob, MO, 106–9
inborn songs. *See* genetically encoded songs and calls
innate songs. *See* genetically encoded songs and calls
intelligence, 91, 131–32, 143, 155–56, 168–69, 261–62, 282

Jackson Hole, WY, 194, 196–97, 240
jay, 252
blue *(Cyanocitta cristata),* 22, 33, 81, 90, 102, 134
eastern Virginia dialect, **7**
learning to call, 22
gray (Canada) *(Perisoreus canadensis),* 160
pinyon *(Gymnorhinus cyanocephalus),* **157**, 157–59, 262
Steller's *(Cyanocitta stelleri),* 227, **229**, 274
mimicry, **258**
John Day Valley, 252–55
Joseph, Chief of the Nez Perce, 219, 235
junco, dark-eyed *(Junco hyemalis),* 160, 164, **193**, 203, 228, 258, 262, 265, 278, **281**
dawn singing, 221, 251–52, **252**, **281**
repertoire, **52**
western variant, 251–52
Juniper Butte, 256
Jurassic period, 185

Kansas
Biking Across Kansas ride, 135, 136–37
Cretaceous era, 125–26
Prairie State Park, MO, to Chanute, 117–24
Chanute to Harvey County East Park, 125–28
Harvey County East Park to Quivira National Wildlife Refuge, 128–31
Quivira National Wildlife Refuge to Ness City, 132–33

Kansas (cont.)
Ness City to Tribune, 133–35
Tribune to Eads, CO, 136–39
Kaufman Field Guide, 142, 183, 194–95, 222, 228, 290 (p. 13)
Kentucky
Rosedale, VA to Virgie, 56–59
Virgie to Pippa Passes, 59–62
Pippa Passes to Booneville, 62–67
Booneville to Irvine, 68–73
Irvine to Lincoln Homestead State Park, 74–81
Bardstown, 82–85
Hodgenville to Utica, 86–90
Utica to Cave-in-Rock, IL, 90–95
kestrel, American *(Falco sparverius)*, 24, 108
killdeer *(Charadrius vociferus)*, 24–25, 68, **103**, 126, 133, 186, 202, 213, 218, 249
kingbird, 126, 256
eastern *(Tyrannus tyrannus)*, 93–94, 108, 122–23, 126, 133, 208, 213, 240
dawn singing, 68–69, **69**, 82, 112, 128–30, **129**, 211
evolution, 246
western *(Tyrannus verticalis)*, 123, 140, 148, 240, 245–46, 256
calls, 140, 245–6
dawn singing, 128–30, **129**, 133, **141**, 143, 146, 148, 213, 244, 254
inborn songs, 245–46, **246**
kingfisher, belted *(Megaceryle alcyon)*, 24–25, **111**, 204, 255
kinglet
golden-crowned *(Regulus satrapa)*, 52, 228–29
dawn singing, **252**
ruby-crowned *(Regulus calendula)*, **159**, 164, 193, **203**, 222
dawn singing, 159, 191

lark, horned *(Eremophila alpestris)*, 138, 169–70, **170**, 176, 182, 208, **209**
dawn singing, 218, **219**
Leading Cloud, Jenny, 289ch 1
learning to sing. *See* song learning
legal protection for birds, 80, 290ch 8
Leopold, Aldo, 254
Lewis, Meriwether, 156, 227, 230
Lewis and Clark expedition, xiii, 116, 221, 226
hardships, 213, 229–30
natural history discoveries, 156, 227
at the Ohio River, 94
at the Pacific Ocean, 279, 282
Sacagawea and Shoshone assistance, 183–84, 210, 213
Lewis River, Falls, and Lake, 199–200, 201
ListeningToAContinentSing.com, vii
Little Salmon River, 237–38
Lloyd, Alice, 61
local dialects. *See* dialects
Lochsa River, 231–32, 235
Lolo Mountain, 226–27
Lolo Pass, 226–29
longspur, McCown's *(Rhynchophanes mccownii)*, 179, **179**
Louisiana Purchase, 116
Love, James, 89–90, **90**
Lower Geyser Basin, 204

Madison River and Range, 201, 204–9
magpie, black-billed *(Pica hudsonial)*, 148, 164, 178, 207, 211, **212**, 256
dawn calling, 168
mob scene, **171**
mate-attraction songs, **31**, **39**, 47, 63, 163, 266, 292 (p. 163). *See also entries for "day (mate) songs" under many warbler species*
mallard *(Anas platyrhynchos)*, 172
Malvern Hill, 16–17
martin, purple *(Progne subis)*, **15**, 15–16, 24, 89, 94, 100, 102, 125, 248
night and dawn singing, 68, 97–98, **97**
mating games, 97
Mary's Peak, 277–78
matched countersinging
cardinal, northern, **21**, 22
meadowlark, eastern, **78**
oriole, Baltimore, 76
titmouse, tufted, 3–4
wren, Carolina, **6**, 22
wren, marsh, 175, **225**, **279**
McKenzie Pass, 243, 261–64
meadowlark
eastern *(Sturnella magna)*, 17–18, 89, 108, 120, **121**, 130–32
calls, **121**, 131
matched countersinging, **78**
hybrid zone birds, 131–33, 142
western *(Sturnella neglecta)*, 78, 132, **133**, 138, 140, 142–43, 167–68, 148, 174, 179, 184, 208, 211, 213, 246, 256, **257**
calls, **131**, **133**, **174**
dueling males, **131**
Medicine Bow Mountains, 171
Midway Geyser Basin, 203
migrating birds, 5, 28–29
calls, 28–29
migration and dispersal, 92–93, 194–95, 271, 273
Migratory Bird Treaty Act, 290 (p. 80)

Mimicry. *See also* song learning
catbird, gray, 23, 211, **211**, **240**
chat, yellow-breasted, 178
finch, Cassin's, **203**
goldfinch, lesser, **155**
jay, Steller's, **258**
mockingbird, northern
eastern songs, 24–26, **24**, 27, 32, 75, **87**, 90–91, 140
western songs, **140**, 140–41
sparrow, fox, 231
starling, European, **80**, 80–81, **249**
thrasher, brown, 27
thrasher, sage, 167–68
vireo, white-eyed, 77, **78**
wren, house, 274
Mississippi River, xiii, 88, 103–5, 106–7, 201, 205, 212
Missouri
Chester, IL to Pilot Knob, 106–9
Pilot Knob to Alley Springs, 109–12
Alley Springs to Hartville, 112–14
Ash Grove to Prairie State Park, 115–17
Prairie State Park to Chanute, KS, 117–24
Missouri-Pacific Railroad, 135
Mitakuye oyasin, 3, 289ch 1
mockingbird, northern *(Mimus polyglottos),* 23–26, **24**, 31, 75, 89, 143
dawn singing, 141
eastern song mimicry, 24–26, 27, 32, 75, 81, 90–91, 140
night singing, 29, 33, 86, **87**, 90, 141
western song mimicry, **140**, 140–41
Montana
West Yellowstone to Madison River Cabins, 204–6
Madison River Cabins to Alder, 207–10
Alder to Dillon, 210–13
Dillon to Wisdom, 213–16
Wisdom to Stevensville, 217–224
Stevensville to Lee Creek Campground, 225–28
Lee Creek Campground to Lowell, ID, 228–32
Mosquito Prince, 51, 65, 84, 154, 190, 265, 274
See also flycatcher, genus *Empidonax*
Mount Lincoln, 160
Mount Moran, 193–94, 198
Mount Rogers, 46–53, 117, 183, 222, 265, 275, 280
Mount Sheridan, 199–200
Mount Washburn, 199–200
Mount Washington, 259, 262–63
Muddy Pass, 167, 171–72
My Old Man and the Sea (Hays and Hays), 45

Napier, Mary Lou, 66, **67**
Native Americans, 183–85
Never Summer Mountains, 172
Nez Perce people, 219, 221, 235
nighthawk, common *(Chordeiles minor),* 155, 168, 254
calls and booming *vroom,* 83, **126**, 220, 242, 244, **256**
nightjars, **98**, 102
night calling or singing
chat, yellow-breasted, 68, **177**, 177–78
chuck-will's-widow, **98**
cuckoo, yellow-billed, 29, 33, 98
flycatcher, Acadian, 96
martin, purple, 97–98, **97**
migrating songbirds, 28–29
mockingbird, northern, 29, 33, 86, **87**, 90, 141
owl, barred, 98, 112
owl, great horned, 86, 213
snipe, Wilson's, 213 **218**
whip-poor-will, eastern, 29, 98
wood-pewee, eastern, 96
wren, marsh, 217–18
northern lights, 218
North Platte River, 174–75
nutcracker, Clark's *(Nucifraga columbiana),* 155–56, **156**, 158, 186, **220**
memory for cached seeds, 227, 262
nuthatch, red-breasted *(Sitta canadensis),* 52, **193**, 258, 262
dawn calling, **52**, 186, 252

Ochoco Divide, 257–59
Ohio River, 87, 93–94, 96
Old Faithful, 201–4
Oregon
Hells Canyon, ID to Baker City, 242–48
Baker City to Dixie Pass, 248–50
Dixie Summit to Dayville, 251–53
Dayville to Ochoco Pass, 253–58
Ochoco Pass to Sisters, 258–60
Sisters to Limberlost Campground, 261–64
Limberlost Campground to Willamette Valley, 264–69
Willamette Valley to Newport, 270–79
Pacific coast, 280–86
Oregon Trail, 180–82, 247
oriole
Baltimore *(Icterus galbula),* 5, 7, 10, 76, 82, 89, 94, 108, 122, **123**, 130, 144
local dialects, **131**
matching songs, 76
Bullock's *(Icterus bullockii),* 122, **144**, 148, 150, 211, **239**, 244
hybrid zone birds, 123, 144
orchard *(Icterus spurius),* 69, **74**, 108

oriole (cont.)
dawn singing, **113**
osprey *(Pandion haliaetus),* 10, 181, **226**, 268
ovenbird *(Seiurus aurocapilla),* **46**, 64, 113
night song, 46
Owens, Terry, 56–57, **56**
owl
asymmetrical ears, 143
barn *(Tyto alba),* 172, 291 (p. 143)
barred *(Strix varia),* 57, 62, 99, 102
mate calls, **11**
night calling, 11, 98, 112
great gray *(Strix nebulosa),* 221
great horned *(Bubo virginianus),* 273, **274**
fledgling, 225, **274**
night calling, 86, 141, 213, 257–58
spotted *(Strix occidentalis),* 264
Ozark Mountains, 106–14

package singers, 53
chat, yellow-breasted, 70–71, 114, 177–78, 181
robin, American, 2
vireo, Bell's, 119
vireo, blue-headed, 53
vireo, Cassin's, 245
vireo, plumbeous, 151
vireo, yellow-throated, 113–14
wren, rock, 168
Painted Hills, 255
parakeet, Carolina *(Conuropsis carolinensis),* 73
Park Range, 172
parula, northern *(Setophaga americana),* 5, 57–58, 63
calls, 99, 112
dawn (aggressive) songs, 63, **99**, 112
day (mate) songs, **42**, 63, 102, 108, 114
eastern and western dialects, 42, 67, 71–72, **72**, 84
pelican, 169
pheasant, ring-necked *(Phasianus colchicus),* 130, 211
phoebe
eastern *(Sayornis phoebe),* 24–26, 30–32, **31**, 102, 114, 122–23, 127
dawn songs, 62, 98–99, 113, **118**
genetically innate songs, 25, 118
Say's *(Sayornis saya),* 122
dawn songs, **147**
physiology. *See* vocal physiology
Picture Gorge, 254–55
pigeon
band-tailed *(Patagioenas fasciata),* 274
passenger *(Ectopistes migratorius),* 73
Pike's Peak, 139, 148
Pioneer Range, 213–15, 227
Pleistocene ice ages, 10, 98, 243, 267
prairie-chicken, greater *(Tympanuchus cupido),* 127–28
prairie dogs, 172, 174
The Precious Present (Johnson), 83

QR Codes, vii
quail, California *(Callipepla californica),* **234**, 240, 276
dawn singing, 254
Quake Lake, 205–6

Rabbit Ears Range, 172
rail, Virginia *(Rallus limicola),* **181**, 226
railroads, 134–35, 140, 143, 176
Ramshorn Peak, 186
raven, common *(Corvus corax),* 42, 148, 174, 184, 203, 204–5, 208
conversations, **205**, 249, **261**
intelligence, 261–62
range, 42
red squirrel, 186, 252
redstart, American *(Setophaga ruticilla),* 108
calls, 42, 63, 112
dawn (aggressive) songs, 41–42, 63, **63**, 112
day (mate) songs, 45, 60, 63, **64**
repertoire estimates, 7–8, 24–26, 70–71, 113–14, 119, 151
Revolutionary War, 5
Rio Grande rift, 164
roadkill, 60, 134, 143, 172
blackbird, red-winged, 239
cardinal, northern, 60
chickadee, black-capped, 238
chipmunk, 60
cow, 172
coyote, 207
deer, 249
flicker, northern, 212
hawk, red-tailed, 172
lark, horned, 170
magpie, black-billed, 207
mallard, 172
nighthawk, common, 256
opossum, 60
oriole, Bullock's, 239
owl, barn, 172
prairie dog, 172
quail, California, 237
scrub-jay, western, 274
shrike, loggerhead, 143
snipe, Wilson's, 239
sora, 239
sparrow, Cassin's, 137–38, 142
starling, European, 108

swallow, bank, 246–48
teal, cinnamon, 172
vulture, black, 60
warbler, yellow, 226, 238
woodcock, American, 38–39, 60
woodpecker, pileated, 60
woodpecker, red-headed, 143
Robbins, Mark, 115
robin, American *(Turdus migratorius),* 2–5, 53, 55, 70, 89, 99, 119, 130, 147, 160, 189, 241, 265, 268, 274
calls, 20, 36, 50, 140, 155, 231, 258
dawn singing (carols and *hissellys*), **2**,12, 20–22, **50**, 59, 62, 68, 86, 112, 125, 155, 163, 167, 186, **190**, 202, 219, **220**, 222, 233, 242, 244, 254, 261, 273,
male improvisation, 20–21
repertoire, 21
Rocky Mountains. *See* names of specific mountain ranges, e.g. Grand Tetons
roosters. *See* fighting cocks
Ruby Range, River, and Basin, 210–13

Sacagawea, 183–85, 210, 213
Salmon River, 235–36, 245
Sand County Almanac (Leopold), 254
Sandpiper, 35
spotted *(Actitis macularia),* 232
upland *(Barbramia longicauda),* 126–27, **127**
Sapphire Mountains, 223, 227
sapsucker, 156, 187, 207, 249
red-breasted *(Sphyrapicus ruber),* 175, **176**
red-naped *(Sphyrapicus nuchalis),* **193**, **221**
science, 45, 266–67, 293 (p. 267)
scrub-jay, western *(Aphelocoma californica),* **240**, 274, **275**
Seeing Is Forgetting the Name of the Thing One Sees (Weschler), 137
Seven Devils Mountains, 233, 236, 238, 245
Sheep Mountain, 158 (WY), 205–6 (MT)
Sheep Rock, 254–55
shortgrass prairie, 136–45
Shoshone Lake, 201
Shoshone National Forest, 188
Shoshone people, 183–85, 210, 213
shrike, loggerhead *(Lanius ludovicianus),* 143, **179**, 277
The Singing Life of Birds (Donald Kroodsma), 161, 287
siskin, pine *(Spinus pinus),* **186**, 194, 203
Smith, Billy Rays, **47**
Snake River, 194–200, 208, 232, 236, 240–41, 243, 245, 259
snipe, Wilson's *(Gallinago delicata)*
winnows, **168**, 210, **211**, 218, 219, 226, **239**, 248
night winnows, **218**, 167, 173, 186, 213
calls, **218**, 226, **239**
Snodgrass, Bert, **54**
solitaire, Townsend's *(Myadestes townsendi),* **222**, 228, 261, **262**
calls, **262**
song improvisation. *See also* song learning
catbird, gray, 23, **23**, 27
chat, yellow-breasted, 178
robin, American, 20–21
thrasher, brown, 27
thrush, wood, 20–21
vireo, red-eyed, 8
wren, sedge, 92
song learning. *See also* dialects; mimicry; song improvisation
bellbird, three-wattled, 267
bunting, indigo, 76–77, 93
bunting, lazuli, 255–56
cardinal, northern, 20
chat, yellow-breasted, 178
chickadee, Carolina, 8
cowbird, brown-headed, 281–82
hummingbird, Anna's, 275
meadowlark, eastern, 131
robin, American, 20–21
songbirds, 5, 123, 244, 262, 267
sparrow, chipping, 13
sparrow, song, **266**
sparrow, white-throated, 14
thrasher, brown, 27
titmouse, tufted, 8, 38
towhee, spotted, 271
vireo, warbling, 276
warbler, chestnut-sided, 39
wren, Bewick's, 271–72
wren, Carolina, 6, 8, 20
wren, marsh, 175, 218, 225
wren, Pacific, 223
songless songbirds
bushtit, **256**
chickadee, chestnut-backed, **281**
jay, blue, 22
grosbeak, evening, **276**
waxwing, cedar, **113**
sora *(Porzana carolina),* 202, **203**, 226, **239**
soundscapes
cattle, **215**
dogs in Kentucky, **65**, 66, **71**
high school students, **79**
listening in Appalachia, **61**
oil pump, **127**
weather forecast, **197**
Yellowstone geysers, **202**

South Pass, 180, 182
sparrow
    Bachman's *(Peucaea aestivalis),* 142
    Brewer's *(Spizella breweri)*
        dawn songs, **167**
        day songs, 164, **168**, 174, 176, 179, 184, 240
    Cassin's *(Peucaea cassinii),* 137–38, **142**, 291 (p. 142)
    chipping *(Spizella passerina),* 94, 100, 108, 193, 202, 252
        calls, 193
        dawn singing, 13–14, **13**, 36–37, 62, **112**, 186, 191, 221, 251
        day singing, 75, **102**, **155**, **258**
        repertoire, 36
    English. *See* sparrow, house
    field *(Spizella pusilla)*
        calls, 37
        dawn songs, **37**
        day songs, **17**, 75–76, **76**, 108
    fox *(Passerella iliaca),* **231**, 261
        calls, 231
        repertoire, **262**
    grasshopper *(Ammodramus savannarum),* 78, 89, 121–22, **122**, 130, 132, 138, 240
        complex song, 121–22, **122**
    Henslow's *(Ammodramus henslowii),* **93**, 121, **122**
    house *(Passer domesticus),* **9**, 55, 80, **108**, 127, 134, 149, 150
        calls, 9
        dawn singing, **82**, 83, 211, **248**
    lark *(Chondestes grammacus),* 142–43, **143**, 148, 211
    Lincoln's *(Melospiza lincolnii)*
        dawn singing, **163**, **221**
    savannah *(Passerculus sandwichensis),* 168–69, **169**, 171, 174, 176, **203**, 213, 248–49
    seaside *(Ammodramus maritimus),* xiii, **7**
    song *(Melospiza melodia),* 5, 47–48, 59, 65, 69, 156, 240, 248, 266, 276, 279
        dawn singing, 62, 219–20, 167, **244**, 254, 281
        learning to sing, **266**
        local dialects, 57
        repertoire, **47**, 76, 244, **266**, 273
    vesper *(Pooecetes gramineus),* 213, 246, 256
        dawn singing, 195–96, **196**, 211
        local dialects, 156, 168, 176, 179, 185, 195–96, **196**, 209, 276
        repertoire, 196, 292 (p. 196)
    white-crowned *(Zonotrichia leucophrys),* 202, 265, 278, 285
        dawn singing, 219–20, 281
        local dialects, **160**, **205**, **220**, 265, 281
    white-throated *(Zonotrichia albicollis)*
        learning to sing, **14**
species and subspecies, 141–42
Split Rock, 180
Spring in Washington (Halle), 27
spring migration, xv–xvi, 5, 28–29
squirrel, red, 186, 252
starling, European *(Sturnus vulgaris),* 16, **80**, 80–81, 89–90, 108, 183
    legal protection, 80, 290 (p. 80)
    mimicry, **80**, 80–81, **249**
States, Carol and Jim, 175
Strawberry Mountains, 253
suboscine. *See* flycatchers
Sumpter Pass, 249
swallow, 15, 62, 68, 75, 112, 167, 186, 203, 225, 233, 241
    bank *(Riparia riparia),* 173, 246–48
    barn *(Hirundo rustica),* 93, 173, 192–93, **193**, 240, 248
        calls, 193
    cliff *(Petrochelidon pyrrhonota),* **133**, 143,155, 173, 248, 256
        calls, **133**
    tree *(Tachycineta bicolor),* 160, 173, 248
        dawn singing, 86–87, 274, **274**
    violet-green *(Tachycineta thalassina),* 160, 248
        dawn singing, 173, 186
Sweetwater River and Hills, 180–82
swift
    chimney *(Chaetura pelagica),* **55**, 69, 94
    white-throated *(Aeronautes saxatalis),* **185**
Table Mountain, 256
tallgrass prairie, 115–35
tanager
    scarlet *(Piranga olivacea),* 5, 7, 10, 31, 61, **73**, 108
        calls, 31, **36**, 191
        dawn singing, **36**, 99, 191
    summer *(Piranga rubra),* 18, **18**, 92–93, 108
        calls, 92, **93**, 99
        dawn singing, **99**
    western *(Piranga ludoviciana),* 99, 148, 149, 155, 191–92,**192**, 229, 244, 249, 265
        calls, 191, **191**, **192**, 220, **233**, 244
        dawn singing, 99, 155, **191**, 220, **233**, 237, 244, 251, 273

Targhee National Forest, 188
teal, cinnamon *(Anas cyanoptera),* 172
Thoreau, Henry David, xvi
thrasher

brown *(Toxostoma rufum)*, 7, 31, 33, 67, 101, 108, 132
dawn singing, 87, 119
learning to sing, 27
mimicry, 27
repertoire, 27, 262
sage *(Oreoscoptes montanus)*, 164, 174, 176, 184, 214, 246
dawn singing, 167–68, **168**, 218
mimicry, 167–68
repertoire, 168
thrush
hermit *(Catharus guttatus)*, 222, 223, 228, 258
dawn singing, 49, 51, **201**, 250, **251**
evening singing, **250**
repertoire, 262, **263**
Swainson's *(Catharus ustulatus)*, 52, 164, **189**, 193, 222, 228, 265, 274, 283, 285
dawn calling, **272**, **280**
dawn singing, 273, **280**
varied *(Ixoreus naevius)*, 222, 229, **230**
repertoire, 229–30, 293 (p. 229)
vocal physiology, 229–30
wood *(Hylocichla mustelina)*, **2**, 4–5, 24–25, 31, 108, 131, 140
calls, 2, 24–25, **35**, 28, 36, 77
dawn singing, 2, 20–22, 62, 113
evening songs, 27
male improvisation, 20–21
repertoire, 21, 25, 48, 49
vocal physiology, 2–3, 289 (p. 3)
time travel, 182–83. *See also* geological events
Tipton Pass, 249
titmouse
juniper *(Baeolophus ridgwayi)*, 150–51, **151**
oak *(Baeolophus inornatus)*, 151
tufted *(Baeolophus bicolor)*, 3–4, 5, 7, 108
dawn songs, 62, 69, 99, 102, 113
matched countersinging, 3–4, **4**
shared repertoires, 8
Tobacco Root Range, 209, 211
Togwotee Pass, 184, 188
Torrey Peak, 213
towhee, 123
eastern *(Pipilo erythrophthalmus)*, **7**, 9, 108, 122
calls, 34–35, 36, 42, 77
dawn singing, 34–35, **35**, 87, 113, 119
green-tailed *(Pipilo chlorurus)*, 148, 164
repertoire, 170–71, **171**
rufous-sided. *See* towhee, eastern; towhee, spotted
spotted *(Pipilo maculatus)*, 123, 148, 149, **181**, 223, 237, 240, 256, 269, 276, **282**
calls, 149, 270, **272**, 280
dawn singing, **148**, 149, 244, 270–71, **272**, **280**, 280–81
far western variant, 227, **227**
learning to sing, 271
mountain variant, 148, **148**
repertoire, 271
TransAmerica Trail, xv–xvi, 5, 32–33, 46, 115, 175–76, 279, 282, 287
Trapper Peak, 223
Triassic period, 185
turkey, wild *(Meleagris gallopavo)*, 47, 102

Union Pacific Railroad, 135, 176
Upper Geyser Basin, 201–3

veery *(Catharus fuscescens)*, **220**, 222
calls, 28–29, **49**, **240**
dawn singing and calling, **49**, 219, **240**
vireo, 277
Bell's *(Vireo bellii)*, 132, 277
dawn singing, 118–19, **119**
repertoire, 119, 291 (p. 119)
blue-headed *(Vireo solitarius)*, 245, 277
dawn singing, 52–53
repertoire, **53**, 151
Cassin's *(Vireo cassinii)*, **245**, 277
Hutton's *(Vireo huttoni)*, **277**
plumbeous *(Vireo plumbeus)*, 245, 277
repertoire, **151**, 168
red-eyed *(Vireo olivaceus)*, 5, 18, 31, 60–61, 102, 113, 234, 245, 277
calls, 61
dawn singing, 233
individual song repertoires, 7–9, **8**
warbling *(Vireo gilvus)*
eastern variant, **95**, 113, 130, 187
western variant,156, 183, **187**, 193, 221, 237, **240**, 245, 274, **276**, 277
white-eyed *(Vireo griseus)*, 77, **78**, 108, 277
calls, 77, **78**
dawn singing, 62
mimicry, 77, **78**
yellow-throated *(Vireo flavifrons)*, 69–70, 108, 277
repertoire, 113–14, **114**
Virginia
northern birds, 49–51
Yorktown to Jamestown, 1–11
Jamestown to Mechanicsville, 12–19
Mechanicsville to Mineral, 20–29
Whippoorwill Lane to Afton, 30–33
Afton to Lexington, Blue Ridge Parkway, 33–43
Wytheville to Mount Rogers, 44–49
Mount Rogers to Rosedale, 49–55
Rosedale to Virgie, KY, 56–59

vocal physiology, 246
    bobolink, 211
    cardinal, northern, 21, 112
    crane, sandhill, 207–8
    flycatcher, great crested, 21–22
    robin, American, 2, 21
    thrush, varied, 229–30
    thrush, wood, 2–3, 21
    two voice boxes, 2–3, 289 (p. 3)
    wren, Carolina, 21
volcanic hotspots, 199–200
    See also Yellowstone National Park
vulture
    turkey *(Cathartes aura)*, 75, 111
    black *(Coragyps atratus)*, 60

Wallowa Mountains, 233, 246
warbler, 73
    Audubon's. *See* warbler, yellow-rumped
    Bachman's *(Vermivora bachmanii)*, 73
    black-and-white *(Mniotilta varia)*
        calls, 63
        dawn (aggressive) songs, 63, **63**
        day (mate) songs, 45, 63, **64**, 108
    Blackburnian *(Setophaga fusca)*, 45
    black-throated blue *(Setophaga caerulescens)*, 231
        day (mate) songs, 45, **46**
    black-throated gray *(Setophaga nigrescens)*, 276, **277**
    black-throated green *(Setophaga virens)*, 5
        calls, **52**
        dawn (aggressive) songs, **52**
        day (mate) songs, 45, **52**
    blue-winged *(Vermivora cyanoptera)*, 5
        calls, 63
        dawn (aggressive) songs, 63, **63**
        day (mate) songs, 46, 63, **64**, 108
    Canada *(Cardellina canadensis)*
        calls, 49–50, 53
        dawn (aggressive) songs, 49–50, **50**
        day (mate) songs, 53
    chestnut-sided *(Setophaga pensylvanica)*, 5, 40–41
        dawn (aggressive) songs, **39**
        day (mate) songs, **39**, 45, 53
        range, 39
    hooded *(Setophaga citrina)*
        calls, 63
        dawn (aggressive) songs, **63**, 64
        day (mate) songs, **41**, **61**, 63
    Kentucky *(Geothlypis formosa)*, 64, **64**, 76, 90, 108
        dawn singing, 100, 113
        pitch-shifting, 64
    MacGillivray's *(Geothlypis tolmiei)*, 64, **193**, **220**, 221
    magnolia *(Setophaga magnolia)*, 52
    male mate-attraction ("day") songs, 46
    mourning *(Geothlypis philadelphia)*, 64
    Nashville *(Oreothlypis ruficapilla)*, **228**
    orange-crowned *(Oreothlypis celata)*, 273, **274**, **276**, 281
    pine *(Setophaga pinus)*, 22, 108
        calls, 109
        dawn (aggressive) songs, 14, **109**
    prairie *(Setophaga discolor)*, 5, 7, 108
        dawn (aggressive) songs, 31, **31**, 40
        day (mate) songs, 31, **31**, 40, 76
    prothonotary *(Protonotaria citrea)*, **90**
    Swainson's *(Limnothlypis swainsonii)*, 56
    Townsend's, 231–232, **232**, 252
    Wilson's *(Cardellina pusilla)*, **160**, **276**
    worm-eating *(Helmitheros vermivorum)*, 102, 108
        calls, 100, 102
        dawn singing, **100**
    yellow *(Setophaga petechia)*, 5, 183, 184, 211, **220**, 226, 238, **239**, 240, 241, 248, 254
        calls, **59**, 112, 147, 244
        dawn (aggressive) songs, **59**, 63, 112, 147, 167, 186, 210, 219, **233**, 236, 244
        day (mate) songs, 63, 76
    yellow-rumped *(Setophaga coronata)*, 202, **203**, 237, 281
    yellow-throated *(Setophaga dominica)*, 40–41, **41**, 65
Washakie, Shoshone chief, 183–85, 292 (p. 185)
waterthrush
    Louisiana *(Parkesia motacilla)*, 53, 90, 108
        dawn singing, 99–100, **100**
    northern *(Parkesia noveboracensis)*, **220**
        dawn singing, 219–21
waxwing, cedar *(Bombycilla cedrorum)*
    calls, **113**, 274
website, *ListeningToAContinent.Sing.com*, vii
Weschler, Lawrence, 137
Western Cascades, 243, 255
West Mountains, 238
whip-poor-will, eastern *(Caprimulgus vociferus)*, 29, **36**, 57, 98–99
White Bird, Nez Perce, 235
Whiteley Peak, 169
Why travel?, 4, 44–45, 77, 83, 285–86
wigeon, American *(Anas americana)*, **226**
Willamette National Forest, 264
Willamette Valley, 265–78
willet *(Tringa semipalmata)*, 249
William L. Finley National Wildlife Refuge, 110, 271–77

Wilson, Alexander, 227
Wind River Basin and Mountains, 181–88
Wolford Mountain, 169
woodcock, American *(Scolapax minor),* 38–39, 60
  call, 35
  whistling wings, 35, **35**, 39
woodpecker, 252
  acorn *(Melanerpes formicivorus),* 276, **277**
  downy *(Picoides pubescens),* 70, 101
  ivory-billed *(Campephilus principalis),* 73
  Lewis's *(Melanerpes lewis),* **227**
  pileated *(Dryocopus pileatus),* 10, **14**, 56, 60, **60**, 70, 66, 75, 102, 265
    announcing sunrise, 14, 60, 70, 102
  red-bellied *(Melanerpes carolinus),* 24, 140
    announcing sunrise, 70, 101, 113
    percussionist, **106**
  red-headed *(Melanerpes erythrocephalus),* 143
wood-pewee, 186
  eastern *(Contopus virens),* 122–23, 127, 153
    dawn prelude, 112
    dawn (and dusk) singing, 102, **112**, 113, 122
    day singing, **102**, 122
    night singing, 96
  genetically inborn songs, 262
  western *(Contopus sordidulus),* 122–23, 140, 148, 183, 240, 249, 262
    dawn prelude, 153, 210
    dawn singing, **153**, **210**, 220, 233, 236–37, 244, 254, 273
    day singing, 153–56, **156**, 256, 258, 261
wren
  Bewick's *(Thryomanes bewickii),* 110, 149, 269, 276, 281
    learning to sing, 271–72
    local dialects, 272–73, **273**
    repertoire, **110**, 148
  canyon *(Catherpes mexicanus),* 236, 244–45, 254, 256, **237**, 273
  Carolina *(Thryothorus ludovicianus),* **6**, 24, 55, 61, 62, 65, 69, 77, 94, 108, 113, 140, 141, 289 (p. 3)
    calls, 77
    dawn singing, 12, 20–22
    learning to sing, 8, 20
    local dialects, 12–13, 20
    matched countersinging, **6**, 20, 22
    repertoire, 6, 8, 21, 289 (p. 12)
    song repetition, 12, 21
  house *(Troglodytes aedon),* 47, 89, 109–10, **110**, 134, 148, 156, 163, 164, 186, 213, 221, 254, 276
    mimicry, 274
  marsh *(Cistothorus palustris),* 276
    matched countersinging, 175, 225–26, **225**, **279**
    night singing, 217–18
    repertoire, 175, **176**
    western variant, 175, 211, **279**
  Pacific *(Troglodytes pacificus),* **222**, 223, **228**, 264–65, 274, 278, 283–85, **284**, 293 (p. 222)
    calls, 264
    dawn singing, 264–65, **265**
  rock *(Salpinctes obsoletus),* 139–41, 156, 168, **170**, **209**, 237, 244, 254
    calls, 140
    dawn singing, 168, 245
    repertoire, **138**, **170**, **209**, **245**
  sedge *(Cistothorus platensis),* **92**, 93
    improvisation vs. imitation, 92
  winter *(Troglodytes hiemalis),* 293(p. 222)
    dawn songs, 51, **52**
    western variant, 222
wrentit *(Chamaea fasciata),* 277, **285**
Wyoming
  Walden, CO to Saratoga, 173–75
  Saratoga to Muddy Gap, 175–78
  Muddy Gap to Lander, 179–83
  Lander to Dubois, 183–86
  Dubois to Jenny Lake, 186–89
  Jenny Lake (rest day), 190–95
  Jenny Lake to Yellowstone National Park, 195–98
  Yellowstone National Park, 199–204

Yellowstone Lake, 200
Yellowstone National Park, xiii, 114, 177, 179, 188, 195, 197–206, 208, 292(p. 199)
  fault zones, 205–6
  formation of the caldera, 199–200
  hydrothermal activity/geysers, 201–4
Yellowstone River, 201
yellowthroat, common *(Geothlypis trichas),* 64, 69, 76, 89, 119–20, 132, 183, 211, 219, 225, **226**, 273, 281
  songs and countersongs, 17, **18**
Yopoah Crater, 263

Zen and the Art of Motorcycle Maintenance (Pirsig), 70, 72, 83, 177